Putting Research into Practice in the Elementary Grades

Readings from Journals of the
National Council of Teachers of Mathematics

Edited by
Donald L. Chambers
Wisconsin Center for Education Research
University of Wisconsin—Madison
Madison, Wisconsin

National Council of Teachers of Mathematics
Reston, Virginia

The special contents of this book are
Copyright © 2002
THE NATIONAL COUNCIL OF TEACHERS OF MATHEMATICS, INC.
1906 Association Drive, Reston, VA 20191-1502
(703) 620-9840; (800) 235-7566; www.nctm.org
All rights reserved

ISBN 0-87353-516-2

> The publications of the National Council of Teachers of Mathematics present a variety of viewpoints. The views expressed or implied in this publication, unless otherwise noted, should not be interpreted as official positions of the Council.

Printed in the United States of America

CONTENTS

Acknowledgments ix

Introduction 1

Part 1: Reasoning, Communicating, Problem Solving 3

Section 1: Reasoning, Student Thinking, and Invented Strategies 5

Constructivist Learning and Teaching 6
  Douglas H. Clements and Michael T. Battista
  Arithmetic Teacher, September 1990

Constructivism and First-Grade Arithmetic 9
  Constance Kamii and Barbara A. Lewis
  Arithmetic Teacher, September 1990

Direct Modeling and Invented Procedures: Building on Students' Informal Strategies 12
  Donald L. Chambers
  Teaching Children Mathematics, October 1996

Invented Strategies Can Develop Meaningful Mathematical Procedures 16
  William M. Carroll and Denise Porter
  Teaching Children Mathematics, March 1997

Story Problems and Students' Strategies 21
  Larry Sowder
  Arithmetic Teacher, May 1989

Connecting Instructional Practice to Student Thinking 24
  Patricia F. Campbell
  Teaching Children Mathematics, October 1997

Children's Understanding of Zero and Infinity 29
  Margariete Montague Wheeler
  Arithmetic Teacher, November 1987

Integrating Assessment and Instruction 33
  Megan M. Loef, Deborah A. Carey, Thomas P. Carpenter, and Elizabeth Fennema
  Arithmetic Teacher, November 1988

Section 2: Communicating 37

Improving Instruction by Listening to Children 38
  Donald L. Chambers
  Teaching Children Mathematics, February 1995

Experience, Problem Solving, and Discourse as Central Aspects of Constructivism 41
  Erna Yackel, Paul Cobb, Terry Wood, and Graceann Merkel
  Arithmetic Teacher, December 1990

Discourse That Promotes Conceptual Understanding 44
  Elham Kazemi
  Teaching Children Mathematics, March 1998

Dialogue and Conceptual Splatter in Mathematics Classes 50
  Jack Easley, Harold A. Taylor, and Judy K. Taylor
  Arithmetic Teacher, March 1990

The Use of Verbal Explanation in Japanese and American Classrooms 55
James W. Stigler
Arithmetic Teacher, October 1988

Section 3: Problem Solving 59

Issues in Problem-Solving Instruction 60
Douglas A. Grouws and Thomas L. Good
Arithmetic Teacher, April 1989

Teaching Mathematics and Thinking 63
Edward A. Silver and Margaret S. Smith
Arithmetic Teacher, April 1990

Arithmetic as Problem Solving 68
Magdalene Lampert
Arithmetic Teacher, March 1989

Creating a Problem-Solving Atmosphere 72
Paul Cobb, Erna Yackel, Terry Wood, Grayson Wheatley, and Graceann Merkel
Arithmetic Teacher, September 1988

Mathematical Problem Solving in and out of School 75
Frank K. Lester Jr.
Arithmetic Teacher, November 1989

Part 2: Children's Thinking in the Content Domains 79

Section 4: Numbers, Number Sense, and Number Operations 81

Number Sense–Making 82
Judith Sowder and Bonnie Schappelle
Arithmetic Teacher, February 1994

Number Sense and Nonsense 87
Zvia Markovits, Rina Hershkowitz, and Maxim Bruckheimer
Arithmetic Teacher, February 1989

Children's Multiplication 90
Kurt Killion and Leslie P. Steffe
Arithmetic Teacher, September 1989

Multiplication and Division: Sense Making and Meaning 93
Vicky L. Kouba and Kathy Franklin
Teaching Children Mathematics, May 1995

Misconceptions about Multiplication and Division 97
Anna O. Graeber
Arithmetic Teacher, March 1993

Developing Understanding of Computational Estimation 101
Judith Sowder
Arithmetic Teacher, January 1989

Section 5: Place Value 104

Place Value for Tens and Ones ... 105
Joseph N. Payne
Arithmetic Teacher, February 1988

Place Value and Addition and Subtraction ... 109
Diana Wearne and James Hiebert
Arithmetic Teacher, January 1994

Place Value as the Key to Teaching Decimal Operations ... 113
Judith Sowder
Teaching Children Mathematics, April 1997

Decimal Fractions ... 119
James Hiebert
Arithmetic Teacher, March 1987

Section 6: Fractions 121

Using Sharing Situations to Help Children Learn Fractions ... 122
Susan B. Empson
Teaching Children Mathematics, October 1995

The Problem of Fractions in the Elementary School ... 128
Leslie P. Steffe and John Olive
Arithmetic Teacher, May 1991

Fractions: In Search of Meaning ... 133
Mary Lou Witherspoon
Arithmetic Teacher, April 1993

Making Connections to Understand Fractions ... 137
Nancy K. Mack
Arithmetic Teacher, February 1993

Children's Strategies in Ordering Rational Numbers ... 141
Thomas Post and Kathleen Cramer
Arithmetic Teacher, October 1987

Building a Foundation for Understanding the Multiplication of Fractions ... 145
Nancy K. Mack
Teaching Children Mathematics, September 1998

How Children Think about Division with Fractions ... 150
Mary Ann Warrington
Teaching Children Mathematics, May 1997

Section 7: Geometry, Geometric Reasoning, and Spatial Sense 155

Concept Learning in Geometry ... 156
David J. Fuys and Amy K. Liebov
Teaching Children Mathematics, January 1997

Spatial Abilities ... 160
Douglas T. Owens
Arithmetic Teacher, February 1990

Enhancing Mathematics Learning through Imagery ... 164
Grayson H. Wheatley
Arithmetic Teacher, September 1991

Conquer Mathematics Concepts by Developing Visual Thinking 168
Rina Hershkowitz and Zvia Markovits
Arithmetic Teacher, May 1992

Using Spatial Imagery in Geometric Reasoning 174
Michael T. Battista and Douglas H. Clements
Arithmetic Teacher, November 1991

Estimation and Mental-Imagery Models in Geometry 179
John Happs and Helen Mansfield
Arithmetic Teacher, September 1992

Developing Spatial Sense through Area Measurement 183
Elizabeth Nitabach and Richard Lehrer
Teaching Children Mathematics, April 1996

Spatial Sense and the Construction of Abstract Units in Tiling 188
Grayson H. Wheatley
Arithmetic Teacher, April 1992

Similarity in the Middle Grades 191
Glenda Lappan and Ruhama Even
Arithmetic Teacher, May 1988

Section 8: Algebra and Algebraic Reasoning 196

A Foundation for Algebraic Reasoning in the Early Grades 197
Erna Yackel
Teaching Children Mathematics, February 1997

Children's Understanding of Equality: A Foundation for Algebra 202
Karen P. Falkner, Linda Levi, and Thomas P. Carpenter
Teaching Children Mathematics, December 1999

Experiences with Patterning 208
Joan Ferrini-Mundy, Glenda Lappan, and Elizabeth Phillips
Teaching Children Mathematics, February 1997

Developing Algebraic Reasoning in the Elementary Grades 216
Jinfa Cai
Teaching Children Mathematics, December 1998

Helping to Make the Transition to Algebra 221
Carolyn Kieran
Arithmetic Teacher, March 1991

Section 9: Statistics and Statistical Reasoning 225

What Do Children Understand about Average? 226
Susan Jo Russell and Jan Mokros
Teaching Children Mathematics, February 1996

Fourth Graders Invent Ways of Computing Averages 232
Constance Kamii, Michele Pritchett, and Kristi Nelson
Teaching Children Mathematics, October 1996

Teaching Statistics: Mean, Median, and Mode 238
Judith S. Zawojewski
Arithmetic Teacher, March 1988

Learning Probability Concepts in Elementary School Mathematics 241
  Albert P. Schulte
  Arithmetic Teacher, January 1987

Part 3: Tools for Representing Mathematical Ideas 243

Section 10: Concrete Materials as Tools 245

Concrete Materials and Teaching for Mathematical Understanding 246
  Patrick W. Thompson
  Arithmetic Teacher, May 1994

Manipulatives Don't Come with Guarantees 250
  Arthur J. Baroody
  Arithmetic Teacher, October 1989

Rethinking "Concrete" Manipulatives 252
  Douglas H. Clements and Sue McMillen
  Teaching Children Mathematics, January 1996

Section 11: Electronic Tools 264

Research on Calculators in Mathematics Education 265
  Ray Hembree and Donald J. Dessart
  Calculators in Mathematics Education, 1992 Yearbook of the
  National Council of Teachers of Mathematics

Calculators and Constructivism 269
  Grayson H. Wheatley and Douglas H. Clements
  Arithmetic Teacher, October 1990

Calculators and Computers: Tools for Mathematical Exploration and Empowerment 272
  Michael T. Battista
  Arithmetic Teacher, October 1994

Constructing Geometric Concepts in Logo 279
  Michael T. Battista and Douglas H. Clements
  Arithmetic Teacher, November 1990

Section 12: Other Representational Tools 282

Giving Pupils Tools for Thinking 283
  Robert B. Davis
  Arithmetic Teacher, January 1991

Students' Use of Symbols 286
  Deborah A. Carey
  Arithmetic Teacher, November 1992

Part 4: Mathematics for All 291

Section 13: Equity 293

Making Equity a Reality in Classrooms 294
  Patricia F. Campbell and Cynthia Langrall
  Arithmetic Teacher, October 1993

Gender and Race Equity in Primary and Middle School Mathematics Classrooms **299**
    Laurie Hart Reyes and George M. A. Stanic
    Arithmetic Teacher, April 1988

Parental Involvement in a Time of Changing Demographics **302**
    Walter G. Secada
    Arithmetic Teacher, December 1989

Part 5: Teaching and Research 307

Section 14: Using Research to Improve Teaching and Learning 309

Teachers Researching Their Mathematics Classrooms 310
    Neil Pateman
    Arithmetic Teacher, October 1989

Putting Research into Practice 313
    Donald L. Chambers

The Role of *JRME* in Advancing Learning and Teaching Elementary School Mathematics 314
    Michael T. Battista and Carol Novillis Larson
    Teaching Children Mathematics, November 1994

Research Resources for Teachers 319
    Donald L. Chambers

ACKNOWLEDGMENTS

Promoting the professional development of its members is an important part of the mission of the National Council of Teachers of Mathematics. An awareness of research on teaching and learning mathematics is an important component of that professional development. Although research is implicit in nearly every article that appears in an NCTM journal, it became explicit when *Arithmetic Teacher* added a new department, "Research Report," during the 1983–84 academic year. Marilyn Suydam served as the editor of "Research Report" through 1986, when Leroy G. Callahan became editor. In 1987, the name of the department was changed to "Research into Practice," with Glenda Lappan as editor. Subsequent editors were Judith Sowder and Larry Sowder; George M. A. Stanic; Michael T. Battista and Douglas H. Clements; Grayson H. Wheatley; Patricia F. Campbell; and Diana V. Lambdin. I became editor in 1994, when *Arithmetic Teacher* became *Teaching Children Mathematics*, and I continued as editor through 1999.

Previous editors and the authors with whom they worked set a very high standard for communicating important research results to elementary school teachers. My goal was to continue to provide articles that would cause teachers to reflect on their practice, to justify practices that were supported by research, and to alter those that were not.

I wish to thank all the authors who contributed articles to the department during my tenure as editor. I especially appreciated their willingness to collaborate with me to shape the articles in ways that met my goals of clarity, interest, and applicability. The careful reading and thoughtful comments of Deborah Carey, Mazie Jenkins, Douglas Edge, and James Barta, my liaisons to the *Teaching Children Mathematics* Editorial Panel, were reflected in the quality of the published articles. Linda Levi of the University of Wisconsin—Madison also assisted by editing an article during my tenure.

Several members of NCTM's Reston Headquarters staff were indispensable in preparing articles for publication. Andy Reeves often helped me think through thorny problems of editorship. Harry Tunis assisted in many decisions related to converting journal articles into a book. Journal staff members Kathleen Chapman Lay and Danny Breidenbach learned to tolerate my idiosyncrasies with good humor.

Mark Driscoll established a high standard for communicating research results to teachers in his landmark volume *Research within Reach: Elementary School Mathematics*, still a valuable resource twenty years later. In my work with elementary school teachers, I have vigorously disseminated Driscoll's work.

Special thanks are due to Elizabeth Fennema and Thomas Carpenter, who helped me understand how classroom-based research can transform the way teachers think about their work, the way they work, and the learning of students in their classrooms.

INTRODUCTION

Whether you are a beginning teacher, a teacher with many years of experience, or a mathematics specialist who supports elementary school teachers, increasing your knowledge of children's mathematical thinking will increase your effectiveness. This book's descriptions of children's mathematical thinking can guide and support you in your reflection on student thinking and on teaching and learning.

The book is organized into five parts.

- In part 1 the focus is on reasoning, student thinking, and invented strategies independent of the mathematics content domain. It addresses the role of discourse in helping students clarify their thinking; communicate their thinking to other students; and communicate their thinking to the teacher, thus permitting access to their thinking.
- In part 2 the focus is on children's thinking in the specific content domains of numbers, number sense, and number operations; place value; fractions; geometry, geometric reasoning, and spatial sense; algebra and algebraic reasoning; and statistics and statistical reasoning.
- Part 3 addresses research on tools for representing mathematical ideas, including concrete materials, electronic tools, and mathematical symbols as tools.
- Part 4 raises equity issues and describes research that can help teachers reduce inequity in their classrooms.
- Part 5 lists additional resources and presents ideas for using research in the classroom. The theme of this book, putting research into practice, is realized when teachers not only inform themselves about students' mathematical thinking but take the next step to use that knowledge in their instruction. The brief section "Putting Research into Practice" suggests how to do so for many of the chapters in this book. Reflective teachers will think of other ways.

All teachers have knowledge and beliefs about mathematics curriculum, instruction, and assessment. Each teacher's knowledge and beliefs guide her or his instructional decisions. Research may contradict something that many teachers believe. Good teachers are careful to distinguish between what they believe and what they know, although doing so can be difficult, especially if colleagues have beliefs similar to their own. Mutual beliefs may be misconstrued as facts. But when colleagues disagree on a question of teaching and learning, teachers can look for evidence that supports one view or another. Evidence allows us to distinguish between knowledge and beliefs. Research provides that evidence.

The chapters in this book originally appeared as articles in NCTM's journals.* Most were in the "Research Report" department of the *Arithmetic Teacher* or the "Research into Practice" department of the *Arithmetic Teacher* or *Teaching Children Mathematics*. The purposes of these articles are to inform elementary school teachers about research related to teaching and learning mathematics in the elementary grades; to help them examine and reflect on their own teaching; and to help them use research results to influence their own instructional decisions, that is, to put research into practice. The purpose of this book is to make those articles readily accessible to all who would use research to improve children's mathematical understanding.

*Some articles may have been modified editorially for the purposes of this collection.

PART 1
Reasoning, Communicating, Problem Solving

Section 1

Reasoning, Student Thinking, and Invented Strategies

Constructivist Learning and Teaching

Douglas H. Clements and Michael T. Battista

In reality, no one can *teach* mathematics. Effective teachers are those who can stimulate students to *learn* mathematics. Educational research offers compelling evidence that students learn mathematics well only when they *construct* their own mathematical understanding. (MSEB and National Research Council 1989, 58)

Radical changes have been advocated in recent reports on mathematics education, such as NCTM's *Curriculum and Evaluation Standards for School Mathematics* (National Council of Teachers of Mathematics 1989) and *Everybody Counts* (MSEB and National Research Council 1989). Unfortunately, many educators are focusing on alterations in content rather than the reports' recommendations for fundamental changes in instructional practices. Many of these instructional changes can best be understood from a *constructivist* perspective. Although references to constructivist approaches are pervasive, practical descriptions of such approaches have not been readily accessible. Therefore, to promote dialogue about instructional change, each "Research into Practice" column this year* will illustrate how a constructivist approach to teaching might be taken for a specific topic in mathematics.

What Is Constructivism?

Most traditional mathematics instruction and curricula are based on the *transmission,* or *absorption,* view of teaching and learning. In this view, students passively "absorb" mathematical structures invented by others and recorded in texts or known by authoritative adults. Teaching consists of transmitting sets of established facts, skills, and concepts to students.

Constructivism offers a sharp contrast to this view. Its basic tenets—which are embraced to a greater or lesser extent by different proponents—are the following:

1. Knowledge is actively created or invented by the child, not passively received from the environment. This idea can be illustrated by the Piagetian position that mathematical ideas are *made* by children, not found like a pebble or accepted from others like a gift (Sinclair, in Steffe and Cobb 1988). For example, the idea "four" cannot be directly detected by a child's senses. It is a relation that the child superimposes on a set of objects. This relation is constructed by the child by reflecting on actions performed on numerous sets of objects, such as contrasting the counting of sets having four units with the counting of sets having three and five units. Although a teacher may have demonstrated and numerically labeled many sets of objects for the student, the mental entity "four" can be created only by the student's thought. In other words, students do not "discover" the way the world works like Columbus found a new continent. Rather they *invent* new ways of thinking about the world.

2. Children create new mathematical knowledge by reflecting on their physical and mental actions. Ideas are constructed or made meaningful when children integrate them into their existing structures of knowledge.

3. No one true reality exists, only individual interpretations of the world. These interpretations are shaped by experience and social interactions. Thus, learning mathematics should be thought of as a process of adapting to and organizing one's quantitative world, not discovering preexisting ideas imposed by others. (This tenet is perhaps the most controversial.)

4. Learning is a social process in which children grow into the intellectual life of those around them (Bruner 1986). Mathematical ideas and truths, both in use and in meaning, are cooperatively established by the members of a culture. Thus, the constructivist classroom is seen as a culture in which students are involved not only in discovery and invention but in a social discourse involving explanation, negotiation, sharing, and evaluation.

*1990–1991

5. When a teacher demands that students use set mathematical methods, the sense-making activity of students is seriously curtailed. Students tend to mimic the methods by rote so that they can appear to achieve the teacher's goals. Their beliefs about the nature of mathematics change from viewing mathematics as sense making to viewing it as learning set procedures that make little sense.

Two Major Goals

Although it has many different interpretations, taking a constructivist perspective appears to imply two major goals for mathematics instruction (Cobb 1988). First, students should develop mathematical structures that are more complex, abstract, and powerful than the ones they currently possess so that they are increasingly capable of solving a wide variety of meaningful problems.

Second, students should become autonomous and self-motivated in their mathematical activity. Such students believe that mathematics is a way of thinking about problems. They believe that they do not "get" mathematical knowledge from their teacher so much as from their own explorations, thinking, and participation in discussions. They see their responsibility in the mathematics classroom not so much as completing assigned tasks but as making sense of, and communicating about, mathematics. Such independent students have the sense of themselves as controlling and creating mathematics.

Teaching and Learning

Constructivist instruction, on the one hand, gives preeminent value to the development of students' personal mathematical ideas. Traditional instruction, on the other hand, values only established mathematical techniques and concepts. For example, even though many teachers consistently use concrete materials to introduce ideas, they use them only for an introduction; the goal is to get to the abstract, symbolic, established mathematics. Inadvertently, students' intuitive thinking about what is meaningful to them is devalued. They come to feel that their intuitive ideas and methods are not related to *real* mathematics. In contrast, in constructivist instruction, students are encouraged to use their own methods for solving problems. They are not asked to adopt someone else's thinking but encouraged to refine their own. Although the teacher presents tasks that promote the invention or adoption of more sophisticated techniques, all methods are valued and supported. Through interaction with mathematical tasks and other students, the student's own intuitive mathematical thinking gradually becomes more abstract and powerful.

Because the role of the constructivist teacher is to guide and support students' invention of viable mathematical ideas rather than transmit "correct" adult ways of doing mathematics, some see the constructivist approach as inefficient, free-for-all discovery. In fact, even in its least directive form, the guidance of the teacher is the feature that distinguishes constructivism from unguided discovery. The constructivist teacher, by offering appropriate tasks and opportunities for dialogue, guides the focus of students' attention, thus unobtrusively directing their learning (Bruner 1986).

Constructivist teachers must be able to pose tasks that bring about appropriate conceptual reorganizations in students. This approach requires knowledge of both the normal developmental sequence in which students learn specific mathematical ideas and the current individual structures of students in the class. Such teachers must also be skilled in structuring the intellectual and social climate of the classroom so that students discuss, reflect on, and make sense of these tasks.

An Invitation

Each article in this year's "Research into Practice" column will present specific examples of the constructivist approach in action. Each will describe how students think about particular mathematical ideas and how instructional environments can be structured to cause students to develop more powerful thinking about those ideas. We invite you to consider the approach and how it relates to your teaching—to try it in your classroom. Which tenets of constructivism might you accept? How might your teaching and classroom environment change if you accept that students must construct their own knowledge? Are the implications different for students of different ages? How do you deal with individual differences? Most important, what instructional methods are consistent with a constructivist view of learning?

References

Bruner, Jerome. *Actual Minds, Possible Worlds.* Cambridge, Mass.: Harvard University Press, 1986.

Cobb, Paul. "The Tension between Theories of Learning and Instruction in Mathematics Education." *Educational Psychologist* 23 (1988): 87–103.

Mathematical Sciences Education Board (MSEB) and National Research Council. *Everybody Counts: A Report to the Nation on the Future of Mathematics Education.* Washington, D.C.: National Academy Press, 1989.

National Council of Teachers of Mathematics (NCTM). *Curriculum and Evaluation Standards for School Mathematics.* Reston, Va.: NCTM, 1989.

Steffe, Leslie, and Paul Cobb. *Construction of Arithmetical Meanings and Strategies.* New York: Springer-Verlag, 1988.

Constructivism and First-Grade Arithmetic

Constance Kamii and Barbara A. Lewis

For arithmetic instruction in the first grade, we advocate the use of games and situations in daily living in contrast to the traditional use of textbooks, workbooks, and worksheets. Our position is supported by the research and theory of Jean Piaget, called *constructivism,* as well as by classroom research (Kamii 1985, 1990).

Piaget's theory shows that children acquire number concepts by constructing them from the inside rather than by internalizing them from the outside. The best way to explain this statement is by describing children's reactions to one of the tasks Piaget developed with Inhelder (Inhelder and Piaget 1963).

The pupil is given one of two identical glasses, and the teacher takes the other one. After putting thirty to fifty chips (or beans, buttons, etc.) on the table, the teacher asks the pupil to drop a chip into his or her glass each time she drops one into hers. When about five chips have thus been dropped into each glass with one-to-one correspondence, the teacher says, "Let's stop now, and you watch what I am going to do." The teacher then drops one chip into her glass and says to the pupil, "Let's get going again." The teacher and the pupil drop about five more chips into each glass with one-to-one correspondence, until the teacher says, "Let's stop." The following is what has happened so far:

Teacher:

$1 + 1 + 1 + 1 + 1 + 1 + 1 + 1 + 1 + 1 + 1$

Pupil:

$1 + 1 + 1 + 1 + 1 \qquad + 1 + 1 + 1 + 1 + 1$

The teacher then asks, "Do we have the same amount, or do *you* have more, or do *I* have more?"

Four-year-olds usually reply that the two glasses have the same amount. When we go on to ask, "How do you know that we have the same amount?" the pupils explain, "Because I can see that we both have the same amount." (Some four-year-olds, however, reply that *they* have more, and when asked how they know that they have more, their usual answer is "Because.")

The teacher goes on to ask, "Do you remember how we dropped the chips?" and four-year-olds usually give all the empirical facts correctly, including the fact that only the teacher put an additional chip into her glass at one point. In other words, four-year-olds remember all the empirical facts correctly and base their judgment of equality on the empirical appearance of the two quantities.

By age five or six, however, most middle-class pupils deduce logically that the teacher has one more. When we ask these pupils how they know that the teacher has one more, they invoke exactly the same empirical facts as the four-year-olds.

No one teaches five- and six-year-olds to give correct answers to these questions. Yet children all over the world become able to give correct answers by constructing numerical relationships through their own natural ability to think. This construction from within can best be explained by reviewing the distinction Piaget made among three kinds of knowledge according to their sources—physical knowledge, logicomathematical knowledge, and social (conventional) knowledge.

Physical knowledge, on the one hand, is knowledge of objects in external reality. The color and weight of a chip are examples of physical properties that are *in* objects in external reality and can be known empirically by observation.

Logicomathematical knowledge, on the other hand, consists of *relationships* created by each individual. For instance, when we are presented with a red chip and a blue one and think that they are *different,* this difference is an example of logicomathematical knowledge. The chips are observable, but the difference between them is not. The difference exists neither *in* the red chip nor *in* the blue one, and if a person did not put the objects into this relationship, the difference would not exist for him or her. Other examples of relationships the individual can create between the chips are *similar, the same* in weight, and *two.*

Physical knowledge is thus empirical in nature because it has its source partly in objects. Logicomathematical knowledge, however, is not empirical knowledge, as its source is in each individual's head.

The ultimate sources of *social knowledge* are conventions worked out by people. Examples of social knowledge are the fact that Christmas comes on 25 December and that a tree is called "tree." Words such as *one, two,* and *three* and numerals such as 1, 2, and 3 belong to social knowledge, but the numerical concepts necessary to understand these numerals belong to logicomathematical knowledge.

Keeping the distinction among the three kinds of knowledge in mind, one can understand why most four-year-olds in the task described earlier said that the two glasses have the same amount. The four-year-olds had not yet constructed the logicomathematical relationship of number and could therefore gain only physical knowledge from the experience. From the appearance of the chips in the glasses, the pupils concluded that the amount was the same despite the fact that they remembered the way in which the chips had been dropped. Once the concept of number has developed, however, pupils will deduce from the same empirical facts that the teacher has one more chip regardless of the physical appearance.

New Goals for Beginning Arithmetic Instruction

If children develop mathematical understanding through their own natural ability to think, the goals of beginning arithmetic must be that pupils think and construct a network of numerical relationships. To add five and four, for example, pupils have to think $(1 + 1 + 1 + 1 + 1) + (1 + 1 + 1 + 1)$. This operation requires pupils to make two wholes (5 and 4) in their heads and then to make a higher-order whole (9) in which the original wholes (5 and 4) become parts. An example of a network of numerical relationships can be seen when pupils think about $5 + 4$ as one more than $4 + 4$ and as one less than $5 + 5$. Addition thus involves a great deal of thinking, that is, the making of relationships rather than mere skills (such as penmanship).

This definition of goals for instruction is very different from traditional instruction that focuses on correct answers and the writing of mathematical symbols. It is also very different from the assumption that pupils have to internalize "addition facts," store them, and retrieve them in computerlike fashion.

New Principles of Teaching

The following principles of teaching flow from constructivism and the preceding goals:

1. Encourage pupils to invent their own ways of adding and subtracting numbers rather than tell them how. For example, if pupils can play a board game with one die, we simply introduce a second die and let them figure out what to do.

2. Encourage pupils to exchange points of view rather than reinforce correct answers and correct wrong ones. For example, if a pupil says that six minus two equals three, we encourage pupils to agree or disagree with each other. Pupils *will* eventually agree on the truth if they debate long enough because, in logicomathematical knowledge, nothing is arbitrary.

3. Encourage pupils to think rather than to compute with paper and pencil. Written computation interferes with pupils' freedom to think and to remember sums and differences.

Classroom Activities

Paper-and-pencil exercises cause social isolation, mechanical repetition, and dependence on the teacher to know if an answer is correct. We, therefore, replace the textbook, workbook, and worksheets with two kinds of activities: games and situations in daily living.

Games, such as a modification of old maid in which pupils try to make a sum of ten with two cards, are well known to be effective. Although games are typically used only as a reward for pupils who have finished their work, we use games as a staple of instruction. Games give rise to compelling reasons for pupils to think and to agree or disagree with each other. When it is useful to know that $5 + 5$, $6 + 4$, $7 + 3$, and so on, all equal ten, pupils are much more likely to remember these combinations than when they write in workbooks to satisfy the teacher.

Situations in daily living also offer meaningful opportunities for pupils to construct mathematical relationships. Taking attendance, voting, collecting money, and sending notes home are examples of situations the teacher can use to encourage pupils to think. If four people brought their lunch, eight ordered the special, and six ordered soup and sandwich, the teacher can ask if everybody present has been accounted for. Pupils care about real-life situations and think much harder about these questions than about those in workbooks.

References

Inhelder, Bärbel, and Jean Piaget, "De l'itération des actions à la récurrence Elémentaire." In *La formation des raisonnements récurrentiels*, edited by Pierre Greco, Bärbel Inhelder, B. Matalon, and Jean Piaget. Paris: Presses Universitaires de France, 1963.

Kamii, Constance. "Constructivism and Beginning Arithmetic (K–2)." In *Teaching and Learning Mathematics in the 1990s*, 1990 Yearbook of the National Council of Teachers of Mathematics (NCTM), edited by Thomas J. Cooney and Christian R. Hirsch, pp. 22–30. Reston, Va.: NCTM, 1990.

———. Young *Children Reinvent Arithmetic*. New York: Teachers College Press, 1985.

Direct Modeling and Invented Procedures: Building on Students' Informal Strategies

Donald L. Chambers

Young children commonly solve mathematics problems by directly modeling the action or relationship described in the problem. They do not need to be taught how to use direct-modeling strategies, nor do they need such often-assumed prerequisite knowledge as number facts or computational algorithms.

Direct Modeling

An example of direct modeling arose when a third-grade teacher posed this problem:

> On our hospital field trip we saw 12 emergency rooms. Each room had 9 beds in it. How many beds were there?

Pamela counted out twelve groups of blocks with 9 blocks in each group. She then rearranged the blocks into groups of 10, replacing each group of 10 with a tens block. When she got 10 tens blocks, she replaced them with a hundreds block. She got 108. Pamela directly modeled the problem situation, using blocks to represent hospital beds. Pamela's strategy is typical. She does not appear to know the standard multiplication algorithm, but she has devised an effective way to solve this "multiplication" problem.

The ability to solve "multiplication" problems is also common among kindergarten students, as the subsequent example illustrates. A kindergarten teacher posed the following problem:

> A bee has 6 legs. How many legs do 5 bees have?

Sean put out 5 cubes to represent the bees. Then he put out 6 cubes with each bee to represent its legs. He then counted all the legs, but not the bees, and got 30. Jeffrey used a number line. He pointed to the 6 and said, "Six for one bee. That's one." He pointed to the 12 and said, "Six for another bee. That's two." He then counted 6 more on the number line, landing at 18, and said, "Six for another bee. That's three." He continued counting by ones on the number line and keeping track of the number of bees mentally until he got 30 legs for 5 bees. Brianna said that she knew that 6 + 6 = 12. Then, using her fingers, she counted on 6 more from 12 to get 18 and repeated the process until she got 30.

The problem-solving power of these students comes from their ability to model a problem directly. In a study of kindergarten students' problem-solving processes (see Carpenter et al. [1993]), 71 percent correctly solved this problem:

> Robin has 3 packages of gum. There are 6 pieces of gum in each package. How many pieces of gum does Robin have altogether?

Direct modeling is a natural strategy used by many children to solve both routine and nonroutine problems. Over half the kindergarten children in the study cited were able to solve this problem:

> 19 children are taking a minibus to the zoo. They will have to sit either 2 or 3 to a seat. The bus has 7 seats. How many children will have to sit 3 to a seat, and how many can sit 2 to a seat?

This problem is almost impossible for children to solve by using addition and subtraction facts. However, it is easy to model, and modeling readily yields a solution. Children who solve this problem usually draw seven seats and systematically assign children to seats, first putting one on each seat, and then a second, and then a third, until all students are seated. They can then count the number sitting two to a seat and the number sitting three to a seat. High school students are taught an algorithm that can be used to solve this problem, but by using direct modeling, younger students clearly can solve the problem long before the procedure is taught.

Invented Strategies

Direct-modeling strategies using counters are sufficient for problems involving small numbers. As numbers become larger, requiring representations of tens and hundreds, students begin to develop invented strategies. Invented strategies involve mental pictures of the blocks that represent the numbers, or mental calculation with the numbers themselves, or invention of pencil-and-paper notation. Three different invented strategies were used by Lauren and Laurel, second graders, and by Joel, a first grader, to solve the following problem:

> Max had 46 comic books. For his birthday his father gave him 37 more comic books. How many comic books does Max have now?

All three students solved the problem mentally. Lauren started with 46 and counted by tens, then by ones: 46, 56, 66, 76, 77, 78, 79, 80, 81, 82, 83. Laurel rounded 37 to 40 to simplify the addition. She then adjusted the result by subtracting the 3 that she added earlier: $46 + 40 = 86$; $86 - 3 = 83$. Joel used a strategy that is very similar to the standard algorithm, but like many children, he worked with the tens first, then the ones: $40 + 30 = 70$; $6 + 7 = 13$; $70 + 13 = 83$.

Standard algorithms typically form the core of the elementary school mathematics program, but instruction of standard algorithms frequently does not build on students' natural ways of thinking. When standard algorithms abruptly displace children's natural direct-modeling strategies and invented procedures, confusion rather than understanding frequently results. In the following example, Gretchen subtracts 23 from 70 by using the standard algorithm. She then uses an invented strategy based on direct modeling. When she gets two different answers, her confusion is apparent.

Teacher: How would you do this problem? [On a piece of paper, she writes:]

$$\begin{array}{r} 70 \\ -23 \\ \hline \end{array}$$

Gretchen: That's easy. [She first writes a 3 below the 0 and the 3, and then writes a 5 below the 7 and the 2.]

Teacher: And your answer is . . . ?

Gretchen: 53.

Teacher: Could you show me that problem with these [base-ten] blocks?

Gretchen: Okay. [She takes 7 tens blocks. Then she takes away 2 tens blocks and counts 3 units on the next tens block. She then counts the remaining units on that block: 1, 2, 3, 4, 5, 6, 7. She looks at the remaining tens blocks. Then she looks back at her previous answer.] Oh, gee! [She rechecks her previous work with pencil and paper.] I don't get it.

Teacher: What did you get for your answer with the blocks?

Gretchen: [Rechecks her work, exactly the same way, with the same result.] Over here [with the blocks] I get 47, but over here . . . [with the pencil and paper]. Okay, 0 take away 3, yeah, that's 3. Okay. 7 take away 2 equals 5. So I put 3 there [pointing to the 3 she wrote earlier] and 5 there [pointing to the 5].

Teacher: And over there [pointing to the blocks] you got what?

Gretchen: 47. I don't get it.

Teacher: Which one do you think is right?

Gretchen: [Taps the 53 with a tens block she still holds in her hand.]

Gretchen has been taught the standard subtraction algorithm. She does not understand it, but she believes in its power. When confronted with her own conceptually based explanation with tens blocks, she rejects it in favor of her algorithmic answer.

If we had seen Gretchen's solution with the blocks but not her paper-and-pencil solution, we would probably believe that Gretchen understands subtraction with regrouping. If we had not seen Gretchen reject the blocks-based solution in favor of her pencil-and-paper solution, we would probably believe that such a blocks-based demonstration would persuade her that her pencil-and-paper solution was wrong.

According to the National Assessment of Educational Progress (Kouba, Carpenter, and Swafford 1989), 70 percent of third graders in the spring semester could solve this subtraction problem:

$$\begin{array}{r} 54 \\ -37 \\ \hline \end{array}$$

Only 65 percent could solve $44 - 6$. A study of beginning second-grade students who had not yet been taught the subtraction algorithm in school found that 76 percent could correctly solve a two-digit-subtraction-with-regrouping problem by using invented strategies. Performance among third-grade students who had one and one-half additional years of instruction was lower. Why?

Young children understand the strategies they invent. They frequently do not understand standard algorithms, even when conceptually based explanations are included as part of the instruction. When students do not understand an algorithm, they may unintentionally modify it. Without understanding, they have no way to know whether their strategy is flawed.

Teaching

Students gain confidence in their ability to do mathematics when they use strategies that they understand. Students understand the strategies they have invented to solve problems in their prior experience. Teachers are naturally eager to see their students use strategies that are efficient, and efficiency is a feature of standard algorithms. But efficiency without understanding leads to errors, and errors lead to lack of confidence.

Teachers can capitalize on children's ability to use direct-modeling strategies and invented algorithms by valuing them as valid strategies, just as they would value strategies using standard algorithms. In a timely manner, teachers may want to try to move individual students away from direct modeling toward invented strategies, as the teacher does in the following example from a third-grade classroom.

Teacher: A room had 327 cavity-filling caramels and 465 invisible chocolate bars. How many pieces of candy were in that room?

Shannon: Here's my 465 [she shows her 4 hundreds blocks, 6 tens blocks, and 5 ones blocks], and here's my 327 [she shows her 3 hundreds blocks, 2 tens blocks, and 7 ones blocks]. I took out the tens counters and the ones counters. I added the hundreds counters first: 100, 200, 300, 400, 500, 600, 700 [she stacks the hundreds blocks as she counts by hundreds]; 710, 720, 730, 740, 750, 760, 770, 780 [she counts on by tens from 700 as she puts each tens block with the hundreds blocks]; 781, 782, 783, 784, 785, 786, 787, 788, 789, 790, 791, 792 [she counts on by ones from 780 as she puts each ones block with the hundreds blocks and tens blocks].

Teacher: Nice. Now, you started with the hundreds first, then you went to the tens, then you went to the ones. Now, did you do this in your head?

Shannon: No.

Teacher: Do you want to try it in your head? Now remember what you did here [pointing to the blocks]. You did the hundreds first, right? Then you did the tens. Then the ones. Let's see what you could do.

Shannon: Well, I could add the 300 and the 400 together. That would be 700. And then I could add the 60 and 20 together: 60, 70, that would be 80. And then 5, 6, 7, 8, 9, 10, 11, 12. And that would be 780, 792.

Teacher: Nice job. See, you really didn't need these [base-ten blocks], did you?

Teachers find that students in their classrooms typically use various strategies for a given problem when allowed to do so. This freedom gives the teacher the opportunity to have students share different strategies with the class. Students listen to other students describe their strategies, and any student may decide to begin using another student's strategy. In this way, all strategies are valued, and students tend to prefer the most efficient strategy they understand. Students typically move from direct-modeling strategies to invented strategies using counting or recalled facts, and eventually, to standard algorithms. In an environment where understanding is valued, few students will use a strategy they do not understand, regardless of its efficiency. Trouble arises when students perceive the use of standard algorithms as being the only strategy the teacher will accept.

Action Research Idea

Think of a computational procedure that your students will learn later this year, or in a later grade. Examples include multiplication or division in grades K–1, subtraction with regrouping in grade 2, multi-digit multiplication or division in grades 3–4, and adding or subtracting integers in grades 5–6. Create a problem that might be solved using that procedure, situated in a context that is familiar to your students. Pose the problem, and encourage the students to find at least two ways to solve it. Make available various manipulative materials and writing materials.

Circulate among the students as they work, noting which are using direct-modeling strategies with manipulatives, which are using mental or written invented strategies, and which are using a standard algorithm. Assess the students' understanding of their strategies by asking them to explain the reasons for each step in their solutions. Then have students present their various solutions to the class. Only students who understand their solution may explain it to the class; this restriction is especially important for students using a standard algorithm. Invite other students to discuss and evaluate each strategy. Try to avoid indicating that you value one strategy more than another.

At spaced intervals over the next several months, pose a similar problem. By attending to the strategies used by individual students, determine whether each student's understanding increases and whether students exhibit a tendency to progress naturally toward more efficient strategies.

References

Carpenter, Thomas P., Ellen Ansell, Megan L. Franke, Elizabeth Fennema, and Linda Weisbeck. "Models of Problem Solving: A Study of Kindergarten Children's Problem-Solving Processes." *Journal for Research in Mathematics Education* 24 (November 1993): 427–40.

Kouba, Vicky L., Thomas P. Carpenter, and Jane O. Swafford. "Numbers and Operations." In *Results from the Fourth Mathematics Assessment of the National Assessment of Educational Progress,* edited by Mary Montgomery Lindquist, pp. 64–93. Reston, Va.: National Council of Teachers of Mathematics, 1989.

Editor's Note: The examples in this chapter were transcribed from videotaped interactions between teachers and students. Most are available as part of the Cognitively Guided Instruction Project videotape series produced by the Wisconsin Center for Education Research, © 1995. Some episodes have been edited for presentation here.

Invented Strategies Can Develop Meaningful Mathematical Procedures

William M. Carroll and Denise Porter

Over the past decade, a growing consensus among educators favors a shift in mathematics instruction from a curriculum in which children learn and practice the standard school algorithms to one in which reasoning, problem solving, and conceptual understanding play a major role. For example, the *Curriculum and Evaluation Standards for School Mathematics* (NCTM 1989, p. 44) states the following:

> Strong evidence suggests that conceptual approaches to computation instruction result in good achievement, good retention, and a reduction in the amount of time children need to master computational skills. Furthermore, many of the errors children typically make are less prevalent.

One way to improve the understanding of numbers and operations is to encourage children to develop computational procedures that are meaningful to them. Our experience observing classrooms where this approach is used suggests that most primary students are capable of developing their own accurate solution procedures for multidigit addition and subtraction as well as for simple multiplication and division. Although these procedures are used flexibly by the students, they are sometimes called "invented algorithms."

This article is based on interviews with, and observations of, students in a mathematics curriculum that encourages the invention of algorithms. Interviews were conducted throughout the year in ten fourth-grade classes and three second-grade classes, and classrooms were observed in a number of first, second, and fourth grades in urban, suburban, and rural school districts. Additionally, interviews with teachers using this curriculum provided important suggestions about how they used this approach as well as some difficulties they encountered. In addition to making mathematics more meaningful for children, many teachers reported that they found teaching to be more exciting when more emphasis is placed on discovering and sharing procedures than on memorizing and practicing traditional algorithms. A number of teachers reported gaining a new understanding of operations and problem solving. This article explores reasons why invented procedures should enhance both number sense and accuracy. It then suggests some ways to support this activity in the classroom. The purpose is not to suggest that algorithm invention is the only way to promote students' understanding or that children should be discouraged from using a standard algorithm if that is their choice.

Why Invented Procedures Promote Understanding

Invented procedures promote the idea of mathematics as a meaningful activity. Although most of us were taught the standard written algorithms as being the correct way to compute, these methods are fairly recent inventions, having been introduced in Europe around 1000 a.d. and not accepted until around the fifteenth century after a certain amount of controversy (Burton 1991). Although they are efficient techniques, the reasons they work are often unclear. For example, think about the standard subtraction algorithm. What does it mean to "borrow 1 from the 11" and bring the "1" over to the 2 (to make it 12) in the calculation that follows:

$$\begin{array}{r} 112 \\ -89 \\ \hline \end{array}$$

Most of us learned this procedure because it produced the correct answer—or at least we were assured that it did—rather than because it made sense to us. However, an alternative subtraction procedure that children often use is "Add 1 to 89 (to make 90), and

---

The research described in this article was supported in part by a grant from the UCSMP Fund for Research in Mathematics education.

10 to 90 (to make 100) and 12 more (to make 112) makes 23."

$$+1 \quad +10 \quad +12$$
$$89 \text{ - - - } 90 \text{ - - - } 100 \text{ - - - } 112$$

This procedure not only works well but also indicates an understanding of the size of the interval between the two numbers as well as the relationship between them, for example, realizing that this particular problem can be easily solved by "adding up." In fact, many of us use the adding-up method when computing change on purchases because adding is often easier and less prone to error than subtraction.

Another method that some children used to solve this problem is mentally to break the problem into two or more easier problems, for example,

$$112 - 89 = (100 - 89) + 12 = 11 + 12 = 23$$
$$\text{or}$$
$$112 - 89 = (100 - 90) + 12 + 1 = 10 + 13 = 23.$$

By encouraging children to invent and use their own procedures, teachers allow them to use a method that makes them focus not simply on practicing computation but also on developing strategies for which computational approach to use.

Different problems are best solved by different methods. The power of standard written algorithms is that they can be applied to all problems. However, they are not always the best method. For example, consider the problem 7000 – 2. In a study conducted during the late 1970s, none of the above-average third or fourth graders tested reached the correct solution because they experienced difficulties in applying the standard subtraction algorithm on such a problem (Fuson 1992a). However, during interviews of beginning fourth graders in ten classrooms in a curriculum that encourages algorithm invention, we found that 67 percent of these children correctly solved the problem in the fall simply by counting down or using a similar method. By the spring, 80 percent solved it correctly. We think that these percents resulted from children's having had the opportunity to invent and to explore various algorithms and strategies in earlier grades and thus to develop more flexibility in their choice of solutions. Our interviews with students verified our conclusion. For example, many children solved 7000 – 25 by counting down by tens and fives but solved 41 – 25 by adding up, or decomposing, the numbers. In other words, these children seemed to have more computational tools available and were more apt to choose one that matched the problem.

Students' natural tendencies do not fit the standard algorithms. A number of studies have found that for addition and subtraction problems, students are more likely to move from left to right (hundreds to tens to units) rather than from right to left as in the standard school algorithms (Madrell 1985; Kamii and Lewis 1993; Kamii, Lewis, and Livingston 1993). For example, in solving 16 + 35 = ——, we have found that many second graders who have not been taught the traditional algorithms explain, as one student did, that "10 and 30 is 40; 6 more is 46, and 5 is 51." This method seems simpler and more meaningful for many children who regroup mentally with fluency.

Notice that this method also reflects and promotes an awareness of place value, saying ten and thirty rather than one and three and not referring to the 10 in 12 as a "1" to be carried. In adding from left to right, the student is also focusing on the tens, which have more bearing on the magnitude of the answer than do the units, and students are less likely to make large computational errors (see, e.g., Kamii, Lewis, and Livingston [1993]). Our interviews also found that many children prefer other methods to either left-to-right-column or right-to-left-column addition and subtraction. Instead, children may decompose numbers to make easier problems or may count up by hundreds, tens, and units. For example, to add 52 + 45, a student may count, "52, 62, 72, 82, 92, plus 5 is 97." Many teachers have noted that these students are much more capable of manipulating numbers mentally than students in the past who had worked primarily with standard algorithms.

Ways to Encourage Algorithm Invention

Although encouraging children to invent their own procedures may sound appealing, teachers who are accustomed to teaching traditional methods are sometimes uncertain how to begin. We observed teachers using several successful techniques. Although this list is not complete, we suggest five items that can help teachers get started.

1. *Allow students time to explore their own methods.* The most important first step is to allow primary children time to explore problems and their own solutions in low-stress situations. Using small

groups also seems to be very successful in involving all children in extended tasks. Successful lessons we have observed often seem to follow a similar pattern: a problem is posed or children are asked to devise their own problem; children try to find some way of solving it, either alone or in groups; and finally the class is brought together and students discuss different methods they use, being allowed the opportunity to present their own method and learn alternative approaches. Rather than solve twenty problems in five minutes, children are often more successfully engaged by larger problems that take more time but are rich in context.

One factor to keep in mind is that some children may not develop successful procedures, especially for multidigit subtraction or multiplication. The teacher may wish to pair a child with another or suggest an algorithm with which the teacher thinks the child will be successful. Also important to remember is that not all children will be at the same levels of computation: some may begin by using more sophisticated methods, like skip counting by adding hundreds, tens, and then units or by regrouping numbers; other children may rely on simple counting for a longer period.

2. *Have manipulatives available to support children's thinking.* Children begin school capable of solving number problems using objects to model the situations long before they have memorized facts or learned to use written symbols (Fuson 1992b; Carpenter and Moser 1983). Encouraging this modeling helps both to build an understanding of the situation, thus leading to the appropriate solution, as well as to help prevent errors. Without using manipulatives, many second graders would have trouble with the following comparison problem: "A giraffe is 19 feet tall and a polar bear is 11 feet tall. How much taller is the giraffe?" However, many kindergartners can represent the problem by using linking cubes or blocks to make a stack of nineteen cubes and a stack of eleven cubes, then compare the two stacks. For many children, the difficulty in solving story problems is often not the arithmetic but the representation of the problem situation. Manipulatives can help to embody the problem situation and suggest a solution.

A second important reason for using manipulatives is to assist children in choosing correct procedures. When counters are available, fewer children will make the typical subtraction error $21 - 19 = 18$. Keep in mind that manipulatives, like calculators, are a tool and that some children may need help using them optimally, whereas others will prefer not to use them but instead to use mental arithmetic.

3. *Have children build fact strategies as well as fact knowledge.* If children are to be accurate when solving multidigit problems, they need to have a good grasp of the basic facts. But children also use counting, doubles facts, and other derived-fact strategies to develop automaticity. In fact, at times many adults continue to use these good strategies, for example, $9 \times 12 = 9 \times 10 + 18$, along with memorized facts.

Fluency with derived facts, especially the use of facts around ten, may be important in helping children develop alternative multidigit procedures. For example, 12 percent of the second graders we interviewed solved the problem $15 - 9 = \square$ by breaking 9 into 5 and 4, getting $15 - 9 = 15 - (5 + 4) = (15 - 5) - 4$, thereby reducing the problem to a simpler subtraction from 10. These answers were so rapid that only through asking children to explain their solution processes did the interviewers learn that the children had not used a memorized fact. The higher-achieving students were more apt to use this facts-around-ten strategy because although they probably knew the fact $9 + 6 = 15$, they found the decomposition approach to be as easy. These children often then extended these decomposition strategies to multidigit problems, such as $41 - 25 = 41 - (20 + 5) = (41 - 20) - 5$. Although fact memorization is important, the beginning of meaningful multidigit procedures for many students may lie in being allowed to explore fact relationships by counting and using derived facts.

4. *Present problems in meaningful contexts.* Many teachers we have seen successfully using algorithm invention typically present problems in contexts that are meaningful to children: the price of toys or candy, the sizes of animals, the distances to favorite places, or the differences between temperatures. Problems posed in meaningful contexts can motivate children and help them to see that mathematics is more than symbol manipulation. Rather, mathematics and problem solving are applied to real situations. In addition to exploring teacher-

created problems, children can make up their own interesting stories and pose these problems to one another. For example, in one second-grade lesson we observed, children were presented with pictures of different animals and their sizes and asked to make up and then to solve a number story using the pictures and heights. During this time, children worked in small groups to record their stories and solutions.

Following this group time, the children were asked to share their stories and solution strategies with the class. Other children then offered alternative methods for solving the same story problem, and the teacher recorded and shared these methods using an overhead projector. This discussion allowed children the opportunity to observe alternative methods that might be easier than their own as well as to correct errors they may have made.

Although the general trend in school is to first present symbolic problems and later story problems, we think that the reverse is often more efficient. Context often builds meaning and constrains errors in the same way that the use of manipulatives does. From kindergarten on, students are intersted in, and capable of, creating and solving simple number stories. The number stories they create can reveal much about their understanding of mathematical situations **(fig. 1)**.

5. *Encourage children to share strategies.* Many teachers have emphasized the importance of children's sharing their strategies both in small groups and in whole-class discussions, allowing children to learn from one another. This scenario may be especially helpful for children who are having difficulty devising their own computational procedures. To be successful, it is important to create a classroom environment in which children feel comfortable taking risks. The following are some ideas that teachers have used to encourage students to explore and to share their strategies:

- Encourage children to share their strategies regularly at the overhead projector or on the chalkboard. Early in the school year, model how to share strategies and procedures.
- Occasionally, have children write a letter to a friend or parent describing their problem-solving strategies, and have them save it in a "math log." **Figure 2** shows a format that one teacher uses to help children keep track of the algorithms they have devised for each operation. These writings can be shared with parents who are unsure of these alternative algorithms or passed on to the teacher the following year.

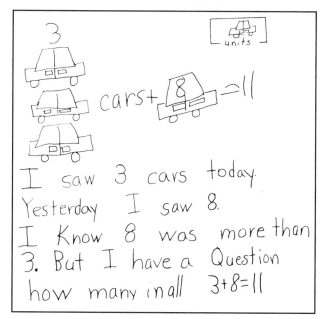

Fig. 1. Number story written by a first grader

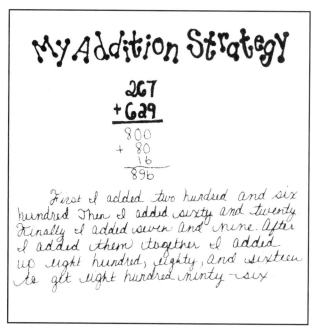

Fig. 2. The format used by a fourth-grade teacher for recording a computation strategy

- Create a "strategy wall" in the classroom to display children's algorithms and strategies. This wall can also be a resource for children who have forgotten or have not developed a procedure.

- Present a rich variety of problem-solving activities, both in context and operation. Include problems that are often neglected, such as comparison stories. Have children create their own problems and produce their own number-story books.

Allowing children time to invent their own algorithms for addition and subtraction initially may appear risky to some teachers, especially if they have not learned mathematics in this manner. Although most children devise accurate procedures, some have difficulty finding a correct or efficient method. Invented algorithms are not always correct. Unfortunately, it is not always clear when to allow a child more time and when to intervene, especially for a teacher used to presenting one algorithm. Parents and other teachers may also sometimes place pressure on a teacher to instruct children in the "correct way" to add and subtract. However, most teachers with whom we have talked learn how to handle these and other questions. The reward of seeing their students make sense of mathematical situations and the resulting appreciation of children's thinking and capabilities more than make up for the difficulties.

References

Burton, David M. *The History of Mathematics: An Introduction.* Dubuque, Iowa: William C. Brown Publishers, 1991.

Carpenter, Tom, and James Moser. "The Acquisition of Addition and Subtraction Concepts." In *Acquisition of Mathematical Concepts and Processes,* edited by Richard Lesh and Marsha Landau, pp. 7–44. New York: Academic Press, 1983.

Fuson, Karen C. "Research on Learning and Teaching Addition and Subtraction of Whole Numbers. In *Analysis of Arithmetic for Mathematics Teaching,* edited by Gaea Leinhardt, Ralph Putnam, and Rosemary A. Hattrup, pp. 53–187. Hillsdale, N.J.: Lawrence Erlbaum Associates, 1992a.

———. "Research on Whole Number Addition and Subtraction." In *Handbook of Research on Mathematics Teaching and Learning,* edited by Douglas A. Grouws, pp. 243–75. New York: MacMillan Publishing, and Reston, Va.: National Council of Teachers of Mathematics, 1992b.

Kamii, Constance, Barbara A. Lewis, and Sally Jones Livingston. "Primary Arithmetic: Children Inventing their Own Procedures." *Arithmetic Teacher* 41 (December 1993): 200–203.

Kamii, Constance, and Barbara A. Lewis. "Viewpoint: The Harmful Effects of Algorithms in Primary Arithmetic." *Teaching Pre-K–8* 23 (January 1993): 36–43.

Madrell, Rob. "Children's Natural Processes." *Arithmetic Teacher* 32 (March 1985): 20–22.

National Council of Teachers of Mathematics (NCTM). *Curriculum and Evaluation Standards for School Mathematics.* Reston, Va.: NCTM, 1989.

Story Problems and Students' Strategies

Larry Sowder

Research . . .

Several researchers have interviewed individual students to find out what the students are thinking as they solve story problems. Entering first graders do seem to "use their heads" as they approach story problems and act out problems in ways consistent with the stories (Carpenter and Moser 1982). Although some encouraging experimental projects in the primary grades show that young pupils will continue to deal with the meaning of a story problem (e.g., Cobb, Yackel, Wood, Wheatley, and Merkel [1988]; Loef, Carey, Carpenter, and Fennema [1988]), the results from our interviews have been discouraging. Most children, even through junior high school, appear to have adopted one or more of the immature strategies in **table 1** (Sowder 1988a, 1988b; Sowder, Threadgill-Sowder, Moyer, and Moyer 1984).

The students who use the first two strategies have a great deal of trouble with story problems, of course, even though by chance they get an occasional answer correct. They are coping the best they can; no doubt story problems are *not* their favorite part of school.

Users of the other strategies are coping somewhat more successfully. Consider this story problem:

> A worker has 78 pounds of candy left to put in boxes. One box holds 3 pounds of candy. How many boxes will the worker fill?

Users of strategy 3 choose division not because they know that 78 ÷ 3 tells how many 3s are in 78, but because, as one said, "That's what I usually think of when I see a big number and a one-digit number. . . . If it's like, 78 and maybe 54, then I'd probably either add or multiply." Fortunately, users of this strategy are not common, but the strategy can be moderately successful at certain points in textbooks, when a par-

---

Many of the interviews described here were conducted with the support of the National Science Foundation under grants SED 8108134, MDR 8550169, and MDR 8696130. Any interpretations or recommendations given here are not necessarily those of the foundation.

Table 1
Immature Stategies Commonly Used by Students

Coping strategies
1. Find the numbers and add.
2. Guess at the operation to use.

Immature, computation-driven strategies
3. Look at the sizes of the numbers; they "tell" you what to try.
4. Try all the operations and choose the most reasonable answer.

Other stategies of limited usefulness
5. Look for isolated key words or phrases that signal the operation.
6. Decide whether the answer should be larger or smaller than the given number. If larger, try addition and multiplication and choose the more reasonable answer. If smaller, make similar computations with subtraction and division.

---

ticular kind of computation is being emphasized.

Strategy 4 is also not a commonly used strategy, but our most striking example of a student using the strategy was a seventh-grader *in the school's gifted program!* Her computational speed made the strategy effective for her on one-step problems. She had great difficulty with the multistep problems in the interview, since her method involved first writing down each occurrence of each number in the problem statement, including those for irrelevant data, and then trying various combinations of operations to get a number that she thought was about the right size.

Here is an excerpt from an interview with a sixth grader who relied on key words (strategy 5); the problem could be solved by multiplying 12 by 36, but the words "in all" appeared in the question: ". . . so you plus 'em up." (He writes 12 + 36.) How do you know? "In all." The key-word strategy is sometimes taught by teachers who are not aware of its limitations or are at their wit's end in looking for some way to help the students with story problems. Indeed, the teachers' *intent* is good: help the students think about what is going on in the situation, and relate that context to the operations. As is well known, however, the strategy often leads to a superficial skimming solely for key words without thought for the total situation; in this problem the student saw "in all" and immediately chose addition without further thought. As with the other immature strategies, the use of isolated key words sometimes gives correct answers, so students may come to adopt the strategy.

Strategy 6 is unfortunately effective on one-step story problems with whole numbers that are familiar in size. Here is a sample of the reasoning: "Well, I would think that you have to subtract, because, er, it'd either be a subtract or, um, division, and the one that sounds right would be the subtraction." The effectiveness of this strategy is unfortunate because (1) strategy 6 is difficult to carry out with multistep problems, problems with large whole numbers, or fractions or decimals for which the student often has little number sense and (2) strategy 6 is short-lived in its reliability: it breaks down when multipliers less than one are involved. For example, consider the problem, "Some cheese costs $2.55 for a pound. Mrs. Duarte bought 0.85 pound. How much did she pay?" Many middle schoolers correctly reason that 0.85 is less than a pound, so under the influence of strategy 6, they consider only subtraction and division. In research studies *they used this reasoning even though the problem immediately before this one involved a whole number of pounds of cheese, and they correctly decided to multiply in that problem!* (The literature on this phenomenon is extensive; see, e.g., Greer [1987].) The many correct answers that can result from the use of strategy 6 conceal from the user its inappropriateness and may hide the narrow nature of the student's thinking from the teacher.

. . . Into Practice

Although some of the immature strategies do involve the students' considering whether an answer is reasonable, it is disappointing that a meaning-based explanation was so rare, even among students who were judged to have high ability in mathematics. One ready explanation is that until recently, textbooks did not offer much exposition on the uses of the operations but depended on the story problems to develop an ability to apply the operations. Thus, students may have adopted the immature strategies by default. Many current texts do give attention to links between settings and operations, as in "to compare two amounts you can subtract" or "1260 ÷ 3.5 tells you how many 3.5s are in 1260." As textbooks and teachers give more attention to these links, it will be interesting to find out whether students will use more reliable strategies in solving story problems.

Nevertheless, how the students arrive at their answers to story problems, not just whether they are getting correct answers, must receive attention. The following are reasonable ways to proceed.

- Routinely ask students for reasons for their choice of operations. "I used division because the 12.8 was to be put into four equal parts" rather than "I divided because I wanted a smaller number" gives valuable positive information that is not revealed by a correct numerical solution to a problem. Including questions about reasons on quizzes also communicates the importance of having acceptable reasons for choices of operations.

- Model such reasoning in your explanations to show the students how to decide on appropriate operations. It may be useful to have a chart giving such descriptions as "When you put known amounts together, addition tells you the total amount," particularly if the phrasing used in the chart varies from that of the textbook so that the students do not absorb the language merely as key words. Talton (1988) gives an example of language that can be used.

- Use more multistep problems and problems with attractive irrelevant information. Such problems make the immature strategies much less effective, since they are relatively unreliable for such problems.

- Some evidence indicates that work in small groups is helpful. Peers can discourage the immature strategies and sometimes illustrate better methods (Noddings 1985). Anecdotal reports also suggest that small-group work with nonroutine problems "liberates" students who previously had not called on their own thinking but relied on immature methods to approach problems.

References

Carpenter, Thomas P., and James Moser. "The Development of Addition and Subtraction Problem-Solving Skills." In *Addition and Subtraction: A Cognitive Perspective,* edited by Thomas Carpenter, James Moser, and Thomas Romberg. Hillsdale, N.J.: Lawrence Erlbaum Associates, 1982.

Cobb, Paul, Erna Yackel, Terry Wood, Grayson Wheatley, and Graceann Merkel. "Creating a Problem-Solving Atmosphere." *Arithmetic Teacher* 36 (September 1988):46–47.

Greer, Brian. "Nonconservation of Multiplication and Division Involving Decimals." *Journal for Research in Mathematics Education* 18 (January 1987):37–45.

Loef, Megan M., Deborah A. Carey, Thomas P. Carpenter, and Elizabeth Fennema. "Integrating Assessment and Instruction." *Arithmetic Teacher* 36 (November 1988):53–55.

Noddings, Nel. "Small Groups as a Setting for Research on Mathematical Problem Solving." In *Teaching and Learning Mathematical Problem Solving: Multiple Research Perspectives,* edited by Edward Silver. Hillsdale, N.J.: Lawrence Erlbaum Associates, 1985.

Sowder, Larry. "Children's Solutions of Story Problems." *Journal of Mathematical Behavior* 7 (1988a): 227–38.

———. *Concept-driven Strategies for Story Problems in Mathematics.* Final report, Grants MDR 8550169 and 8696130. Washington, D.C.: National Science Foundation, 1988a. (ERIC Document Reproduction Service no. ED 290 629)

Sowder, Larry, Judith Threadgill-Sowder, John Moyer, and Margaret Moyer. *Format Variables and Learner Characteristics in Mathematical Problem Solving.* Final report, Grant SED 8108134. Washington, D.C.: National Science Foundation, 1984. (ERIC Document Reproduction Service no. ED 238 735).

Talton, Carolyn F. "Let's Solve the Problem Before We Find the Answer." *Arithmetic Teacher* 36 (September 1988):40–45.

Connecting Instructional Practice to Student Thinking

Patricia F. Campbell

Over the last fifteen years, much research has investigated children's learning of mathematics. This research indicates that when permitted, children frequently devise approaches to solve problems that are distinct from those typically used by adults. The approaches and strategies used by children when solving mathematics problems are rational and reasoned when interpreted in terms of the children's way of thinking and in terms of the children's current level of understanding (Campbell and Johnson 1995). Research indicates that instructional practice that supports and builds on children's thinking is meaningful to the children and will foster their mathematical understanding (Fennema et al. 1993; Hiebert and Wearne 1993).

However, as any classroom teacher can attest, children's initial mathematical responses do not always meet society's standards of mathematical practice. That is, the solution strategies and answers offered by children may be reasoned when interpreted from the perspective of a child's thinking, but sometimes during instruction those strategies and answers might be incorrect, illogical, incomplete, or inefficient when examined in terms of adult society's perception of mathematics. The challenge that constantly faces teachers is how to use children's initial understandings to proactively support their *continued* mathematical development. Instruction must build on children's existing ideas, so that the children will construct progressively more advanced understandings and simultaneously perceive mathematics as "making sense." It is important for each child to confidently think, reason, and explain mathematically. Eventually this understanding should be consistent with established standards for mathematics. Therefore, instructional practice cannot proceed without considering both mathematical content and children's current understandings.

Traditional mathematics teaching has addressed this challenge by focusing on teacher demonstration and explanation, perhaps even using manipulatives to emphasize intended meaning, as children watch, listen, and practice. This method generally emphasizes adult ways of doing mathematics. However, instruction that is defined primarily in terms of the teacher's actions and assigned student practice ignores an essential component. What do the children think? Are the children's ways of addressing the mathematics just as accurate and perhaps more meaningful to them? Are the children's ways mathematically incorrect or inefficient?

Teachers must also ask, *Why* do the children think the way they do? Unless students' thinking is considered, teachers will make instructional decisions without crucial information that could offer important insight. With this idea in mind, the National Science Foundation–funded Project IMPACT, Increasing the Mathematical Power of All Children and Teachers, implemented and evaluated a reform model for elementary mathematics instruction in predominantly minority, urban schools. The results of Project IMPACT reinforce the notion that as teachers come to think more deeply about how children understand and "construct" mathematical meanings, teachers can make instructional decisions and organize their classrooms in ways that support and encourage more meaningful mathematics learning, resulting in significant gains in student achievement. Further, teaching for understanding yields growth for children at all ability levels. The following examples of children's thinking are from Project IMPACT.

Building upon Emerging Understandings

Place value is an essential mathematical concept in the primary grades, and substantial research addresses children's emerging understanding of place value. In particular, Steffe and his colleagues (Steffe, Cobb,

---

The research reported in this article was supported by the National Science Foundation under grant numbers MDR 8954652 and ESI 9454187. The opinions, conclusions, or recommendations expressed in these materials are those of the author and do not necessarily reflect the views of the National Science Foundation.

and von Glasersfeld 1988) have considered what it means for children's place-value understandings to become more sophisticated. This research maintains that children's place-value ideas gradually develop from *(a)* perceiving numbers in terms of a unit of one and counting by ones or counting on by ones; to *(b)* counting by groups or counting on by groups, such as groups of twos, fives, or tens; to *(c)* working separately with the tens and ones that compose numbers; to *(d)* finally coordinating and flexibly adjusting between tens and ones within numbers.

After considering the complexity involved in understanding place-value ideas, many of the teachers in Project IMPACT decided to first support children's counting and ways of representing two-digit numbers. Subsequently, they would use word problems and problematic investigations as settings for having the children examine how two-digit numbers might be composed, decomposed, and related. Finally, as the children standardized their approaches, first for adding and then for subtracting with regrouping, written symbolic representations would be expected. The teachers anticipated that student-invented approaches for solving problems involving regrouping could differ; consequently, written representations for those approaches would also differ.

The teachers did not simply present problems and wait for a miracle to occur. Research results told them how children's counting and thinking about two-digit numbers might change over time. The district mathematics curriculum contained specific objectives for place value and regrouping that implied that the children's algorithms had to be mathematically correct, efficient, and generalizable, but the children were not required to use a particular algorithm. The challenge facing teachers was to listen to their students' thinking and to interpret their explanations. How did individual children solve problems involving two-digit numbers? How far had a child's place-value ideas progressed? Were the children moving toward the intended mathematics objectives? What was missing? What was errant? What problem or activity or question could be used to probe understanding further, to increase abstraction, or to cause a reexamination of an existing idea? The following examples illustrate this crucial interplay between teachers' instructional decisions and children's thinking.

Desmond's second-grade class had been using Unifix cubes, base-ten blocks, and hundreds boards to represent two-digit numbers and to characterize both number patterns and numerical relationships across two-digit numbers. They had not yet investigated written algorithms for addition with regrouping. When presented with the problem

$$25 + 37 = \underline{},$$

Desmond offered the explanation shown in **figure 1**.

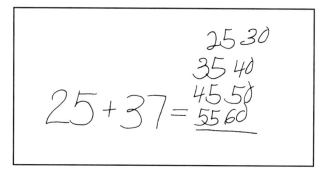

Fig. 1. Desmond began by counting on by fives.

Desmond: [Writing as he speaks] 25, 30. Let's see. 35, 40, 45. Let's see. 50. I'm gonna. . . . [Taps with his pencil on each written value from 30 to 50, counting silently. Then he continues to speak and write.] 55, 60. [See **fig. 1.**] Okay, and plus that 7.

Teacher: Plus 7?

Desmond: Sí. It's going to be 62. 'Cause I already took . . . I took 5 from 7 to make it 35.

Teacher: Uh huh.

Desmond: Then it was 60. So I add the 2, 62. [Rewrites the problem vertically and horizontally, recording the solution of 62. See **fig. 2.**]

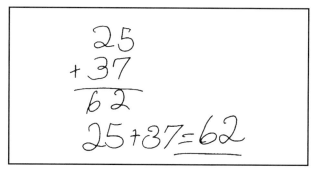

Fig. 2. Desmond rewrote the problem vertically and horizontally.

Desmond's strategy indicates that his interpretation of this problem involves counting up thirty-seven more from twenty-five. However, Desmond did not count up from twenty-five by ones. Rather he clustered or counted on by fives until a total of thirty-five (seven counts of five) had been added. This process then left two more to be added (" 'Cause I already took . . . I took 5 from 7. . . .").

Reflecting on the implications of place-value research, Desmond's teacher did not demand that her students begin by using groups of ten, and she did not expect regrouping initially. Indeed, when the idea of making groups to count a large collection of items was first suggested by a child, the class hypothesized that the ideal group size would be seven! Desmond had moved beyond groups of seven at the time of this interview. However, he was neither regrouping nor using a unit of ten, even though the base-ten blocks were available in his classroom. At the time of this interview, Desmond's teacher was considering removing the Unifix cubes from the classroom to encourage grouping by tens. Many of the children had stopped displaying or counting by ones with the Unifix cubes, perceiving that approach as too slow and too prone to error. The teacher noted that when solving addition-with-regrouping problems presented either as horizontal-computation problems or within word problems, some of the children in the class were beginning to work with tens and ones separately. These children determined the total number of tens and then adjusted for the remaining ones. The teacher indicated that she would begin to focus her "Why?" and "How?" questioning to highlight these approaches so that more children would consider more closely explanations such as Corrine's.

Corrine: That's 50 [pointing to the 2 and the 3 in the numbers 25 and 37]; 57 [pointing to the 7 in 37]; 58, 59, 60, 61, 62.

In these two episodes, the children's explanations were mathematically correct. The teacher recognized that the place-value understandings of the children in her class were not yet complete but were advancing steadily. Her instructional decisions in terms of problem definition, availability of materials, and questioning were aimed at supporting and stimulating that development.

Confronting Mathematical Misconceptions

Sometimes children's explanations are not mathematically correct. Indeed, student explanations may reveal either a total lack of understanding or perceptions that contradict accepted mathematical practice. Teachers in Project IMPACT noted that when first confronted with incorrect explanations, they explicitly directed the children's thinking. However, when the children were subsequently asked to solve other similar problems, the teachers noted that incomplete explanations were frequent and misconceptions were often persistent. For example, consider this problem:

> The Yummy Frozen Yogurt Company sells six kinds of frozen yogurt and four kinds of sprinkles. A yogurt cup has one kind of yogurt and one kind of sprinkles. How many different kinds of yogurt cups could the company make?

In an interview setting, Sheena expressed a misconception common among fourth-grade children.

Sheena: [Reads the problem three times, then rereads the final question.] "How many different kinds of yogurt cups could the company make?" Six kinds of frozen yogurt and four kinds of sprinkles. Okay, here's one cup [draws a U-shaped figure as shown in **fig. 3**] another cup [draws a second U-shaped figure]. [Silently draws four more figures to represent the yogurt cups.] Okay, four kinds of sprinkles. [Reading again.] "A yogurt cup has one kind of yogurt and one kind of sprinkles." Oh, "How many different kinds of yogurt cups could the

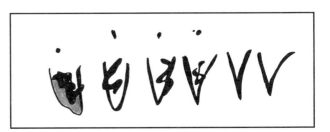

Fig. 3. Sheena's misconception

company make?" Oh, only four kinds of yogurt and sprinkles.

Teacher: So what's your answer?

Sheena: Four.

Teacher: Okay. And, and would you show me why? You've got your six kinds of yogurt here [indicating the drawing], and what happens?

Sheena: No. I was gonna do something like.... Since they said right here, "A yogurt cup has one kind of yogurt and one kind of sprinkles," [indicating where to read in the text], I was gonna make one. And this is the yogurt [makes squiggle marks in the first cup] and sprinkles [draws a dot above the first cup]. That, that means like a lot of sprinkles [indicating the dot]. This is one [indicating the first squiggle cup with a dot] And the yogurt [makes squiggle marks in the second cup] and this [draws a dot above the second cup] and that and this, and that and this [drawing squiggles in two more cups with a dot above each cup]. But it says that "[a] yogurt cup has one kind of yogurt and one kind of sprinkles." And it's only four kinds of sprinkles, so I just said four.

Sheena's teacher knew that simply telling Sheena to change her drawing would not be sufficient. Sheena still believed that only enough toppings were available for four frozen-yogurt cups because only four kinds of sprinkles were involved. A team of fourth-grade teachers decided to address the common misconception expressed by Sheena by having their classes actually make frozen-yogurt cups. The children readily saw that each kind of sprinkles could be associated with more than one kind of frozen yogurt. Then the teachers had their classes investigate how to draw figures or tables to represent all possible choices because, as one child said, "You don't get to make real frozen-yogurt cups everyday!" Eventually, after solving a number of problems of this type with continued questioning and discussion, the children saw a pattern and made a connection with the meaning of multiplication.

The Challenge of Instruction

Instructors who expect all students to explain and justify their strategies or solutions do not presume that every response that a child offers will be correct, complete, or easily interpreted. Deviating from the traditions of mathematics teaching does not mean abandoning standards for what is mathematically correct, generalizable, or efficient. Teachers constantly make instructional decisions. The challenge is to listen to children's thinking and to consider both the children's existing understanding and society's mathematical expectations.

At the same time, teachers must ensure that all children are participating in classroom discussions and investigations. Some children are eager to reveal their ideas, whereas others are reluctant. All children must be expected to contribute and to explain. When children offer their strategies, differences will be evident in their levels of sophistication or mathematical rigor. The challenge is to promote each child's mathematical growth and to maintain high expectations without devaluing any child's current level of understanding.

Action Research Idea 1

Put a small tape recorder in a pocket. Record a typical mathematics lesson. When listening to the recording, ask yourself the following:

- Do I ask the children to explain how they solved a problem without first telling them whether their approach is correct or incorrect?

- Do I ask follow-up questions to help me interpret their thinking or to emphasize or elaborate on ideas?

- If a child explains an incorrect answer and does not self-correct, do I ask a question to encourage re-examination of some part of the incorrect explanation, or do I generally move on, calling on someone else to explain another method?

- Do I ask the children to explain why they used a particular strategy or to justify their approaches or solutions?

- Do I seek different methods and ask the children to compare or contrast approaches to generalize key ideas?

- Do I expect all children to solve problems in the same way? When I ask for the children's ideas about how to solve a problem, am I really only waiting until I get the answer I want so that I can model one correct way?
- Do I give enough time for the children to think before I expect an answer? How long, in seconds, do I really wait?
- Do I encourage the children to question one another?

Choose one or two of these behaviors for your focus. Try to reflect on these patterns in your teaching over the next six weeks. Then record another lesson. Compare the two tapes. What changes do you notice?

Action Research Idea 2

Make a seating chart for your class with a box around each child's name. For two consecutive mathematics classes, carry the chart on a clipboard and place an l in a child's box whenever that child is called on to answer a factual or low-level question or to demonstrate a previously developed procedure. Place an h in a box whenever that child answers a high-level question, shares a hypothesis or possible approach, explains a strategy, or justifies a solution. Place a q in a child's box whenever that child asks a question, either of you or of a peer, during whole-class discussion.

- Analyze your record by making a table that lists all the girls' names and then lists all the boys' names. For each child, total the number of ls, hs, and qs. Then total the number of ls, hs, and qs for the boys and for the girls.
- Make another table that groups your class by perceived ability. List all the names of the children that you would say are in the top third of your mathematics class, then the names for the middle third, and the names for the bottom third. Note the number of ls, hs, and qs for each name. Then total the ls, hs, and qs for each grouping.
- If your class has children of differing ethnic groups, make another table that groups the children's names according to ethnicity. As before, record the number of ls, hs, and qs for each name. Then total the number of ls, hs, and qs within groups.

Look at each of these three tables separately. Were the totals of ls, hs, and qs the same for each group in a table? Did one group in a table have more ls, hs, or qs? Did one group in a table have fewer ls, hs, or qs? What does that result tell you? Are you taking steps in your questioning to be sure that each student has a real chance to become thoughtfully engaged, to be challenged, and to learn?

Next take the table grouped by gender, and total the number of ls, hs, and qs in the class. Consider the pattern of your questions. Are more ls or more hs evident? What does that occurrence mean? How does the total of qs compare? If instructional-communication patterns emerge that you would like to address, focus on them for three weeks. Then rechart your class for two consecutive mathematics periods. Compare your charts. What changes do you notice?

Bibliography

Campbell, Patricia F., and Martin L. Johnson. "How Primary Students Think and Learn." In *Seventy-five Years of Progress: Prospects for School Mathematics*, edited by Iris M. Carl, pp. 21–42. Reston, Va.: National Council of Teachers of Mathematics, 1995.

Fennema, Elizabeth, Megan L. Franke, Thomas P. Carpenter, and Deborah A. Carey. "Using Children's Mathematical Knowledge in Instruction." *American Educational Research Journal* 30 (fall 1993): 555–83.

Hiebert, James, and Diana Wearne. "Instructional Tasks, Classroom Discourse, and Students' Learning in Second-Grade Arithmetic." *American Educational Research Journal* 30 (summer 1993): 393–425.

Steffe, Leslie P., Paul Cobb, and Ernst von Glasersfeld. *Construction of Arithmetical Meanings and Strategies*. New York: Springer-Verlag, 1988.

Children's Understanding of Zero and Infinity

Margariete Montague Wheeler

> 8, 80, and 800 are all the same. Zero doesn't make any difference.
> (Davis and Greenstein 1973, 77)

> [Zero] isn't really a number... it is just nothing.
> (Reys and Grouws 1975, 601)

> Zero has no amount; therefore it is not a number.
> (Wheeler and Feghali 1983, 151)

These erroneous responses given by third-grade, eighth-grade, and university students highlight some of the complexities of dealing with zero. Zero concepts are complex, in part, because zero is something—it is a number. If zero is not a number, how might exercises such as $8 - 8 = \square$, $0 + 8 = \square$, $0 \times 8 = \square$, or $0 \div 8 = \square$ be explained and solved? Without a deepening understanding of zero and the operations of arithmetic, how might intermediate-grade students meaningfully internalize that for all whole numbers N different from zero, $N + 0 = N$ and $N - 0 = N$, whereas $0 \times N = 0$ and $0 \div N = 0$?

Zero answers the same question any counting number answers: How many? India gave the world zero and with it the last of the ten digits used today. But it was not until the 1800s that these symbols were completely accepted throughout Europe. Even in this century, the place of zero in the elementary school mathematics curriculum is controversial.

1937: "Through persistent effort, the children finally learn to set down the approved answers when the zero is used. Since the actual use and meaning of the zero are neglected, what the children learn to do is necessarily barren of meaning. Their human minds are turned into machines." (Harry G. Wheat, *The Psychology and Teaching of Arithmetic*, p. 80)

1954: "It should be obvious that the teaching of the zero facts and the one facts should be delayed until the pupil has experienced a need for them in the multiplication of two-figured numbers." (Herbert F. Spitzer, *The Teaching of Arithmetic*, p. 149)

1984: "Although the notion of zero will continue to be expanded and developed throughout the study of elementary school mathematics, the role of zero in place value should be experienced early and often." (Robert E. Reys, Marilyn N. Suydam, and Mary M. Lindquist, *Helping Children Learn Mathematics*, p. 75)

Zero difficulties permeate whole-number computation. A survey of computational skills listed the seven most frequent whole-number errors made by students. "Errors with zero for *each* operation" was one of the seven found (Suydam and Dessart 1976). This finding is consistent with Lepore's (1979) work with educable mentally handicapped and learning disabled children.

Division involving zero seems to cause more than ordinary difficulty for students of today as it did for mathematicians of the eighteenth and ninteenth centuries (Romig 1924). This is true when the division problem has zero as the dividend, the divisor, or the quotient (Reys and Grouws 1975) or when the division problem has zero as one or more digits in the dividend, the divisor, or the quotient. An example involves Jennifer, age eight. When asked to solve $8\overline{)4808}$, Jennifer began, "8 doesn't go into 4, so you have to say that 8 goes into 48, 6 times" and later continued, "the 0 doesn't count for anything, so you say '8 goes into 8 once.'" She wrote

$$8\overline{)4808}^{61}$$

(Davis and Greenstein 1973).

In the realm of basic fact division involving zero, third-grade and sixth-grade students (Reys and Grouws 1975) as well as preservice teacher candidates (Wheeler and Feghali 1983) found it very difficult to explain that a unique whole-number solution exists for $0 \div 6$, that no whole-number solution exists for $6 \div 0$, and why an infinite number of solutions appear to exist for $0 \div 0$. Such understandings are preparatory to understanding that $0 \div 6$ is 0, that $6 \div 0$ is undefined, and that $0 \div 0$ is indeterminate.

National Assessment of Educational Progress (NAEP) results indicate lower performance on all decimal exercises involving a zero (Carpenter et al. 1981). About 50 percent of the thirteen-year-olds could change decimals such as 0.2 or 0.76 to fractions, whereas for decimals involving zero (e.g., 0.09), the performance was weaker. Asked to identify the greatest decimal among four choices (0.19, 0.036, 0.195, and 0.2), more students selected the third choice (0.195) than the correct fourth choice, which could be solved by thinking of 0.2 as 0.200. This error pattern persists throughout the high school years and beyond (Grossman 1983).

Infinity, like zero, is a meaningful, precise mathematical concept that has been accepted relatively recently in the history of mathematics. About a century ago, contemporaries Georg Cantor (1845–1918) and Richard Dedekind (1831–1916) independently established infinity as a noncontradictory concept, consistent with all other mathematical concepts.

Preschool and young elementary school children show intuitions of infinity when the questions are incorporated into a game setting or are phrased as "Is there a biggest number?" or "Can you successively halve a line segment?" These understandings seem to cluster in developmental levels (Falk et al. 1986; Gelman 1980; Langford 1974; Piaget and Inhelder 1967), to be related to knowledge of numbers greater than 100 and the ability to add systematically (Evans 1984), and to be sensitive to the conceptual or figural context of the question posed (Fischbein, Tirosh, and Hess 1979). Knowledge of infinite sets and infinite processes interacts with rather stable intuitive notions of infinity and may be resistant to the effects of schooling (Fischbein, Tirosh, and Hess 1979; Langford 1974).

The developmental character of infinity is suggested when three siblings, ages eight, eleven, and thirteen, were questioned about the counting numbers, lines, and halving fractions. The youngest gave evidence of appreciating the infinity of the counting numbers but rejected the possibility that you could draw a line on and on and never come to the end, commenting, "You'd get fed up." The eleven-year-old accepted the unbounded infinity of the natural numbers and a line but would not consider the possibility of halving fractions. Only the 13-year-old related infinity to the counting numbers and to lines and also appreciated the implications that when halving fractions, the process could be continued indefinitely and would approach zero but never quite get there (Hives 1971).

An understanding of infinity includes issues of cardinality. When a pair of finite or infinite sets can be placed in one-to-one correspondence, the sets are said are to have the same cardinality or to be equivalent. Before ninth grade, few students give evidence of being able to deal with conflicts arising from equivalent sets and questions of infinity. When a conflict such as the correspondence between the infinite set of counting numbers and a proper subset such as the evens is posed to preadolescents, the finitist interpretation dominates over an infinitist interpretation. That is, most students will agree the sets to be equivalent yet argue the set of natural numbers to be "bigger" (Fischbein, Tirosh, and Hess 1979; Langford 1974). Infinity explanations are sometimes triggered by external symbolism. Recent interviews with university students showed that students distinguished between the infinity of points in a line and a segment because "the arrows tell you the line is infinite." These same students distinguished between 0.999 . . . and 1 because "the three dots tell you the first number is an infinite decimal."

Instruction in infinity may begin as early as third grade in a geometric setting involving symmetries of a circle or in the fourth grade with the closure property of addition for the whole numbers. More often the context is geometric and includes the properties of circles, rays, lines, and angles. This pattern has curricular implications because Langford (1974) found numerical questions about infinity to be easier than spatial questions, probably because numerical situations have fewer concrete connotations. Concrete distractions include "I cannot mark another midpoint because my pencil isn't fine enough" and "when halving a segment, you cannot reach an endpoint of the segment because you'll always be off by a point."

Our decimal numeration system provides two perspectives of infinity: counting leads to a consideration of numbers that continue to get large without bound, whereas successive partitionings of an interval lead to a consideration of numbers that continue to get smaller and are bounded. Does a hierarchical relationship exist between understanding the decimal system and understanding infinity, or does each concept increasingly contribute to an understanding of the other? Piaget raised these issues, but the questions, along with the instructional implications of these questions, remain unanswered today (Piaget and Inhelder 1967).

More middle school and junior high school students (grades 5–9) have a finitist rather than a nonfinitist or an infinitist point of view when questioned as to successive halvings of a segment (Fischbein, Tirosh, and Hess 1979).

Research into Practice

The research on zero and infinity reported here has focused on the products of children's learning. The evidence can be used to infer something about the effectiveness of the conditions under which the concepts were learned. The evidence should not be used to infer what occurs inside the learner's head. Points of difficulty have been identified to help teachers be more sensitive to students' conceptions and misconceptions. Thus far, the research on students' understanding of zero and infinity has not led to the design of teaching units. However, the following suggestions may help students construct a better understanding of zero and infinity.

- Use zero in such social contexts as street addresses, telephone numbers, and automobile license plates when it arises naturally. Is the school telephone number, 753-0567, verbalized as "seven-five-three, *oh*-five-six-seven" or as "seven-five-three, *zero*-five-six-seven"?
- Can your students explain the difficulties of dividing by zero, or do they claim that "you cannot divide by zero because that is what my teacher said"?
- Increase the attention given to the role of zero in decimal computation when using a calculator or measuring with metric units.
- Do your students explain the equivalence of 0.3 and 0.300 by saying, "It's nothing; you just added some zeros"? The use of physical and pictorial models of decimal equivalents can facilitate students' understanding.
- Introduce infinity with such questions as "Can you name a bigger number?" and "Can you count forever?" with children who have mastered decade counting.
- Videotape "Square One" segments that consider infinity by means of the "and add one" or the additive-successor function to stimulate "Is there a biggest number?" discussions. "Square One" is the Public Broadcasting television series that began January 1987.
- Challenge students to explain the mathematics of Aesop's tale of the hare and the tortoise. Discuss differences among infinity concepts held by members of the class.
- Encourage students from the middle grades on to read the classic volume *From Zero to Infinity* by Constance Reid (1964).
- Discuss infinity in geometrical as well as numerical settings with discussions of the number of lines of symmetry for a circle or the properties of a line.
- Decimal numeration focuses on tenths, hundredths, and thousandths. Raise "what if . . . " questions about successive partitionings of segments of various lengths (lengths such as one centimeter or the width of this page) and different partition-intervals (include halves and eighths as well as tenths) to contribute to a richer understanding of infinity.

Children show considerable understanding about infinity—a topic not in the mainstream of the elementary school mathematics curriculum. They also show considerable misunderstanding about zero—a topic that is in the mainstream of the curriculum. Their understandings (and misunderstandings) are sometimes contradictory (e.g., zero is not a number or a biggest counting number exists) and are remarkably stable across time as shown when university students agree that $0 \div 8$ and $8 \div 0$ have the same quotient or that the limit of 0.999 . . . is never actually equal to 1. Even if the relationship between instruction and understandings is unclear, a goal of mathematics instruction is helping students modify their existing conceptual frameworks. With time they may develop more sophisticated understandings of zero and infinity. Growth in elementary school children's understandings of zero and infinity is needed.

References

Carpenter, Thomas P., Mary Kay Corbitt, Henry S. Kepner, Jr., Mary Montgomery Lindquist; and Robert E. Reys. "Decimals: Results and Implications from National Assessments." *Arithmetic Teacher* 28 (April 1981): 34–37.

Davis, Robert B., and Rhonda Greenstein. "Jennifer." In *Teaching Mathematics: Psychological Foundations,* edited by F. Joe Crosswhite, Jon L. Higgins, Alan R. Osborne, and Richard J. Shumway. Belmont, Calif.: Wadsworth/Jones, 1973.

Evans, Diane Wilkinson. "Understanding Zero and Infinity in the Early School Years." *Dissertation Abstracts International* 44 (January 1984): 2265B. (University Microfilms No. DA8326, 285)

Falk, Ruma, Drorit Gassner, Francoise Benzoor, and Karin Ben-Simon. "How Do Children Cope with the Infinity of Numbers?" In *Proceedings of the Tenth International Conference for Psychology of Mathematics Education*, pp. 13–18 London, 1986.

Fischbein, E., D. Tirosh, and P. Hess. "The Intuition of Infinity." *Educational Studies in Mathematics* 10 (February 1979): 3–40.

Gelman, Rochel. "What Young Children Know about Numbers." *Educational Psychologist* 15 (spring 1980): 54–68.

Gelman, Rochel, and C. R. Gallistell. *The Child's Understanding of Number.* Cambridge, Mass.: Harvard University Press, 1978.

Grossman, Anne S. "Decimal Notation: An Important Research Finding." *Arithmetic Teacher* 30 (May 1983): 32–33.

Hives, B. M. "A Question of Infinity." *Mathematics Teaching* 54 (spring 1971): 6–7.

Langford, P. E. "Development of Concepts of Infinity and Limit in Mathematics." *Archives of Psychology* 42 (1974): 311–22.

Lepore, Angela V. "A Comparison of Computational Errors between Educable Mentally Handicapped and Learning Disability Children." *Focus on Learning Problems in Mathematics* 1 (January 1979): 12–33.

Piaget, Jean, and Barbel Inhelder. *A Child's Conception of Space.* London: Routledge & Kegan Paul, 1967.

Reid, Constance. *From Zero to Infinity.* New York: Thomas Y. Crowell Co., 1964.

Reys, Robert E., and Douglas A. Grouws. "Division Involving Zero: Some Revealing Thoughts from Interviewing Children." *School Science and Mathematics* 75 (November 1975): 593–605.

Reys, Robert E., Marilyn N. Suydam, and Mary M. Lindquist. *Helping Children Learn Mathematics.* Englewood Cliffs, N. J.: Prentice-Hall, 1984.

Romig, H. G. "Early History of Division by Zero." *American Mathematical Monthly* 31 (October 1924): 387–89.

Spitzer, Herbert F. *The Teaching of Arithmetic.* 2d ed. Boston: Houghton Mifflin Co., 1954.

Suydam, Marilyn N., and Donald J. Dessart. *Classroom Ideas from Research on Computational Skills.* Reston, Va.: National Council of Teachers of Mathematics, 1976.

Wheat, Harry G. *The Psychology and Teaching of Arithmetic.* 1937. Reprint. Boston, Mass.: D. C. Heath & Co., 1983.

Wheeler, Margariete Montague, and Issa Feghali. "Much Ado about Nothing: Preservice Elementary School Teachers' Concept of Zero." *Journal for Research in Mathematics Education* 14 (May 1983): 147–55.

Integrating Assessment and Instruction

Megan M. Loef, Deborah A. Carey, Thomas P. Carpenter, and Elizabeth Fennema

Research ...

Research indicates that young children can solve a variety of story problems before any formal schooling occurs (Carpenter and Moser 1983). Children also use a variety of solution strategies in solving story problems. These strategies range from modeling the action in the problem with counters or fingers, through counting techniques, to derived facts and recall of facts (Carpenter, Carey, and Kouba 1990). Children's abilities to solve story problems appear to change naturally as they systematically move through a hierarchy of problem types (Carpenter and Moser 1983; Riley, Greeno, and Heller 1983).

Within this hierarchy of problem types, a story problem's difficulty is determined by its underlying structure. Problem structure refers to the kind of action or the relationship among the quantities described in the problem and the nature of the unknown in the problem. For example, all three of the problems shown in **figure 1** can be represented by "8 − 3 = ?" and are typically described as subtraction problems. Yet in each of these problems the structure differs, and the pupil is being asked to find a different unknown. The natural strategies that children use to solve each of these three problems reflect the problem structure and consequently differ noticeably. In solving problem A, a pupil may count out an initial group of eight objects, remove a group of three objects, and then count the remaining objects. In problem B, a pupil may count out a set of three objects and then continue to add on to that set, keeping the added objects separate until a total of eight is reached. Then the pupil will count the objects in the added-on set. In problem C, a pupil may count out a group of eight objects, count out a group of three objects, line the two groups up next to each other, and count the objects without partners. Although each problem can be solved by subtracting 3 from 8, the solution strategies are different and reflect the structures of the story problems being solved. In effect, the structure of a story problem determines not only the difficulty of the problem but also the solution strategies that pupils are likely to use.

A knowledge of problem types and of solution strategies enables the teacher to design instruction that builds on children's knowledge. In a recent research project, teachers first participated in an in-service program in which they learned about problem types and solution strategies (Carpenter, Fennema, Peterson, and Carey, 1988). The teachers then used this knowledge to design lessons that focused on problem solving, to plan assessment activities, and to develop a curriculum that could be adapted to their pupils.

At the end of the year, the teachers from the in-service program showed different classroom behaviors and more extensive knowledge of their pupils than teachers who were not given the in-service training. Teachers in the in-service group posed problems, listened to pupils' processes, and worked in smaller groups more often than the non-in-service teachers. The in-service teachers were better able to predict the strategies that their pupils would use to solve a variety of story problems and number facts. They spent 20 percent of the mathematics class time working with number facts, in contrast with the non-in-service teachers, who spent 40 percent of their mathematics time on number facts. Nonetheless, results of an

---

A. Separate, result unknown
Connie has 8 marbles. She gives 3 marbles to Jim. How many marbles does she have left?

B. Join, change unknown
Connie has 3 marbles. How many more marbles does she need to collect to have 8 marbles all together?

C. Compare
Connie has 8 marbles. Jim has 3 marbles. How many more marbles does Connie have than Jim?

Fig. 1. Examples of three problem types for subtraction (Carpenter and Moser 1983)

achievement test on number facts showed no difference between the performance of the pupils of the two groups of teachers! In interviews, in fact, pupils from the in-service teachers' classrooms demonstrated a higher level of recall of number facts than did pupils from the other teachers' classrooms (Carpenter, Fennema, Peterson, Chiang, and Loef 1988).

Although knowledge of problem types and children's solution strategies is not usually taken into account when teachers make instructional decisions, a pupil's progress through the mathematical sequence can be facilitated by the teacher's structuring of the learning environment to match the child's natural development (Secada, Fuson, and Hall 1983; Carpenter and Moser 1984). The knowledge of a variety of problem types and of children's strategies can thus influence children's growth in the ability to solve addition and subtraction problems.

. . . Into Practice

Assessing children's thinking and knowing the processes that individual children use to solve problems are essential aspects of a program that focuses on problem solving. Once teachers know *how* pupils solve problems, they can then use this information to determine the sequence of their instruction. The assessment of pupils' cognitions is a formidable task for a teacher faced with a classroom of twenty-five to thirty pupils. Individual interviews may be the most effective way to gather information on pupils, but a more practical approach is to integrate assessment into the regular instruction. Activities can be designed for a whole-class setting and can be used for both instruction and assessment purposes.

The following vignette portrays Ms. Silver assessing her pupils' knowledge early in the school year. Ms. Silver was not so much concerned with knowing which pupils could respond quickly or accurately to story problems as she was interested in the strategies that the children used to solve particular types of problems. Her participation in the research project had given her content knowledge of addition and subtraction word problems. As a result, Ms. Silver expected her pupils to use a variety of successful strategies to solve different problems, even though neither the problems nor the strategies had been presented to the pupils in a formal mathematics lesson.

Ms. Silver's knowledge of the content also helped her decide which problems would give her the best information about the majority of her pupils. For example, problem B in figure 1 will give Ms. Silver information about the different levels of problem-solving abilities among her pupils because it is more difficult to solve than problem A, yet it can still be directly modeled. Problem A would probably be solved by many of the pupils and as a result would not offer an efficient assessment (Carpenter and Moser 1984; Riley, Greeno, and Heller 1983). In this lesson segment, the teacher gave the pupils counters and began the activity with a story problem.

Ms. S.: This morning I collected five dimes for milk money. Bob gave me some more dimes, then I had eight dimes all together. How many dimes did Bob give me?

(The teacher gave the pupils time to work on the problem until most of the class had a solution. As the pupils worked, she observed and recorded the names of those who were using a modeling or a counting strategy, planning to ask them to explain how they figured out their answers. If they responded with only the answer, she would ask them to tell the class what numbers they were thinking about when they were solving the problem.)

Ann: I know, 3. I counted 1, 2, 3, 4, 5, . . . [on her desk she had made a group of 5, then added on, keeping the second group separate] 6, 7, 8. See [pointing to the second set], 3.

Nick: No, it's 8. I counted 1, 2, 3, 4, 5 [pause], 6, 7, 8.

(He made a group of 5 with the counters, then added 3 more, and identified the total quantity as the solution.)

Ms. S.: Did anyone else solve the problem like Ann or Nick?

(For the next minute or so, the children discussed both strategies. The class agreed that they were similar but that Ann had kept the second set of three separate from the first set of five as she added on her counters. Ms. Silver listened as the children discussed the problem.)

Ms. S.: Did anyone solve the problem a different way than Nick or Ann? [Pause] Jan, how did you solve the problem?

(It was not apparent, from observing, how Jan had solved the problem.)

Jan: Well, I know that 4 plus 4 is 8. If you change one of the 4s to a 5, then the other number is changed to a 3, so 5 plus 3 is 8. Bob gave you three dimes.

Ms. S.: Everyone is doing such a good job of thinking about how to solve this problem. Would anyone else like to share another solution?

Pat: Well, I thought . . . you had 5 to start with, so I counted 6, 7, 8. [For each count, Pat extended a finger to keep track of the number that was added to 5 as he counted up to 8.] So you got 3 more.

Ms. S.: Nice job. Let's try another problem.

During this activity the teacher gained information about the problem-solving abilities and strategies of a number of her pupils. From her observations, Ms. Silver decided that the majority of the pupils could solve the problem. The discussion gave her information about the strategies that specific pupils used to arrive at the correct solution. For example, Ann's solution involved direct modeling with physical objects; Pat was able to use a more advanced counting strategy; and Jan figured out the answer using a known fact. From the responses of these pupils, Ms. Silver was confident that they could solve easier join-and-separate problems in which the result set is unknown. She noted that Nick successfully performed the counting procedure but was unable to identify the quantity of the second set. She decided that it would be necessary to check Nick's performance on the easier join-and-separate problems. The information gathered through observation and discussion could be used for instructional decisions on grouping, problem selection, and activities to encourage the development of more advanced solution strategies.

Key Aspects of Assessment

- Effective assessment can be informal and should be continual. Because a great deal of variability occurs in children's problem-solving strategies, it is necessary to integrate assessment into each lesson. Finding out how a few pupils are thinking about one or two story problems each day gives the teacher information on individual pupils' thought processes that can immediately influence instructional decisions.

- The decision to select certain problems for instruction is influenced by the teacher's knowledge of pupils' thinking. For example, if some pupils are solving story problems by direct modeling with counters, then the decision can be made to select problems that facilitate progress toward more advanced solution strategies. Pupils who usually count all the objects in both sets in addition story problems, say, may be encouraged to count on from the larger number when the second addend is 1, 2, or 3.

- Instruction that focuses on problem solving emphasizes the process rather than the product of the problem-solving activity. Mathematics becomes a subject that is shared and discussed among pupils. As a result, teachers gain new insights about their pupils and the pupils become aware of their peers' different ways of thinking about problems. Although pupils arrive at the same solution to a problem, it becomes apparent that no one correct procedure exists for solving the problem. All pupils can share in the discussion at their own level of problem-solving development.

Teachers can design an individual curriculum focusing on problem solving by basing it on their knowledge of the subject matter and their knowledge of their pupils. Curricular decisions are then based on what the learners know and how they think about the mathematics. This article comes from our research in the primary grades, but focusing on children's thinking and understanding and listening to students can be implemented with any subject matter in any grade. Teachers, as decision makers, can find effective ways of assessing children's mathematical knowledge and of understanding how that knowledge changes.

References

Carpenter, Thomas P., Deborah A. Carey, and Vicky L. Kouba. "A Problem-Solving Approach to the Operations." In *Mathematics for the Young Child*, edited by Joseph Payne. Reston, Va.: National Council of Teachers of Mathematics, 1990.

Carpenter, Thomas P., Elizabeth Fennema, Penelope L. Peterson, and Deborah A. Carey. "Teachers'

Pedagogical Content Knowledge of Students' Problem Solving in Elementary Arithmetic." *Journal for Research in Mathematics Education* 19 (November 1988):385–401.

Carpenter, Thomas P., Elizabeth Fennema, Penelope L. Peterson, Chi-Pang Chiang, and Megan Loef. "Using Knowledge of Children's Mathematics Thinking in Classroom Teaching: An Experimental Study." Paper presented to the American Educational Research Association, New Orleans, Louisiana, April 1988.

Carpenter, Thomas P., and James M. Moser, "The Acquisition of Addition and Subtraction Concepts." In *Acquisition of Mathematics Concepts and Processes,* edited by Richard Lesh and Marsha Landau. New York: Academic Press, 1983.

———. "The Acquisition of Addition and Subtraction Concepts in Grades One through Three." *Journal for Research in Mathematics Education* 15 (May 1984):179–202.

Riley, Mary S., James G. Greeno, and Joan G. Heller. "Development of Children's Problem-Solving Ability in Arithmetic." In *The Development of Mathematical Thinking,* edited by Herbert Ginsburg. New York: Academic Press, 1983.

Secada, Walter G., Karen C. Fuson, and James W. Hall. "The Transition from Counting All to Counting On in Addition." *Journal for Research in Mathematics Education* 14 (January 1983):47–57.

Section 2

Communicating

Improving Instruction by Listening to Children

Donald L. Chambers

In the vignette that follows, identify the ways in which the teacher's behavior is or is not consistent with the NCTM's *Standards*. The first-grade teacher presents the following problem to her students:

> "I made eight pies for my husband, and he ate two of them. How many were left? Everyone get out your counters, and we will do it together." She pauses briefly. "Lay down eight counters." She pauses again. "Picture them as pies. My husband eats two of them. How many pies do I have left? Write your answer beneath the line."

After making sure that each child has used his or her counters to model the eight pies, removed two of the counters, counted the remainder, and written 6 in the appropriate spot, the teacher continues, "Nan has five candy hearts. Dan gave her six more. How many candy hearts does she have now?" She asks Jim to show the class how he solved the problem. Jim demonstrates his solution on the overhead projector. First he counts out five hearts, then he counts out six into another pile. Then he counts all the hearts to get eleven. The teacher points out that Jim should have started with the larger pile and added on to get the answer. She says that they should always start with the larger pile when adding on.

Some features of this vignette seem consistent with the recommendations of the NCTM's *Professional Standards for Teaching Mathematics* (1991). The children are engaged in a problem-solving activity. They are using manipulative materials to help them solve the problem. They have an opportunity to describe how they solved the problem, and the manipulative materials serve as tools for discourse.

But a closer look reveals less desirable features. The teacher seems unreceptive to a student's solution unless it conforms to principles that she has formulated. In this example, the student must start with the larger pile. Furthermore, students' communicating to negotiate good mathematics is not valued.

In traditional mathematics teaching, communication by teachers consists largely of explanations; teachers explain and demonstrate mathematical procedures to the children. Students watch, listen, and practice. Even when teachers emphasize understanding by explaining reasons for the procedures, the pattern of communication remains the same. The teacher talks and the students listen.

During the last decade, our understanding of cognition in learning mathematics has increased, and curriculum developers and teachers are being encouraged to use this knowledge. Studies indicate that the behavior of teachers changes as their understanding of children's thinking increases and that this change in behavior results in changes in patterns of classroom communication and in better mathematics learning by their students.

Indeed, this teacher did change. Consider the ways in which a subsequent episode, recorded in the same teacher's classroom one year later, reflects a changed pattern of communication.

Teacher: Alice collects stamps. She had twenty-five stamps and she wants to have forty stamps. How many more does she need?

This example is a change-unknown problem, which is more complex than the simple result-unknown problem in the first episode. The numbers are also much larger. The teacher wanders around the room looking at individual papers. One child has written 46 rather than 40, so she prompts the student to correct the number by repeating, "She's got 25 and needs 40." She notices that Morgan is puzzled, so she suggests that he use counters. She restates the problem for Tom and says, "How are we going to find the answer?" She then calls on various children to explain how they arrived at a solution.

Tom: 20 plus 20 is 40, so I just took 5 away.

Teacher: Why did you take away 5? (Long wait time)

Tom: Because he had 25 so that would be 15.

This article is based on "Learning to Use Children's Mathematics Thinking: A Case Study" by Elizabeth Fennema, Thomas P. Carpenter, Megan L. Franke, and Deborah Carey. It appears in *Schools, Mathematics and the World of Reality*, edited by Robert Davis and Carolyn Maher (Needham Heights, Mass: Allyn & Bacon. 1992). The authors have given their permission for its use.

Teacher: Good! Judy?

Judy: I counted by nickels.

Teacher: I didn't think of it that way. Ben, would you show us on the overhead projector how you solved it?

Ben counts by ones from 25 to 40 and puts a counter on the projector for each number. He then counts the counters he placed on the projector.

Teacher: Did anyone solve it a different way?

Alice: I thought in my head; one 5 to get to 30; one 10 to get to 40; 5 and 10 is 15.

Teacher: Susan?

Susan shows how she counted from 25 to 40 with her fingers. She counts the fingers on one hand three times and says it is 15.

If the strategy that a child uses is unclear to the teacher, she responds with a comment such as, "I don't understand. Could you tell me another way?" She offers hints only when a child appears confused. After about five problems have been solved, the children separate into groups to continue their problem solving. The teacher works directly with a group of several students who appear not to understand. She gives them problems with slightly smaller numbers, and she has them think through the problems step by step. She does not tell them explicitly what to do but gives such hints as "Can you use your fingers?"

In this episode, the teacher asked children to explain their thinking and listened to what they said. If she did not understand, she continued to question them. The textbook was not evident; the teacher had decided what to do on the basis of what she knew about her students' thinking. She was not explicitly directing the children's thinking. Instead, she had selected problems that were appropriate to their abilities and was encouraging them both to report and to think about their own problem solutions. Instead of asking children to follow procedures that she supplied, she was asking them to engage in problem-solving behavior.

These instructional episodes represent teaching by the same teacher at different times. The classroom looked much the same and the teacher and children were comfortable at both times, but major differences were apparent in the teacher's instructional behavior.

The role of the teacher had changed. She was no longer totally directing the children's learning. Instead, she listened carefully to children and then used their thinking in making instructional decisions. The children were learning mathematics differently as they solved more challenging problems.

From the beginning of the project, this teacher strongly believed that it was her responsibility to establish an environment conducive to learning, but she no longer believed that learning would occur just because she clearly told the children what they were to learn. She described the change in her teaching in the following way:

> I listen to kids more. I am so much more aware of what the kids can really do that before I never would have thought about—first grade trying multiplication and division—that just comes from [my knowing] they can do it.... Then when I gave the children the opportunity and saw how they really do ... solve those problems, now I realize that they can do them and I give them the opportunity.

She also reported that she did less direct teaching. When asked if she told the children how to solve problems, she replied in the following manner:

> I don't really share strategies. I mean, they come up with more ideas than I could ever think of.... If a child is struggling and really has no idea where to begin, I might ... give a starting place, like "Why don't you start with three counters?" but I don't give them the strategies.

She also reported that she became more aware that children learned from each other:

> [When the kids listen to each other they] understand better than when they hear directly from me. It makes more sense to them.... I really see a lot of learning going on when [children] listen to the other children. I really do.

These examples reveal a classroom in which the pattern of communication changed dramatically. Initially, the communication was teacher directed. Students listened to the teacher's explanations, moving their counters according to the teacher's instructions. Students had an opportunity to explain how they got the answer to the practice problems, but their explanations were expected to conform to the explained method. Innovation was discouraged. The teacher had a very limited knowledge of each student's thinking and so had no way to use that knowledge to inform her instructional decisions.

The first-grade teacher in the foregoing episodes had seven years of teaching experience at the time the study began. She is representative of many teachers who have learned to modify their practice. As a result of her access to structured knowledge about children's thinking within well-defined content domains, she was able to understand her students' thinking (Carpenter 1990). Over a period of several years she learned how to use her knowledge of children's thinking to make more informed instructional decisions.

To understand the thinking of children, teachers need to spend more time listening to children describe how they think and less time explaining to the children how the teacher thinks. Although this transition is very difficult for most teachers, they can incorporate their knowledge of children's thinking into their decision making and change their instruction as their ability to assess their students' thought processes increases. These changes in instruction involve major shifts in the patterns of communication in the classroom.

Action Research Idea 1

Record a mathematics lesson without doing any special preparation. Analyze the lesson by listening to the tape and asking such questions as the following:

1. Do I hear or see evidence that I encourage students to—
   - engage in problem solving throughout the lesson?
   - invent their own solution strategies?
   - share their strategies with other students?
   - listen to other students as they describe their strategies?
   - negotiate and validate correct mathematics?

2. Do I hear or see evidence that I—
   - listen to students as they describe their strategies?
   - make an effort to understand their strategies by asking questions to clarify their thinking?
   - avoid imposing my way of thinking as the correct way or the best way?

Reflect on patterns of classroom communication over two months. Then record and analyze another lesson. What changes do you notice in your communication? What changes do you notice in the students' communication?

Action Research Idea 2

Ask your students to solve a problem in at least two ways. As the students work on the problem, identify three students who each use a different correct strategy. Call on these three students to present their different strategies to the class.

- Which strategies seem more advanced? Why?
- Think about what each student's strategy tells about what that student knows. How does the knowledge of these three students differ?
- Think about ways to build on the knowledge of each student. What problems might be appropriate for each of them? Why?

After two months, see how these same students think about the same problem.

- Are any students using a strategy different from their earlier strategy?
- Are any students using more advanced strategies?
- How has the knowledge of these three students changed?
- Think about new ways to build on the knowledge of each student. What problems might be appropriate for each of them? Why?

References

Carpenter, Thomas P., Deborah A. Carey, and Vicky L. Kouba. "A Problem-Solving Approach to the Operations." In *Mathematics for the Young Child,* edited by Joseph N. Payne, pp. 111–31. Reston, Va.: National Council of Teachers of Mathematics. 1990.

National Council of Teachers of Mathematics (NCTM). *Professional Standards for Teaching Mathematics.* Reston, Va.: NCTM, 1991.

Experience, Problem Solving, and Discourse as Central Aspects of Constructivism

Erna Yackel, Paul Cobb, Terry Wood, and Graceann Merkel

Over the past five years, we have collaborated with teachers to develop forms of instructional practice in elementary school mathematics that are compatible with a constructivist view of teaching and learning. Two key aspects of our work form the basis for this discussion: first, the process of developing instructional activities, and second, the importance of engaging students in mathematical discussion.

Serious attempts to develop instructional practices compatible with a constructivist perspective require confronting taken-for-granted assumptions that underlie current educational practices. Few, if any, would disagree with the view that learning occurs not as students take in mathematical knowledge in readymade pieces but as they build up mathematical meaning on the basis of their experiences in the classroom. This aspect of the constructivist perspective is easily embraced by anyone interested in meaningful learning and is reflected in the NCTM's *Curriculum and Evaluation Standards for School Mathematics (Standards)* (1989). Von Glasersfeld (1988) describes the instructional implications of this perspective as follows:

> If you believe that knowledge has to be constructed by each individual knower, . . . teaching becomes a very different proposition from the traditional notion where the knowledge is in the head of the teacher and the teacher has to find ways of conveying it or transferring in to the student. . . . I'm primarily interested in developing ways of thinking in the student. And if you want to do that, you are constantly working with conjectures . . . about what goes on in the student's head, and on those . . . you base your strategies. . . . What you present is never something that you expect the student to adopt as it is, but what you present is something that you think will make it possible for the student to find his or her own way of constructing.

The research reported in this article was supported by the National Science Foundation under grants nos. MDR 847-0400 and MDR 885-0560. All opinions and recommendations expressed are, of course, solely those of the authors and do not reflect the position of the foundation.

In line with von Glasersfeld's recommendations, we have attempted to use constructivism as a guiding framework within which to develop instructional situations that facilitate students' progressive construction of increasingly abstract mathematical conceptions and procedures. In concert with the guided-discovery approach, our goal is to have students develop mathematical concepts and relationships in ways that are personally meaningful. Important differenecs can be cited, however, between guided discovery and our own approach in terms of the basic assumptions that guide the development of instructional activities and the corresponding instructional strategies. The guided-discovery approach typically starts with a mathematical analysis of the relationships that students are supposed to discover and attempts to embody them in manipulative materials or in activities in a readily apprehensible form (Treffers 1987). In contrast, a constructivist approach acknowledges that individual students interpret instructional situations in profoundly different ways. In this approach, our understanding of students' mathematical experiences as informed by cognitive models (Steffe, von Glasersfeld, and Cobb 1988) rather than a mathematical analysis constitutes the starting point. Consequently, the instructional activities were not designed to ensure that every student make the same preselected mathematical constructions or apprehend the same relationships. Instead, we attempted to develop instructional activities so that the interpretations of students at different conceptual levels would be productive for their individual learning. In particular, because learning often occurs as students attempt to resolve cognitive conflicts, we used the cognitive models to anticipate which types of situations might potentially be problematic for students at different conceptual levels (Cobb, Wood, and Yackel, 1991). As an example, consider the pair of problems shown in **figure 1** that were used in a second-grade classroom in February. The following solutions for the second task were suggested by three students working together.

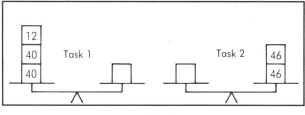

Fig. 1

Brenda: I'd say it's the same thing as this (the first task) because there's twelve—and six and six—and six and six make twelve.

Consuela: Wait a minute. Let me do it. Forty and forty makes eighty. Take these sixes away makes eighty, eighty-one, eighty-two, eighty-three, eighty-four, eighty-five, . . . , ninety-two (counting by ones).

Dai-soon: Forty-seven, forty-eight, forty-nine, ... (While listening to the girls, Dai-soon counts by ones starting with forty-seven, but he loses track of the number of times he has counted. Later he gets a hundred[s] board to use on the subsequent tasks.)

Brenda's problem was relating the two tasks. Consuela and Dai-soon both solved the problem of adding 46 and 46, but their solutions indicate differences in their interpretations and in their personal mathematical conceptions. Consuela's solution shows that she is capable of operating on two-digit numbers by thinking of them as made up of tens and ones. Dai-soon's solution of attempting to count on forty-six ones is more primitive.

Although students construct their own mathematical understandings, they do not do so in isolation. Interactions with both other students and the teacher give rise to crucial learning opportunities. For this reason, the students typically work in pairs or small groups and are expected to collaborate in their solution attempts. Collaboration involves much more than combining solution procedures to develop a joint solution. It involves developing explanations that are meaningful to someone else and trying to interpret and make sense of another's ideas and solution attempts as they evolve. As part of this process, students attempt to verbalize and interpret only partially formed ideas, thereby engaging in what Barnes (1976) calls *exploratory talk* as opposed to *final-draft talk*.

During subsequent whole-class discussion, students are expected to give coherent explanations of their problems, interpretations, and solutions and to respond to questions and challenges posed by their peers. They are also expected to listen to, and try to make sense of, explanations given by others, to pose appropriate questions, and to ask for clarifications.

When students engage in this type of discourse, not only is the amount of time they spend participating in problem-solving activities increased, but the nature of their problem-solving activity is itself extended to encompass learning opportunities that rarely arise in traditional instructional settings (Yackel, Cobb, and Wood 1991). These opportunities have their genesis in social interaction. As students participate in the discourse of what Richards (1991) calls *inquiry mathematics,* they learn to develop mathematical explanations and justifications and engage in what is typically termed *analytical reasoning.* Consider, for example, the following explanation of Jerome's method for solving $56 - 10 = \underline{}$: "Ten. Take away the six; that'd be fifty. And if six and four is ten, if you take away four that'd leave forty-six." On this occasion, unlike on previous ones, Jerome is describing and explaining his method, not just using it to get an answer. He does not say, "*Since* six and four is ten . . ." but "*If* six and four is ten. . . ." Thus he is explaining his methodology to justify it; his result is secondary.

The teacher's role is to facilitate the development of this type of mathematical discourse (Wood and Yackel 1990) by helping students in their attempts to express their mathematical thinking while also encouraging them to conceptualize situations in alternative ways. More generally, the teacher facilitates the students' mathematical development by subtly highlighting selected aspects of their mathematical contributions. In the process, he or she initiates and guides in the classroom the evolution of taken-to-be-shared mathematical understandings that are compatible with those of the wider society.

In conclusion, we note that our emphasis on mathematical discourse, with its resultant development of explanation and justification, is consistent with the National Council of Teachers of Mathematics's standards on communication, reasoning, and connections. Our experience in the classroom suggests that these standards are both realistic and attainable. Furthermore, both students and teachers find learning mathematics in this manner to be an exciting adventure.

References

Barnes, Douglas. *From Communication to Curriculum.* London: Penguin, 1976.

Cobb, Paul, Terry Wood, and Erna Yackel. "A Constructivist Approach to Second Grade Mathematics." In *Radical Constructivism in Mathematics Education,* edited by Ernst von Glasersfeld. Dordrecht, Netherlands: Kluwer Academic Publishers, 1991.

National Council of Teachers of Mathematics (NCTM), Commission on Standards for School Mathematics. *Curriculum and Evaluation Standards for School Mathematics.* Reston, Va.: NCTM, 1989.

Richards, John. "Mathematical Discussions." In *Radical Constructivism in Mathematics Education,* edited by Ernst von Glasersfeld. Dordrecht, Netherlands: Kluwer Academic Publishers, 1991

Steffe, Leslie P., Ernst von Glasersfeld, and Paul Cobb. *Construction of Arithmetical Meanings and Strategies.* New York: Springer-Verlag, 1988.

Treffers, A. *Three Dimensions: A Model of Goal and Theory Description in Mathematics Instruction—the Wiskobas Project.* Dordrecht, Netherlands: Reidel, 1987.

von Glasersfeld, Ernst. *Using Mathematical Thinking.* Tape 1. BBC audiotape of an interview by Barbara Jaworski at the Open University, produced by John Jaworski. Recorded at the International Congress of Mathematics Education, Budapest, Hungary, July 1988.

Wood, Terry, and Erna Yackel. "The Development of Collaborative Dialogue within Small Group Interactions." In *Transforming Children's Mathematics Education: An International Perspective.* Hillsdale, N.J.: Lawrence Erlbaum Associates, 1990.

Yackel, Erna, Paul Cobb, and Terry Wood. "Small-Group Interactions as a Source of Learning Opportunities in Second-Grade Mathematics." *Journal for Research in Mathematics Education* 22 (November 1991): 390–408.

Discourse That Promotes Conceptual Understanding

Elham Kazemi

As mathematics teachers, we want students to understand mathematics, not just to recite facts and execute computational procedures. We also know that allowing students to explore and have fun with mathematics may not necessarily stimulate deep thinking and promote greater conceptual understanding. Tasks that are aligned with the NCTM's curriculum standards (NCTM 1989) and that are connected to students' lives still may not challenge students to build more sophisticated understandings of mathematics. The actions of the teacher play a crucial role.

This article presents highlights from a study that demonstrates what it means to "press" students to think conceptually about mathematics (Kazemi and Stipek 1997), that is, to require reasoning that justifies procedures rather than statements of the procedures themselves. This study assessed the extent to which twenty-three upper elementary teachers supported learning and understanding during whole-class and small-group discussions. "Press for learning" was measured by the degree to which teachers (1) emphasized students' effort, (2) focused on learning and understanding, (3) supported students' autonomy, and (4) emphasized reasoning more than producing correct answers. Quantitative analyses indicated that the higher the press in the classroom, the more the students learned.

Like researchers in other studies (e.g, Fennema et al. [1996]), we observed that when teachers helped students build on their thinking, student achievement in problem solving and conceptual understanding increased. To understand what press for learning looks like in classrooms, we studied in depth two classes with higher scores for press and two classes with lower scores, and we looked closely at mathematical activity and discourse in the classes. The high-press classroom of Ms. Carter is contrasted with the low-press classroom of Ms. Andrew.

The data for this article are from the Integrating Mathematics Assessment project; Maryl Gearhart, Geoffrey Saxe, and Deborah Stipek, principal investigators; funded by grant MDR-9154512 from the National Science Foundation. The opinions expressed here do not necessarily reflect those of the Foundation.

Students in Ms. Carter's and Ms. Andrew's classes were exploring the concept of equivalence and the addition of fractions. They worked on fair-share problems, such as the following:

> I invited 8 people to a party (including me), and I had 12 brownies. How much did each person get if everyone got a fair share? Later my mother got home with 9 more brownies. We can always eat more brownies, so we shared these out equally too. This time how much brownie did each person get? How much brownie did each person eat altogether? (Corwin, Russell, and Tierney 1990, p. 76)

Similarities between Classrooms: Social Norms

In both Ms. Carter's and Ms. Andrew's classes, we saw students huddled in groups, materials scattered about them, figuring out how to share a batch of brownies equally among a group of people. The students seemed to be engaged in and enjoying their work. Often each group found a slightly different strategy to solve the problem. After moving from group to group, listening to and joining student conversations, both teachers stopped group activity to ask students to share their work and explain how they solved the problem.

The NCTM *Standards* documents support the view that social norms—practices such as explaining thinking, sharing strategies, and collaborating that we see in both classrooms—afford opportunities for students to engage in conceptual thinking. Many teachers establish those social norms in their classrooms quite readily. But social norms alone may not advance students' conceptual thinking.

Differences between Classrooms: Sociomathematical Norms

Although Ms. Andrew and Ms. Carter both valued problem solving and established the same social norms in their classrooms, important differences were seen in the quality of their students' engagement with the mathematics. To understand those differences, we looked more closely at the norms that guide the quality of mathematical discourse, the *sociomathematical* norms (Yackel and Cobb 1996). Teachers and students actively negotiate the sociomathematical norms that develop in any classroom. Sociomathematical norms identify what kind of talk in valued in the classroom, what counts as a mathematical explanation, and what counts as a mathematically different strategy. In the brownie problem, for example, students grapple with ideas of equivalence, part-whole relations, and the addition of fractional parts. Sociomathematical norms help us understand the ways in which fraction concepts are supported within the context of sharing and explaining strategies.

Through our study of the four classrooms, we identified four sociomathematical norms that guided students' mathematical activity and helped create a high press for conceptual thinking:

- Explanations consisted of mathematical arguments, not simply procedural summaries of the steps taken to solve the problem.
- Errors offered opportunities to reconceptualize a problem and explore contradictions and alternative strategies.
- Mathematical thinking involved understanding relations among multiple strategies.
- Collaborative work involved individual accountability and reaching consensus through mathematical argumentation.

Other norms may also contribute to a high press, but these norms captured the major differences in the way that mathematics was treated by the high- and low-press teachers.

Explaining strategies

The following examples illustrate some of the differences in the two classrooms. First, in Ms. Carter's class, explanations were not limited to descriptions of steps taken to solve a problem. They were always linked to mathematical reasons. In the following example, Ms. Carter asked Sarah and Jasmine to describe their actions and to *explain why* they chose particular partitioning strategies.

Sarah: The first four we cut them in half. [Jasmine divides squares in half on an overhead transparency. See **fig. 1**.]

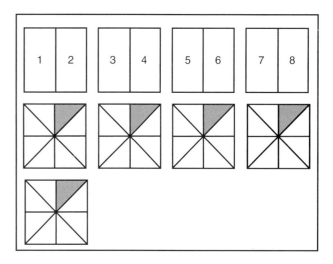

Fig. 1. Sharing nine brownies among eight people

Ms. Carter: Now as you explain, could you explain why you did it in half?

Sarah: Because when you put it in half, it becomes four . . . four . . . eight halves.

Ms. Carter: Eight halves. What does that mean if there are eight halves?

Sarah: Then each person gets a half.

Ms. Carter: Okay, that each person gets a half. [Jasmine labels halves 1 through 8 for each of the eight people.]

Sarah: Then there were five boxes [brownies] left. We put them in eighths.

Ms. Carter: Okay, so they divided them into eighths. Could you tell us why you chose eighths?

Sarah: It's easiest. Because then everyone will get . . . each person will get a half and [addresses Jasmine] "How many eighths?"

Jasmine: [Quietly] Five-eighths.

Ms. Carter: I didn't know why you did it in eighths. That's the reason. I just wanted to know why you chose eighths.

Jasmine: We did eighths because then if we did eighths, each person would get each eighth, I mean one-eighth out of each brownie.

Ms. Carter: Okay, one-eighth out of each brownie. Can you just, don't have to number, but just show us what you mean by that? I heard the words, but . . .

[Jasmine shades in one-eighth of each of the five brownies that were divided into eighths.]

Jasmine: Person one would get this . . . [points to one-eighth].

Ms. Carter: Oh, cut of each brownie.

Sarah: Out of each brownie, one person will get one-eighth.

Ms. Carter: One-eighth. Okay. So how much did they get if they got their fair share?

Jasmine and Sarah: They got a half and five-eighths.

Ms. Carter: Do you want to write that down at the top, so I can see what you did?

[Jasmine writes 1/2 + 1/8 + 1/8 + 1/8 + 1/8 + 1/8 at the top of the overhead transparency.]

The exchange among Sarah, Jasmine, and Ms. Carter highlighted the conceptual focus of the lesson on fair share. Ms. Carter asked Sarah to explain the importance of having eight halves and why the partitioning strategy using eighths made sense. After Jasmine gave a verbal justification, Ms. Carter continued to press her to link her verbal response to the appropriate pictorial representation—by shading the pieces—and to the symbolic representation—by writing the sum of the fractions.

The same degree of press did not exist in Ms. Andrew's classroom. Ms. Andrew's students engaged in the same social practice of sharing their strategies with the class, but the mathematical content of classroom conversations was different. Students shared solutions by giving procedural summaries of the steps they took to solve the problem, as demonstrated by the following exchange, in which Raymond described his solution for sharing twelve brownies among eight people. Ms. Andrew had drawn twelve squares on the chalkboard.

[Raymond divides four of the brownies in half.]

Ms. Andrew: Okay, now would you like to explain to us what . . . loud . . .

Raymond: Each one gets one, and I give them a half.

Ms. Andrew: So each person got how much?

Raymond: One and one-half.

Ms. Andrew: One-half?

Raymond: No, one and one-half.

Ms. Andrew: So you're saying that each one gets one and one-half. Does that make sense?

[After a chorus of "yeahs" comes from students, Ms. Andrew moves on to another problem.]

Unlike Ms. Carter, Ms. Andrew did not ask her students to justify why they chose a particular partitioning strategy. Instead, Ms. Andrew often asked questions that required a show of hands or yes-no responses, such as "How many people agree?" "Does this make sense?" or "Do you think that was a good answer?" Ms. Andrew wanted to engage her students in the activity and to see if they understood, but the questions she asked yielded general responses without revealing specific information about the students' thinking.

Reacting to mathematical errors

By emphasizing mathematical reasons for actions, Ms. Carter created opportunities for her students to prove that their solutions were correct. She resisted telling students that an answer or reason was wrong, and she invited others to respond to incorrect solutions. Ms. Carter modeled the kinds of questions that may help students think through their own confusion by using their existing knowledge. Those questions usually involved graphical representations of the fractions. In small groups, students challenged one another when they disagreed on a solution and helped one another find errors.

The interaction among Ms. Carter, Jasmine, and Sarah continued with the following conversation.

[Jasmine writes 1/2 + 1/8 + 1/8 + 1/8 + 1/8 + 1/8 at the top of the overhead transparency.]

Ms. Carter: Okay, so that's what you did. So how much was that in all?

Jasmine: It equals 1 1/8 or 6/8.

Ms. Carter: So she says it can equal 6 and 6/8? [She misheard Jasmine.]

Jasmine: No, it can equal 6/8 or it can equal 1 1/8.

Ms. Carter: Okay, so you have two different answers. Could you write them down so people can see it? And boys and girls, I'd like you to respond to what they've written up here. She says it either could equal 6/8 or 1 1/8.

Ms. Carter: Matthew had his hand up and was thinking about it. Someone from team 5. Anybody from team 6 that has a response? Right now. I'm just going to let you look a minute. See if anyone has a response. Andrew, you had your hand up, is that right? [waits] Still only have four hands up. I wonder if you're all looking up here and seeing? She's given us two answers here, 6/8 or it can equal 1 1/8. Okay, could those four people right now . . . do you agree with both answers?

Students: No. . . .

Ms. Carter: Do you have a reason why you don't agree? Don't explain it to me, but do you have a reason? Raise your hand if you have a reason why you don't agree. [hands] One, two, three, four, five, six . . . okay. Would those six people please stand? Okay. Would you please, you're going to be in charge of explaining why you don't agree to your team. [She assigns those six students to teams.] Right now, if you don't agree, would you please tell them what you think the answer is and why you don't agree. Go ahead. Explain.

Ms. Carter could have stepped in and pointed out why 6/8 and 1 1/8 are not equal. Instead, her response to this mistake was to encourage her students to explore the error by providing the conceptual reasons for why 6/8 and 1 1/8 are not equal. She engaged the entire class in thinking about which solution was correct instead of talking with only the two presenters or correcting their mistake herself, and she created an opportunity for her students to practice articulating their thinking.

The mistake also created an opportunity for the entire class to explore contradictions in the solution and to build an understanding of fractional equivalence and the addition of fractions by using an area model. This type of activity and discourse was typical in Ms. Carter's classroom. In a whole-class discussion, each group shared its proof that 1 1/8 was correct. Neither the students nor Ms. Carter belittled, penalized, or discredited anyone who made a mistake. The atmosphere of mutual respect between the students and Ms. Carter allowed the class to think about and build conceptual understandings eagerly.

Ms. Andrew treated errors differently. Note how she provides the mathematical reasoning when three boys explained their solution for sharing five brownies among six people.

Ms. Andrew: They got 1/2, you already said that. And then 1/6 and then another sixth. So, how many sixths did they get?

Anthony: One, two.

Ryan: One, two.

Joe: 1/12.

Ms. Andrew: What did you say? [to Joe] They got two . . .

Ryan: Sixths.

Anthony: 2/12.

Joe: 2/6.

Ms. Andrew: 2/6 [confirming the right answer].

Why did you say 2/12? Because there are twelve parts altogether?

Anthony: Yeah.

Ms. Andrew: Okay, be sure not to get confused. Because there are two brownies not one. Perfect. Good, good job.

At first, the boys appeared to be guessing the answer to Ms. Andrew's question. She focused on Joe once he stated the right answer. Although she predicted accurately why Anthony said 2/12, she did not ask him to think about why his answer did not work. Instead, she asked and answered the question herself and did not press Anthony to sort out his confusion. Her statement "Because there are two brownies, not one" was left unexplained. As this example illustrates, limited opportunity was available for the members of the group to engage in conceptual thinking about what 1/6 and 1/12 signify and how the graphical representation is linked to the numeric representation.

Both Ms. Carter and Ms. Andrew allowed students to make mistakes. That social norm, however, was not enough to press students to examine their work conceptually. Both teachers wanted their students to learn from their mistakes, but Ms. Andrew often supplied the conceptual thinking for her students. In Ms. Carter's class, inadequate solutions served as entry points for further mathematical discussion.

Comparing strategies

Students in both classrooms worked together, shared their strategies, and were praised for their efforts. Students in both classrooms attended to nonmathematical similarities between shared solutions, such as the layout of the paper or the uses of color. In Ms. Andrew's class, strategies were typically offered one after the other, with discussion limited to nonmathematical aspects of students' work. For example, a pair of students noted that they cut the paper brownies and pasted the pieces under stick-figure illustrations. Another pair had drawn lines from the fractional parts of the brownies to the individuals that received them. Although the partitioning strategy in both was the same, students viewed the strategies as different because the representations were different. Ms. Carter, however, pressed her students to go beyond their initial observations and reflect on the mathematical similarities and differences between strategies.

Accountability and consensus

In inquiry-based classrooms, students often work together to share interpretations and solutions and construct new understandings. Important differences arose between Ms. Andrew's and Ms. Carter's classes in the way in which they emphasized individual accountability and consensus. Ms. Carter required her students to make sure that each person contributed to, and understood the mathematics involved in, the group's solution. If students disagreed about an answer, she encouraged them to prove their answers mathematically and to work until they arrived at a consensus. If she noticed that students were not listening to others during an activity, she reminded them that they had to prove their solutions and that each group member must be prepared to discuss the reasons for the solution in front of the class. As a result, the distribution of work was more equitable. Students listened to one another's ideas and evaluated their appropriateness before using them.

Ms. Andrew did not describe and discuss collaboration beyond the general directive to "work with a partner" or "remember to work together." Neither individual accountability nor consensus emerged as topics of discussion in whole-class activity. Typically, only one person would be in control of group work at any particular time and would complete most of the work.

Conclusion

We saw a consistently higher press for conceptual thinking in Ms. Carter's class. She took her students' ideas seriously as they engaged in building mathematical concepts. In both whole-class discussions and small-group work, all students were accountable for participating in an intellectual climate characterized by argument and justification. Four sociomathematical norms governed mathematical discourse in Ms. Carter's classroom: explanations were supported by mathematical reasons, mistakes created opportunities to engage further with mathematical ideas, students drew mathematical connections between strategies, and each student was accountable for the work of the group.

When teachers create a high press for conceptual thinking, mathematics drives not only the activities but the students' explanations as well. As a result, student achievement in problem solving and conceptual understanding increases.

Action Research Ideas

- Over time, listen for differences in the number of times that you that interrupt a student's explanation, restate a student's explanation, or provide a solution strategy. By keeping a daily log, notice any changes in the nature and quantity of your responses.
- *(a)* Identify the social norms and the sociomathematical norms that characterize your classroom.

(b) Discuss the issue of sociomathematical norms with a colleague. Share your goals and the problems that you expect to encounter. Continue to discuss your progress with your colleague over time. Encourage your colleague to engage in a similar program to create a higher press. *(c)* Observe and discuss each other's teaching.

- *(a)* Reflect on the discourse associated with a problem recently discussed in your classroom. Using a four-point scale from 0 (low press) to 4 (high press), rate the discourse according to each of the sociomathematical norms that characterize Ms. Carter's classroom. *(b)* Set personal goals for each of the sociomathematical norms. Use such questions as the following to help establish a high press: "How can you prove that your answer is right? Can you prove it in more than one way? How is your strategy mathematically different from, or mathematically like, that of [another student]? Do you agree or disagree with [another student's] solution? Why? Why does [strategy *x*] work? Why does [strategy *y*] not work?" *(c)* After four weeks, reevaluate your classroom, using the same scale and the same sociomathematical norms. Note your areas of improvement, and set new goals for the next four weeks.

References

Corwin, R. B., S. J. Russell, and C. C. Tierney. *Seeing Fractions: Representations of Wholes and Parts. A Unit for the Upper Elementary Grades.* Sacramento, Calif.: Technical Education Research Center, California Department of Education, 1990.

Fennema, Elizabeth, Thomas P. Carpenter, Megan L. Franke, Linda Levi, Victoria R. Jacobs, and Susan B. Empson. "A Longitudinal Study of Learning to Use Children's Thinking in Mathematics Instruction." *Journal for Research in Mathematics Education* 27 (July 1996): 403–34.

Kazemi, E., and D. Stipek. "Pressing Students to Be Thoughtful: Promoting Conceptual Thinking in Mathematics." Paper presented at the annual meeting of the American Educational Research Association, Chicago, 1997.

National Council of Teachers of Mathematics (NCTM). *Curriculum and Evaluation Standards for School Mathematics.* Reston, Va.: NCTM, 1989.

Yackel, Ema, and Paul Cobb. "Sociomathematical Norms, Argumentation, and Autonomy in Mathematics." *Journal for Research in Mathematics Education* 27 (July 1996): 458–77.

Dialogue and Conceptual Splatter in Mathematics Classes

Jack Easley, Harold A. Taylor and Judy K. Taylor

We have studied first- and second-grade mathematics classes in Japan and the United States in which story problems and problems about mathematical concepts were used to promote higher-level thinking. Our research leads us to question the common assumption that teachers should aim each lesson at a particular conceptual goal for all pupils and expect nearly all of them to succeed and to understand the same concept. Instead, a teacher's aim should be as close as possible to accepting conceptual diversity or "splatter" among pupils. We believe that steady work on such an aim through problem solving and dialogue will lead to outstanding conceptual power in pupils. In this article, we present three examples from lessons that reflect this aim, as well as suggestions that we have drawn from these and other lessons.

As an example of the splatter of ideas a teacher is likely to encounter as pupils are presented with a problem, consider the following lesson taught by one of the authors (Easley) to a class of first and second graders in the United States. The author was asked to talk about the experience he had in a Japanese school and to demonstrate something of the process of teaching and learning he had observed in Japanese classrooms. He told the class that the Japanese government had decided that teachers in Japan should not encourage counting to add or subtract (Hatano 1982; Easley 1983), and he asked the pupils about different ways to add. Their preferred method was to add each column separately, although they knew people who made a lot of dots and counted them. During this discussion, one of the pupils asked how far away Japan was, so the author tried to remember the number of hours he had flown and guess how fast he flew. One pupil then said that he had flown to Florida, which was 900 miles away. When asked how long he flew, he said it was hard to figure out because he flew in one plane for one hour, another plane for one-half hour, and a third plane for two and one-half hours. The author wrote these times on the board vertically, without keeping the fractions in a separate column. He added a plus sign and drew a line:

$$\begin{array}{r} 1\,hr \\ \tfrac{1}{2}\,hr \\ + \,2\tfrac{1}{2}\,hr \\ \hline \end{array}$$

He let the pupils work on the sum briefly and then organized them into groups of three to five to share their work. Two groups came to a consensus quickly; one argued in favor of four hours and the other one, five hours. The author allowed two more minutes of group time for the other four groups that were still struggling with the problem. One group with more than one answer split in two. The answers that the class produced included four hours, four and one-half hours, five hours, seven hours, eight hours, eleven hours, and twenty-seven hours. The spokesperson from the group with the answer of four hours reported first. Then the next group showed how, by counting the numerators of the two fractions as 1 each, they got five hours for the answer.

The next group had four and one-half hours for the answer. The spokesperson drew a diagram that consisted of a pizza divided with ten radial lines, without explaining why the group had chosen a pizza diagram or why they had divided it in that way. He said that they removed four of the pieces to represent four hours. As he spoke, he erased all five radial lines bounding these four pieces; that left five radial lines defining four more pieces:

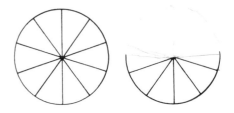

The spokesperson then said, "So four were left, which made a half; therefore, the answer was four and one-half."

No questions were posed from the floor about this solution, nor were any asked when the spokesperson for the group with an answer of eleven hours explained, by pointing to the numerals on the board, that they had added all the numerators and denominators together with the integers; when he worked the problem for the class, however, he got an answer of nine hours. The next group got eight hours for the answer, but while the spokesperson for the group was explaining the solution, she realized that they had forgotten to add in one of the numerators, so she changed the answer to nine hours, with the approval of the rest of her group.

Seven hours was the answer from the next group. Their argument was that 1 and 2 are 3, and 1/2 plus 1/2 is 4. The author asked whether putting together two half apples made a whole. Yes, the spokesperson said, but that "wasn't the same as one." The spokesperson then drew a pie diagram and divided it in half twice with two perpendicular lines to show how they had obtained 4 for 1/2 plus 1/2. He then concluded with, "Altogether than makes seven."

The group with twenty-seven hours for an answer was never able to explain the solution. The spokesperson at the board drew a hexomino and changed it step by step into a cross-shaped pentomino:

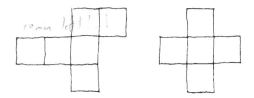

Even with help from another member of the group, no explanation came forth. They were sent back to their seats to think again about their solution, and a member of another group volunteered to help them. They then came up with an answer of five hours.

By this time, four or five pupils were asking questions of their peers, and the author was asked what the winning answer was. Taking a poll was considered fair. The results showed nine people for four hours, five people for five hours, and the rest of the people not voting. The boy who had made the trip announced that he was sure his answer was right—four hours. The author told him that he had a lot of people to convince, and the boy quickly rose and explained to the class that 2 1/2 plus 1/2 makes 3, and 1 more makes 4. Many classmates objected, but the class time had run out. The classroom teacher said they "could work on it tomorrow."

This kind of activity was, in fact, typical in the Japanese school that the author had hoped to describe to these American students. In addition, if this specific interaction had occurred in a Japanese classroom, the teacher probably would have given time the next day for the pupils to discuss further and resolve their differences about the total flying time. The Japanese teachers almost always allowed the pupils to resolve conflicts about mathematical ideas. The teachers were not reluctant to introduce ideas or alternatives, but they understood that pupils often drop resistance they may feel toward a teacher's point when the teacher stops actively pressing it. In the Japanese first-grade classroom from which the following example was taken (Easley and Easley 1983), the issue was whether adding zero could be considered a case of adding, since no net increase resulted. The teacher had dramatized the point with a demonstration of "pouring" into two deciliters of orange juice the contents of a similar, but empty, container. After the pouring demonstration, the resulting two deciliters of orange juice were read off and recorded. Then adding a basket of nothing to a basket containing three apples led to a controversial number sentence: $3 + 0 = 3$. One pupil said that there was no answer, since adding zero really wasn't adding; another said that since zero meant nothing, zero should be the answer instead of three. The Japanese teacher was obviously fascinated by such minority opinions and did not press the majority case for three apples as the answer. After a few minutes of encouraging more dialogue by the pupils, all opposition to the meaning of adding zero ceased. In this discussion and others like it that were observed, the pupils clearly enjoyed the mathematical arguments.

A third example of encouraging conceptual splatter comes from a first-grade class in the United States, again taught by one of the authors (Harold Taylor). Near the end of the school year, the author gave the pupils written story problems to work on in groups of five. Each group presented its solution, in a skit, to the whole class. The following excerpt from the lesson is part of the presentation of Sarah's group. This example does not fully demonstrate the conceptual splatter that could be seen across all the presentations during the lesson, but it does show how a pupil can clarify her ideas when given the chance to express them. Sarah was fairly directive with her fellow pupils, and she perhaps reaped the greatest benefit from the activity. But the crucial point is that Sarah (S) had a chance to speak about mathematics with the class, as did all the pupils during the course of the week. (R= Respondent.)

S: Here's our problem. A baker made fourteen cookies. She gave all of them to four kids. How many cookies and pieces of cookies did each kid get? (Sarah quizzes the class on the problem, calling on those with hands raised for answers.) How many cookies did the baker make?

R: Fourteen.

S: How many kids were there?

R: Five.

S: Nuh-uh, I'm the baker. What were we supposed to do with the fourteen cookies?

R: Give 'em to all the four girls. (Four pupils correct the word *girls* to *kids*.)

S: What were we supposed to find out?

R: How many they got.

S: How many together? How many did each one get? Let's start out with passing them [out]. (Sarah passes out one paper cookie to each of the other four pupils in her group.) How many cookies do they have?

R: One.

S: I still have all these cookies. Maybe I can do it again. (She begins to distribute more cookies.) Cookie number five, six, seven, eight. How many cookies do they each have?

R: Two.

S: I still have all these cookies. I still have enough to go around again. (She begins to distribute more cookies.) Okay, cookie number nine, ten, eleven, twelve. Now how many does each have?

R: Three.

S: With four kids, I have only two left. Either two kids get four or two kids get none (meaning, presumably, that only two pupils could get another whole cookie). (Sarah then gets a pair of scissors from a boy in the audience.) Let's see, what if I just cut em in . . . . Well, I think there's just (pointing to the children) one, two, three, four left, so cut these cookies into fourths, I think. (She begins to cut one of the last two paper cookies.) For cookie number one, two slices. (She cuts it in half.) Cut 'em again—three slices. (She cuts one of the halves in half.) Fourths. (She cuts the other half in half and then takes the last cookie to cut it.) Cut one again! (She cuts the last paper cookie in half.) Now, how many cookies? (She momentarily slips into counting each fraction of a cookie as a whole cookie.)

R: Two.

S: Cut again, and how many cookies do I have? (She cuts one of the halves in half.)

R: Three.

S: Cut them again, and we have four. (She cuts the other half in half.) Let's pass 'em out. You get three cookies and a fourth. (She directs each pupil to put the whole cookies in one hand and the fractions of cookies in the other hand.) How many do they have now?

R: Four.

S: No! Three and one-fourth, three and a fourth. . . . What *is* a fourth?

R: When you cut a cookie into four pieces, one of those pieces is a fourth. When you cut anything into four pieces . . .

S: (Sarah refers to the last paper cookie, which has been cut up but has not yet been distributed.) If I give this cookie to one kid, they would get four and one-fourth. That wouldn't be fair because everyone else would get three and one-fourth. (Sarah distributes the four parts of the last cookie, one part to each of the four pupils in the group.) Three cookies and two-fourths.

R: So I have four cookies.

S: No you don't, three cookies and two-fourths. (Sarah goes to Jimmy and counts each cookie and each fourth of a cookie.)

Several days later, when the class saw the videotape of the lesson, Sarah developed her concept of fractions even further, saying that two-fourths was the same as one-half, so she could have just cut each of the two remaining cookies in half. Sarah's directive style during the activity clearly reflects the fact that the classroom teacher (Harold Taylor) was also directive at times but that the pupils had the major voice in the classroom. The main point is that teachers are likely to overlook what pupils are actually thinking if they do not use problem-solving activities and dialogue to encourage the expression of pupils' conceptual splatter.

Suggestions for Practice

The examples of classroom dialogues have embedded in them suggestions for classroom practice, but we offer some specific guidelines for teachers who would like to include more mathematical dialogue and encourage the expression of conceptual splatter in their classrooms. The following suggestions come from one of the authors (Judy Taylor) and are based on her experiences as a first-grade teacher:

1. Try to maintain an atmosphere of freedom in the classroom, that is, an atmosphere that encourages pupils to feel free to express themselves.

2. When in doubt, remain silent! Give the pupils a chance to work through mathematical difficulties.

3. Don't paraphrase, because paraphrasing is usually a technique for pushing the pupils too quickly toward the teacher's point of view. Let the pupils' words stand by themselves. Ask pupils to clarify the expression of their own thoughts rather than comply with your thoughts.

4. Encourage constructive arguments. For example, ask, "Do you believe that?" "Do you agree or disagree?" "How do you know?" Or say, "Tell him what you mean."

5. Play dumb but avoid condescension. Let the pupils teach you, as well as each other.

6. Don't focus on "right" answers. Alternative answers can create a kind of tension that leads to lively and fruitful interaction. In addition, many good problems have more than one right answer, or at least more than one possible solution technique.

7. Let pupils have time to work on a problem as individuals before you ask them to share their thoughts in a small or large group, thereby encouraging each pupil to get involved in the problem. Furthermore, tell the pupils that when groups are formed, the entire class should be brought into the conversation.

8. No best way exists to form small groups. For example, you can group pupils at random or on the basis of differing ability levels or of their conflicting opinions.

9. Dialogue groups, whether small or large, can be called "meetings." Take time to model expected etiquette for the meetings.

10. Save the written work of the pupils to evaluate and substantiate learning. Record the classroom dialogues on videotapes, audiotapes, and in your own journal of the activities; share these recordings with the pupils. And finally, bring continuity to the dialogues by basing future sessions on your evaluation of the conceptual splatter that has been expressed.

Conclusion

Teachers often consciously or unconsciously "paper over" the diversity of ideas and concepts held by pupils. But the strategy of encouraging dialogue and conceptual splatter can lead to a more solid mathematical foundation for pupils; children's ideas must be viewed as genuine intellectual proposals deserving every consideration. Once teachers realize that the papered-over surface is smooth but weak and the alternative ideas still remain under the surface, waiting to sidetrack understandings of the next lesson, they can see that the paper needs to be ripped off so that all ideas can emerge and be evaluated.

Besides the obstacle of our own ingrained views of what mathematics classes should look like, we must contend with the obstacle of trying to explain to colleagues, administrators, and parents that dialogue and conceptual splatter constitute serious mathematics and that brainstorming and the simple airing of ideas often lead to self-correction (Easley and Zwoyer 1975). If, however, we want pupils to learn and experience mathematics, we have no choice but to try to overcome these obstacles.

References

Easley, Jack. "A Japanese Approach to Arithmetic." *For the Learning of Mathematics* 3 (March 1983):8–14.

Easley, Jack, and Elizabeth Easley. "What's There to Talk about in Arithmetic?" *Problem Solving* 5 (March 1983) (Newsletter published by the Franklin Institute Press, Philadelphia).

Easley, Jack, and Russell Zwoyer. "Teaching by Listening." *Contemporary Education* 47 (Fall 1975):19–25.

Hatano, Giyoo. "Learning to Add and Subtract: A Japanese Perspective." In *Addition and Subtraction: A Developmental Perspective,* edited by Thomas P. Carpenter, James M. Moser, and Thomas A. Romberg, pp. 211–23. Hillsdale, N.J.: Lawrence Erlbaum Associates, 1982.

The Use of Verbal Explanation in Japanese and American Classrooms

James W. Stigler

Mathematics represents a universal system for communicating quantitative ideas, and the Hindu-Arabic system of numeration is used throughout the world. Although the concepts of mathematics are universal, the beliefs and practices that underlie mathematics instruction are not. In ten years of work comparing Japanese, Chinese, and American elementary schools, I have repeatedly been impressed that beliefs and practices taken for granted by American teachers are not necessarily adhered to by their Asian counterparts. In mathematics we should be particularly motivated to examine and reflect on these differences because American children fall far behind Asian students in their knowledge of mathematics.

How real are differences in mathematics achievement between Asian and American children? For at least twenty years we have known that American middle and secondary school students compare poorly on tests of mathematics achievement with students from many other countries, and especially with students from Japan (Husen 1967; McKnight et al. 1987; Travers et al. 1985). The interpretation of these results, however, is not straightforward. A great deal happens between elementary school and secondary school, and most of it is different in Japan and the United States. For example, Japanese students take more mathematics courses than do American students. If all that has been shown is that people who take more courses learn more, then nothing is surprising in the results. Furthermore, international tests of mathematics achievement may not measure the creativity and problem-solving skills that we as Americans value most. Americans often cite the American advantage over Japan in winning Nobel prizes.

Unfortunately, our most recent research suggests that such rationalizations do not apply. First, we find that the Japanese superiority in mathematics exists as early as kindergarten, and it is dramatic by the time children reach fifth grade. In comparing the mean achievement test scores of children from representative samples of fifth-grade classrooms in Sendai, Japan; Taipei, Taiwan; and Minneapolis, U.S.A., we found that the highest scoring American classroom did not perform as well as the lowest scoring Japanese classroom and scored higher than only one of the twenty classrooms in Taipei (Stevenson, Lee, and Stigler 1986). Second, in data as yet unpublished, we find that the superiority of Japanese elementary school students is not limited to basic computational skills but extends to nearly every mathematics-related area we tested, including novel problem solving, estimation, visualization, graphing, measurement, complex mental calculation, conceptual knowledge about mathematics, and so on. The differences are real, they are large, they are broad, and they appear early.

Explaining these dramatic differences presents a challenge to researchers and also to educators who must grapple with the problem of declining mathematical competence in American society. In the remainder of this article I discuss one of the cross-cultural differences that we have identified from observations of mathematics teaching in first- and fifth-grade classrooms in Japan and the United States. (For more detailed descriptions of the two observational studies we have conducted, see Stigler, Lee, and Stevenson [1987]; Stigler and Perry [1988].)

This article was written while the author was supported by a Spencer Fellowship from the National Academy of Education. The research discussed in this paper was conducted in collaboration with Harold Stevenson at the Center for Human Growth and Development, University of Michigan, and funded by Grant BNS8409372 from the National Science Foundation. The author would like to acknowledge the contribution of Michelle Perry to this article, both conceptually and in terms of data analysis; and of Kevin F. Miller for comments on an earlier draft.

Some Findings on Verbal Explanations

In our most recent observational study, we collected detailed narrative observations from 120 first- and fifth-grade classrooms in Sendai, Japan, and the Chicago metropolitan area. Each classroom was visited four times during mathematics instruction, for a total of 480 narrative accounts. The observations were coded in Chicago by a team of researchers representing the two cultures. The description of each class was divided into segments based on changes in topic, activity, or materials; summarized into English; and entered into a computer data base. Thus, we were able to maintain a sense of what was happening during mathematics instruction, at the same time facilitating systematic analysis and quantification of cultural differences. Numerous differences emerged. Here I shall focus only on differences in the use of verbal explanation.

One of the most striking aspects of Japanese classrooms, especially at the first-grade level, was the amount of verbal explanation that occurred during mathematics class. We were able to identify segments that contained explanations by either the teacher, a student, or both the teacher and a student. The percents of segments containing these three categories of verbal explanation in Japanese and American classrooms are presented in **figure 1**. A segment is a unit of time in which observations are made. Nearly 50 percent of all Japanese first-grade segments contained verbal explanations, compared with only about 20 percent of the American segments. By the fifth grade the gap has narrowed somewhat, but still more explanation is given in Japanese classrooms than in American classrooms.

Many differences between Japanese and American classrooms help to explain differences in verbal explanation. Whereas the American teachers were more likely to stress participation in nonverbal activities or the asking of short-answer questions to lead students into a new topic, Japanese teachers would give, and ask students to give, lengthy verbal explanations of mathematical concepts and algorithms. The narrative records suggest that Japanese teachers not only explained more but also produced more complicated and abstract explanations than did American teachers, especially in the first grade. For example, one Japanese first-grade teacher started a lesson by asking a child, "Would you explain the difference between what we learned in the previous lesson and what you came across in preparing for today's lessons?" That question surprised us as Americans. What surprised us more, however, was that the six-year-old student was able to reply with a coherent explanation.

Another difference apparently related to the difference in verbal explanations is that Japanese classrooms appeared to move at a more relaxed pace than American classrooms. Only teachers in Japan were ever observed to spend an entire forty-minute lesson on one or two problems. To illustrate this point, we coded the number of mathematical problems covered during all fifth-grade instructional segments with durations of five minutes. The distribution of segments according to number of problems covered is presented in **figure 2**. Whereas attention was paid to only one problem in 17 percent of American segments, attention focused on only one problem in fully 75 percent of Japanese segments (see Stigler and Perry [1988]). Japanese teachers seem not to rush through material but rather are constantly pausing to discuss and explain. Perhaps Japanese teachers and students talk more because they proceed more slowly; alternatively, they may proceed more slowly because they devote so much time to verbal exchanges. Regardless, the pace of instruction does seem to be an important difference.

One final difference that leads to more verbalization in the Japanese classroom is in the methods used for evaluating students' work. The most common method employed by the Japanese teachers in our study was to have students who had produced incorrect solutions present their solutions to the whole

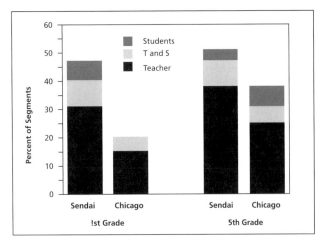

Fig.1. The percent of segments containing explanations by teachers, teachers and students, or students in Sendai (Japan) and Chicago (USA)

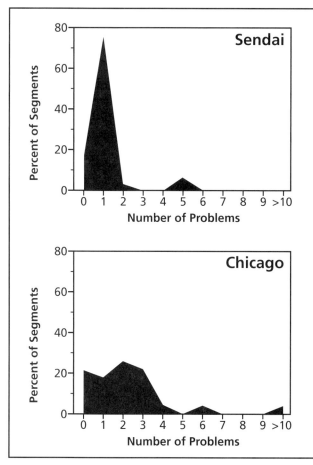

Fig. 2. Distribution of fifth-grade segments with durations of five minutes according to the number of problems covered in the segment

class, usually on the chalkboard, for discussion and correction. American teachers, by contrast, were more likely to evaluate work privately than were the Japanese teachers. Also, American teachers were more likely to restrict public evaluations simply to reporting how many problems were answered correctly (e.g., "Raise your hand if you got them all right") or, most commonly, praising students' correct solutions. American teachers probably steer away from having a child display his or her errors in public because they fear it may damage the child's self-esteem. One clearly, however, can conduct a lot more discussion about errors than about correct answers.

Conclusion

Differences in the use of verbal explanation compose just one among many differences we are finding between Japanese and American mathematics classrooms. I have presented it not as a prescription but as a mirror that American teachers can use to question in a concrete way their own beliefs and practices about mathematics learning and teaching. In particular, we need to reexamine the role of direct explanation in teaching mathematics to young children. A widespread belief in our society maintains that concrete experiences are the best way to teach young children and that language will either go over their young heads or lead to learning of a superficial kind. We need to ponder two important findings from the Japanese observations: (1) young children are capable of responding to, and apparently understanding, complex verbal explanations, and (2) as the Japanese demonstrate, it is possible to stress *both* concrete experiences *and* verbal explanations at the same time. It is possible, in fact, that both are necessary to promote high levels of learning.

A second assumption we should question is the one that ties greater learning to more repetition and practice, that conceives of "covering the material" in a quantitative rather than a qualitative way. The relaxed pace of learnning in Japanese classrooms, combined with the high level of achievement, is a fact worthy of more consideration and more research. Perhaps it is our hurry to get through all the problems in the textbook that makes the "spiral" curriculum necessary. Understanding takes time, and perhaps spending that time at an early stage will lead to future benefits.

Specific techniques may or may not be able to be lifted directly from Japanese classrooms. Far more important, though, is the thinking process among educators that such comparisons bring about. I leave it to teachers to ponder and test the validity of these ideas.

References

Husen, Torsten. *International Study of Achievement in Mathematics.* New York: John Wiley & Sons, 1967.

McKnight, Curtis C., F. Joe Crosswhite, John A. Dossey, Edward Kifer, Jane O. Swafford, Kenneth J. Travers, and Thomas J. Cooney. *The Underachieving Curriculum: Assessing U.S. School Mathematics from an International Perspective.* Champaign, Ill.: Stipes Publishing Co., 1987.

Stevenson, Harold W., S. Y. Lee, and James W. Stigler. "Mathematics Achievement of Chinese, Japanese, and American Children." *Science* 231 (1986):693–99.

Stigler, James W., S. Y. Lee, and Harold. W. Stevenson. "Mathematics Classrooms in Japan, Taiwan, and the United States." *Child Development* 58 (1987):1272–85.

Stigler, James W., and Michelle Perry. "Mathematics Learning in Japanese, Chinese, and American Classrooms." In *Developmental and Cultural Perspectives on Mathematics Learning,* edited by G. Saxe and M. Gearhardt. San Francisco: Jossey-Bass, 1988.

Travers, Kenneth J., F. Joe Crosswhite, John A. Dossey, Jane O. Swafford, Curtis C. McKnight, and Thomas J. Cooney. *Second International Mathematics Study Summary Report for the United States.* Champaign, Ill.: Stipes Publishing Co., 1985.

Section 3

Problem Solving

Issues in Problem-Solving Instruction

Douglas A. Grouws and Thomas L. Good

Research on problem solving can assist teachers by raising issues to consider as instructional decisions are made. Research that focuses on the study of current classroom practice is one good way of identifying such issues. This article first characterizes some results from our recent research on classroom teaching of problem solving and then summarizes the related issues that teachers should consider as they plan and teach problem-solving lessons.

Research . . .

As part of an ongoing study of problem solving, we have interviewed and regularly observed a group of twenty-four seventh- and eighth-grade teachers over a three-year period. The following are some of the important findings from this work:

- *Lessons that focus on problem solving as a topic did not occur frequently.* In the initial round of observations of teachers, we did not observe a single lesson with a problem-solving focus, even under a liberal definition of what constituted a lesson devoted to problem solving. Since our intent was to study naturally occurring problem-solving lessons, we then asked teachers to teach problem-solving lessons during subsequent observations, although we did not elaborate on what we meant by a problem-solving lesson.

- *When asked to teach a problem-solving lesson, most teachers based their lessons on the textbook and chose a section of the textbook that dealt with verbal problems.* Lessons typically were based on textbook story problems in which solving the problem frequently involved just selecting a computational operation and applying it to the only two numbers mentioned in the problem. The problems were also relatively superficial and offered little apparent challenge or interest to the students. Hence, neither the mathematical content nor the instructional context was particularly conducive to the development of problem-solving ability.

- *The use of time in problem-solving lessons was substantially different from teacher to teacher.* Particularly noticeable were differences in the amount of time spent discussing, illustrating, and explaining—that is, in what we have called the developmental phase of the lesson (Good, Grouws, and Ebmeier 1983). Some teachers spent the majority of the class period on this part of the lesson, whereas others gave it little attention and instead allowed extended time for independent practice or for going over seatwork or homework.

- *Teachers' conceptions of problem solving varied widely.* In an interview situation (Grouws, Good, and Dougherty 1988), we asked teachers to define problem solving to help us gain an understanding of teaching practices. Most teachers (60%) focused their response on the type of situations that would be involved in problem solving, that is, on the nature of the problems. These responses were quite easily subdivided into the type of problem mentioned: word problems, practical problems, and problems requiring higher-order thinking. The remaining teachers' descriptions centered on the solution process rather than on the problems themselves. Thus an interesting dichotomy emerged in which teachers narrowly interpreted problem solving by considering it either only in terms of problems or only in terms of solution methods. Even the teachers that conceptualized problem solving by focusing on solution methods, however, seldom allowed opportunities for students actively to construct their own solution strategies.

- *Some teachers were relatively successful in consistently fostering growth in problem-solving ability across classes and school years, but others were quite unsuccessful.* We examined changes in students' problem-solving performance by giving

---

The research reported in this article was supported by the National Science Foundation under Grant No. MDR-847-0265. All opinions and recommendations are, of course, solely those of the authors and do not necessarily reflect the position of the foundation.

a problem-solving test to each class at the beginning and at the end of the school year. The ten-item test was oriented toward requiring some critical thinking rather than mere selection of an operation (e.g., "How many times must the number keys on a typewriter be hit in order to type page numbers on a paper that has 124 pages?"). On this test substantial differences occurred in the average amount of student growth among classes across the school year, but interestingly some of these junior high school teachers facilitated significant growth in every one of their classes. In fact, an examination of the five most successful teachers reveals that 100 percent of their classes gained more than would have been predicted statistically (Grouws and Good 1988). However, 75 percent of the classes of the five least successful teachers gained less than would have been predicted from pretest scores. Hence, a relationship appears to exist between he teaching process and students' performance on relevant problem-solving tasks.

. . . Into Practice

These results raise issues that are helpful for teachers to consider as they reflect on how to present appropriate instruction in problem solving.

It is important to conceive of problem solving not as just a topic in the curriculum but as a process that can permeate almost every aspect of school mathematics. Appropriate time allocations for problem solving are thus not easy to measure or to recommend; reflection on our personal allocations relative to our beliefs and to professional recommendations about the importance of problem solving, however, is important. Every teacher should consider, especially in view of the dearth of problem-solving lessons in some classrooms, these two questions: Is the amount of time devoted to problem solving consistent with my beliefs about the importance of problem solving? Is it consonant with the NCTM's recommendation to make problem solving the focus of mathematics instruction?

The nature of problem-solving lessons obviously affects immediate learning, but it also has subtle long-term effects on the kinds of problems that students can solve, and perhaps just as important, it affects the types of problems that students are willing to attempt and are interested in solving. Given the heavy reliance on the textbook in many classrooms, one can also profitably consider these questions: Do the problem-solving lessons I conduct offer sufficient variety and depth? Are enough of my lessons based on real-life, naturally occurring problems? Should more of my lessons use data that my students have collected? What portion of my lessons should use problems that my students pose? Although a textbook is a valuable *resource* for teaching problem solving, it is not useful for many aspects of problem solving (e.g., providing student-posed problems), yet the textbook is the only resource used for problem solving in too many classrooms.

Considerable research supports the importance of allowing substantial amounts of time for development in mathematics lessons, but we found that many teachers allow little time for development during problem-solving lessons. Although no magic formula exists for determining how much time should be spent on discussing, illustrating, and explaining during problem-solving lessons, these lessons do call for more than working several problems on the chalkboard and then assigning seatwork. Problem solving is a complex endeavor that requires critical thinking and therefore, on a logical basis, *more* development than most other types of lessons, such as those directed to the development of computational skills. Our experience, based on classroom observation, suggests that it is likely that most teachers are allowing too little time for development. As one aspect of increasing the time devoted to development, we believe that teachers should offer students more opportunities to interact among themselves. Structuring activities for students to discuss problems with one another and to describe their own attempts at solving a problem is, in our view, an important part of good problem-solving instruction. Opportunities for active teaching that focus on meaning (Good, Grouws, and Ebmeier 1983), as well as opportunities for students actively to construct knowledge (Confrey 1987; Romberg and Carpenter 1986), are necessary for a successful and systematic approach to mathematical problem solving.

Teachers' beliefs about problem solving no doubt affect instruction, but the relationship between them is not simple and probably varies with the context. Nonetheless, the differences we found in teachers' conceptions about problem solving make it only prudent for teachers to reflect on their beliefs and to ascertain both the depth of these beliefs and the extent to which they are comfortable with the beliefs in light of alternative conceptions. In particular, one might analyze the degree to which one's current belief structures tend to limit the vision of problem solving to story problems and answers or alternatively to problem solving as a process. Simply put, if a teacher conceives of mathematics solely in terms of speed, accuracy, and one right answer, then it is unlikely that such a teacher will stimulate students to monitor their solution processes, estimate answers, search for alternative solution methods, pose problems, or engage in similar worthwhile activities.

Finally, some teachers clearly promote more learning of problem solving than do others. This fact should act as an incentive for all teachers to try to improve their teaching of problem solving. Major contributions to our knowledge of how to foster problem solving are beginning to emerge from the research on learning (see, e.g., Cobb et al. [1988] and Loef et al. [1988]), and we can profit from those studies in our attempts to make problem solving indeed the focus of the curriculum.

References

Cobb, Paul, Erna Yackel, Terry Wood, Grayson Wheatley, and Graceann Merkel. "Creating a Problem-solving Atmosphere." *Arithmetic Teacher* 36 (September 1988):46–47.

Confrey, Jere. "Mathematics Learning and Teaching." In *The Educator's Handbook: A Research Perspective,* edited by Virginia Richardson-Koehler. White Plains, N.Y.: Longman, 1987.

Good, Thomas L., Douglas A. Grouws, and Howard Ebmeier. *Active Mathematics Teaching.* New York: Longman, 1983.

Grouws, Douglas A., and Thomas L. Good. "Teaching Mathematical Problem Solving: Consistency and Variation in Student Performance in the Classes of Junior High Teachers." Paper presented at the annual meeting of the American Educational Research Association, New Orleans, April 1988.

Grouws, Douglas A., Thomas L. Good, and Barbara Dougherty. "Junior High School Teachers' Conceptions of Problem Solving and Beliefs about Problem-Solving Instruction." Technical Report, Center for Research in Social Behavior, University of Missouri, Columbia, Mo., 1988.

Loef, Megan M., Deborah A. Carey, Thomas P. Carpenter, and Elizabeth Fennema. "Integrating Assessment and Instruction." *Arithmetic Teacher* 36 (November 1988):53–55.

Romberg, Thomas A., and Thomas P. Carpenter. "Research on Teaching and Learning Mathematics: Two Disciplines of Scientific Inquiry." In *Handbook of Research on Teaching,* 3d ed., edited by Merlin C. Wittrock. New York: Macmillan, 1986.

Teaching Mathematics and Thinking

Edward A. Silver and Margaret S. Smith

One afternoon, eleven-year-old Michael stopped by the home of an adult friend and found her nearly buried in decorating books, charts, and samples. Michael's sudden appearance at the door was a welcome sight, and he was asked to assist his friend in the process of decorating her home office. Although quite sure he knew little about interior decorating, Michael agreed to lend a hand. The friends began to measure the room and calculate the areas of the ceiling, walls, and floor. They looked at wallpaper swatch books, paint-color charts, and rug samples from various manufacturers and discussed the differences in price, the expected coverage per gallon of paint and roll of wallpaper, and the relative quality and ease of upkeep for different products. One style of wallpaper they liked had a horizontal stripe and required matching. After approximating how many rolls of striped wallpaper they would need to buy, they performed a similar approximation for a design that did not require matching. They looked at carpet samples, discussed the relative merits of light and dark colors and various styles, and considered the issues of cost versus quality with respect to durability and stain resistance. Finally, they selected a set of materials to complete the decorating project on the basis of cost, product quality, maintenance required, and personal preference. After helping his friend with this task for several hours, Michael departed for his home and his dinner.

In a conversation with Michael the next day, it became clear to the adult friend that Michael saw little relationship between the decorating task and the concept of area as he had studied it in school. In school, he had memorized the appropriate vocabulary and the rules for calculating areas of certain figures, and he had practiced applying the rules to a large number of problems presented in the textbook and on dittoed sheets. His school experience, however, had not revealed a connection between the concept of area and the kinds of interesting and challenging questions that were being considered in the decorating task.

Research on Teaching Mathematics and Thinking

Unfortunately, Michael's view of mathematics—as a subject not connected to interesting thinking—is all too common. On the basis of their school experience, too many students adopt the view that mathematics is a dry and dusty subject. Results from the fourth mathematics assessment of the National Assessment of Educational Progress (NAEP) indicate that the majority of seventh-grade students view mathematics as rule based and noncreative, and many feel that learning mathematics is mostly memorizing (Kouba et al. 1988). Given these responses, it is not surprising that these students performed so poorly on those NAEP problems involving mathematical reasoning, nonroutine situations, or multiple steps in the solution. Carpenter and others (1988) argued that the NAEP results demonstrate that a greater emphasis must be placed on helping students become better problem solvers who can communicate and reason about mathematics.

Many would argue that the poor performance of students on mathematics problems requiring more than the routine application of a simple procedure is the direct result of excessive instructional attention to low-level knowledge and rules to be memorized. Students are taught to calculate with numbers but not to think in numerical quantities. This emphasis on basic skills is due to a pervasive educational belief that low-level knowledge and skills, those that require little or no independent thinking and judgment, must be taught and learned before high-level thinking can be developed.

The accumulated mass of evidence from the past few decades of cognitively oriented research on mathematics learning and problem solving challenges this belief. For example, evidence (e.g., Carpenter [1985])

Preparation of this article was supported in part by a grant from the Ford Foundation (grant number 890–0572) to the first author for the QUASAR (Quantitative Understanding Amplifying Students' Achievement and Reasoning) project. Any opinions expressed herein are those of the authors and do not necessarily reflect the views of the Ford Foundation.

has demonstrated that children who lack certain presumed prerequisite low-level skills are often capable of performing high-level tasks successfully using reasoning and problem solving. Furthermore, the process of acquiring and using skills presumed to be low level is itself quite complex (e.g., Wearne and Hiebert [1988]). In fact, complete mastery of basic skills probably requires an understanding of fundamental concepts sufficient to judge when it is appropriate to use a particular skill (e.g., Schoenfeld [1986]). This research base makes it possible to envision a new mathematics curriculum in which the learning of basic skills is integrated with thinking, problem solving, and reasoning.

Thinking skills

In recent years, reform has been called for in many school subjects, not just mathematics, to make the teaching of thinking a central part of the curriculum. The impetus for this movement comes in large part from two different bodies of research: *(a)* research that has revealed and analyzed poor performance by students on complex tasks and *(b)* research that has documented children's capabilities for complex thinking and reasoning on which current curricula are not building. Although interest in teaching thinking is not new, the current focus on teaching *all* students to think, not just a privileged elite, is especially promising. Several programs (e.g., Nickerson, Perkins, and Smith [1985]) have been developed in recent years to respond to the apparent need to upgrade students' reasoning and problem-solving skills (i.e., high-level thinking skills) in many disciplines.

Despite the widespread interest in teaching thinking, no simple, clear, universally accepted definition of what is meant by "high-level thinking" has emerged. Nevertheless, general agreement is found on certain hallmarks of high-level thinking. Resnick's (1987) review of research and scholarship on learning to think identified several of the features that characterize high-level thinking. In particular, high-level thinking is nonalgorithmic and complex; it often yields multiple solutions; and it involves nuanced judgment, uncertainty, and self-regulation of the thinking process. Considering the redecorating task in light of this description, we see that it presented Michael with an opportunity to engage in high-level thinking because *(a)* although some procedural knowledge was required, *no algorithm* was available that would solve the entire problem and no clear path of action had been laid out (e.g., what do you do after you find the areas?); *(b)* the reasoning about the task was *complex,* including many components (e.g., finding the areas, determining the number of gallons of paint needed, weighing cost against quality); *(c)* the problem had *multiple solutions* rather than a single "right answer"; *(d) nuanced judgment* was required (e.g., stain-resistant carpet is probably the better choice even though it costs more money, because it will wear better and require less maintenance); *(e)* the reasoning involved *uncertainty* (e.g., how many coats of paint will really be required to cover an existing darker color?); and *(f)* the solution process required *self-regulation* (e.g., no checklist of things to consider was available for determining the materials to purchase). Although the redecorating task might have been approached rather algorithmically by an experienced interior decorator, it served as an opportunity for complex quantitative thinking for Michael and his adult friend.

General versus specific approaches

Much attention has been given to the development of programs for teaching high-level thinking. *General approaches* to teaching thinking skills, such as the Productive Thinking Program (Covington 1985) and CoRT (de Bono 1985), focus instruction on helping students develop a variety of strategies for planning, managing, and monitoring their own cognitive activity. As Adams (1989) noted, the fundamental assumption underlying these general approaches is that . . . certain . . . processes are common to all high-level thinking, regardless of the domain in which they will be applied. *Content-specific approaches,* in contrast, infuse thinking skills and strategies into the teaching of a particular subject area. In elementary school mathematics, the teaching of heuristic problem-solving strategies, such as drawing a diagram or restating the problem, can be part of a content-specific approach, and evidence suggests that students can benefit from such instruction. For example, Lee (1982) was able to teach fourth graders a set of heuristic strategies, and the students who received the instruction were more successful in using the strategies to solve a variety of mathematics problems than students who had not received the instruction. Putt (1979) found that fifth-grade students could benefit from instruction in heuristics, so that they were able to use many of the strategies effectively, were able to develop an appropriate vocabulary for discussing the strategies, and were able to suggest many questions appropriate for understanding the problems they were asked to solve. Heuristics are not synonymous with high-level thinking skills, however, and are sometimes inappropriately presented to students as a memorizable list of steps that must be followed.

Resnick (1987) suggests that although evidence shows that elements of high-level thinking can be taught, little evidence can be cited to support a particular program or approach. Perkins and Salomon (1989) argue that if our goal is to teach skills that can be generalized and transferred to other domains, we need to give students instruction in the principles of both general and content-specific high-level thinking. They claim that domain-specific knowledge without general heuristics is inert and that general heuristics that are detached from a rich domain-specific knowledge base are weak. In the next section, we turn our attention to some ways to teach high-level thinking in elementary school mathematics.

Suggestions for Classroom Practice

Evidence suggests that children are capable of high-level thinking and that they often demonstrate this ability naturally in nonschool settings. Helping students to be more thoughtful about the mathematics they learn in school does not necessarily require nonroutine problems or special materials. Opportunities can be found in the topics commonly taught in elementary school. For example, any division computation exercise involving a remainder can be used to model and solve several different story problems. Therefore, any division exercise can serve as an instructional opportunity for rich thinking and discussion about the application of mathematics to real-world situations and the nature of mathematical models (Silver 1988). Moreover, routine computational procedures can in some ways be a source of high-level thinking and reasoning about mathematics (Lampert 1989). For example, having students explore alternative algorithms for standard operations offers an opportunity for a discussion of why a particular algorithm works and why it might be preferable to others. Such opportunities to emphasize high-level thinking are fairly easy to integrate with standard textbooks and instructional practice.

Problems with multiple methods of solution

Opportunities for students to think about mathematics are often associated with their talking about mathematics with one another and with their teacher. One tactic that some teachers have found effective to stimulate students' thinking and discussion is to present problems that have multiple methods of solution. Consider, for example, the following problem that one teacher recently presented to her seventh-grade class:

> A farmer puts his chickens into cages and finds that he has 2 cages left over if he puts 6 chickens in a cage, but he has 2 chickens left over if he puts 4 chickens in a cage. How many chickens and how many cages does the farmer have?

After a brief time in which the teacher presented the problem and discussed it with the class, the students were asked to solve the problem independently or in small groups, using any method they wished. During the solution period, the teacher circulated around the room, taking note of the various approaches being used by the students, asking clarifying questions, and giving other help to those who needed it. During her travels around the room, the teacher designated some of the students to present their solutions to the class. When the time for work on the solution of the problem was over, the designated students, who had written their solutions on large sheets of butcher paper, presented and explained their solutions to the entire class. The multiple methods—some involving clever counting, some based on a guess-and-test strategy, others with a geometric flavor, and even one that was fairly algebraic in nature—were displayed for subsequent summary and discussion, during which the similarities and differences, unique features, and other characteristics of the solutions were identified and examined by the teacher and her students.

Opportunities for students to create and discuss multiple solution methods for interesting mathematical problems constitute invitations to engage in high-level thinking in mathematics class, and they represent important opportunities for teachers to learn about and enhance their students' mathematical thinking and reasoning. Although textbooks do not typically emphasize multiple solution procedures, many textbook problems could be used for this kind of activity.

Open-ended problems

Another technique that some teachers have used to foster opportunities for high-level thinking in mathematics classes is the posing of open-ended problems. Although mathematics textbooks almost always present problems that have a well-specified goal, interesting activities can result from modifying existing problems to make them more open ended. Silver and Adams (1987) have suggested a few ways to use open-ended problems in elementary school classes, and Brown and Walter (1983) have suggested many examples appropriate for older students. Consider the following two problems, which illustrate the difference between a typical school mathematics problem and a more open-ended version:

Problem 1

> John has 34 marbles, Bill has 27 marbles, and Mike has 23 marbles. How many marbles do they have together?

Problem 2

> John has 34 marbles, Bill has 27 marbles, and Mike has 23 marbles. Write and solve as many problems as you can using this information.

Problem 1, which is a typical elementary school problem, is not likely to encourage high-level thinking if the students are simply expected to apply a particular algorithm they have learned. However, the open-ended nature of problem 2 invites students to consider alternatives, to analyze a simple situation and relate it to the mathematics they know, and to propose problems that are interesting and complex.

Teachers who have used such activities report that their students often propose many interesting problems that are more challenging than the problems they are typically given to solve. For example, problem 2 would likely lead not only to the question posed in problem 1 and other typical questions, such as, How many more marbles than Bill does John have? It would also lead to more interesting and challenging questions, such as, How many more marbles would they need in order to have as many marbles as Sammy, who has 103 marbles? or How many marbles would Bill need to give Mike in order for them to have the same number of marbles? Moreover, a question like the following could lead to an interesting discussion of the reasonableness of problems and solutions: How many marbles would John need to give Mike in order for them to have the same number of marbles?

Situational problems

Many teachers have also found that it is possible to bring into the mathematics classroom problems like the decorating task with which Michael was helping his friend. For example, consider the following:

> Rebecca has a pocketful of change. She would like to buy a soda, which costs $0.55. How could she pay for the soda so that she would eliminate the most change from her pocket?

This applied situational problem has features of the classes of problems previously discussed. It is open with respect to interpretation. (Is Rebecca buying the soda from a machine? If so, even if she had fifty-five pennies, they would not be useful in this situation. But if Rebecca is at a store, fifty-five pennies is a valid solution.) The problem is also open with respect to solution. (If she is buying the soda from a machine, she might first look for eleven nickels in her pocket to eliminate a maximum number of coins. If she does not have eleven nickels, then using nine nickels and one dime might be her next attempt to solve the problem.) Furthermore, the problem is even open with respect to goal. (Should Rebecca use volume, weight, or number of coins in her pocket as the criterion for the "best" solution?) Problems such as this one give students experience in using mathematics and offer opportunities for high-level thinking.

Conclusion

All the activities we have suggested could be used in classrooms by teachers to encourage the kind of high-level thinking described by Resnick (1987)—thinking that is complex, that is nonalgorithmic, and that involves some judgment and uncertainty. Using such activities will not only help students develop their ability to think at a high level but also help them see that such thinking is valued in the subject area of mathematics and in your classroom.

References

Adams, Marilyn Jager. "Thinking Skills Curricula: Their Promise and Progress." *Educational Psychologist* 24 (Winter 1989):25–77.

Brown, Stephen I., and Marion I. Walter. *The Art of Problem Posing.* Hillsdale, N.J.: Lawrence Erlbaum Associates, 1983.

Carpenter, Thomas P. "Learning to Add and Subtract: An Exercise in Problem Solving." In *Teaching and Learning Mathematical Problem Solving: Multiple Research Perspectives,* edited by Edward A. Silver, pp. 17–40. Hillsdale, N.J.: Lawrence Erlbaum Associates, 1985.

Carpenter, Thomas P., Mary M. Lindquist, Catherine A. Brown, Vicky L. Kouba, Edward A. Silver, and Jane O. Swafford. "Results of the Fourth NAEP Assessment of Mathematics: Trends and Conclusions." *Arithmetic Teacher* 36 (December 1988):38–41.

Covington, Martin. "Strategic Thinking and the Fear of Failure." In *Thinking and Learning Skills,* vol. 1, *Relating Instruction to Research,* edited by Judith W. Segal, Susan F. Chipman, and Robert Glaser, pp. 389–416. Hillsdale, N.J.: Lawrence Erlbaum Associates, 1985.

de Bono, Edward. "The CoRT Thinking Program." In *Thinking and Learning Skills,* vol. 1, *Relating Instruction to Research,* edited by Judith W. Segal, Susan F. Chipman, and Robert Glaser, pp. 363–88. Hillsdale, N.J.: Lawrence Erlbaum Associates, 1985.

Kouba, Vicky L., Catherine A. Brown, Thomas P. Carpenter, Mary M. Lindquist, Edward A. Silver, and Jane O. Swafford. "Results of the Fourth NAEP Assessment of Mathematics: Measurement, Geometry, Data Interpretation, Attitudes, and Other Topics." *Arithmetic Teacher* 35 (May 1988):10–16.

Lampert, Magdalene. "Arithmetic as Problem Solving." *Arithmetic Teacher* 36 (March 1989):34–36.

Lee, Kil S. "Fourth Graders' Heuristic Problem-Solving Behavior." *Journal for Research in Mathematics Education* 13 (March 1982):110–23.

Nickerson, Raymond, David Perkins, and Edward Smith. *The Teaching of Thinking.* Hillsdale, N.J.: Lawrence Erlbaum Associates, 1985.

Perkins, David, and Gavriel Salomon. "Are Cognitive Skills Context-Bound?" *Educational Researcher* 18 (January-February 1989): 16–25.

Putt, Ian J. "An Exploratory Investigation of Two Methods of Instruction in Mathematical Problem Solving at the Fifth-Grade Level." Ph.D. diss., Indiana University, 1979.

Resnick, Lauren. *Education and Learning to Think.* Washington, D.C.: National Academy Press, 1987.

Schoenfeld, Alan H. "On Having and Using Geometric Knowledge." In *Conceptual and Procedural Knowledge: The Case of Mathematics,* edited by James Hiebert, pp. 225–64. Hillsdale, N. J.: Lawrence Erlbaum Associates, 1986.

Silver, Edward A. "Solving Story Problems Involving Division with Remainders: The Importance of Semantic Processing and Referential Mapping." In *Proceedings of the Tenth Annual Meeting of the North American Chapter of the IGPME,* edited by Merlyn J. Behr, Carol B. Lacampagne, and Margariete M. Wheeler, pp. 127–33. DeKalb, Ill.: n.p., 1988.

Silver, Edward A., and Verna M. Adams. "Problem Solving: Tips for Teachers: Using Open-ended Problems." *Arithmetic Teacher* 34 (May 1987):34–35.

Wearne, Diana, and James Hiebert. "A Cognitive Approach to Meaningful Mathematics Instruction: Testing a Local Theory Using Decimal Numbers." *Journal for Research in Mathematics Education* 19 (November 1988):371–84.

Arithmetic as Problem Solving

Magdalene Lampert

Teachers often feel torn, especially in upper elementary school mathematics classes, between spending time on problem-solving work that will get students to understand mathematics and spending time on developing computational skills. The conflict is especially strong because the computational skills in the curriculum at this level are complex: "long" multiplication and division, relating fractions to decimals and percents, and operations on fractions. The procedures involved in doing these computations involve many steps, and students often have difficulty remembering what to do and in what order.

The four-year teaching experiment described here was aimed at determining whether it is possible to redefine the problem that students must solve in upper elementary arithmetic—to change it from a problem of *remembering* what to do and in what order to do it, to a problem of *figuring out* why arithmetic rules make sense in the first place. The purpose of teaching this way is not only to have students understand arithmetic procedures but also to communicate the idea that mathematics is a way of thinking and doing things that should make sense to the persons using it, not just to the teacher or the textbook author. I have conducted these experiments in my own fourth- and fifth- grade classrooms with students from many different backgrounds. They are not grouped by ability, and many speak English as a second language.

The Mathematics behind the Arithmetic

For each topic in the curriculum, the research began with a careful analysis of the basic mathematical ideas behind the arithmetic I wanted students to master. The first topic to be studied was the multiplication of large numbers. The essential idea behind the way we do this kind of computation is that we can take numbers apart, operate on the pieces, and then put them back together again. The problem to be solved is that most people cannot do a multiplication like 182×357 all at once. In conventional arithmetic teaching and learning, we solve this problem by using rules for doing the computation that have been passed down from one generation to another, making arithmetic more a cultural tradition than a process of mathematical problem solving.

To multiply 182×357, we usually take the 182 apart into 2, 80, and 100 and multiply 357 by each of these numbers. But we do even that computation in pieces, first multiplying 2×7, then 2×50, then 2×300, then 80×7, 80×50, and so on. The rules that we follow tell us where to place the numbers and how to keep track of all these separate multiplications and put them together appropriately at the end.

Multiplying large numbers does not require using this conventional grouping procedure, however, to get the correct product. If we found the "six times" table easier to manage than the "eights," for example, we could calculate 182×357 by doing 60×357 three times (because 180 is 3×60), adding up the results, and then adding 357 and then adding it again (i.e., adding 2×357). (Try it!) We could even multiply 200×357, and then *take away* both 10×357 and 8×357, because $182 = 200 - 10 - 8$. The reason all these procedures work is the same, and it has to do with what multiplication means at this stage of the curriculum: counting the number of objects in a large group that is formed by putting together some collection of smaller groups that are all the same size. (This is only one interpretation of multiplication in mathematics, but it is the one most accessible to beginners. Other interpretations are outlined in James Kaput, *Multiplicative Word Problems and Intensive Quantities: An Integrated Software Response* [Cambridge, Mass.: Educational Technology Center, Harvard University, 1985].)

Given this understanding of the mathematical ideas behind how we do "long" multiplication, the next part of the experiment was to find learning activities that students could do that would involve them in grouping as a strategy for solving the problem of how to multiply large numbers. In all the activities in the experiment, students were able to refer to familiar situations as the basis for their mathematical reasoning. They could create different ways of taking the numbers apart, and they could competently recombine them when appropriate.

Multiplication Stories and Drawings

In these activities, students learned to use familiar situations to figure out how to operate on numbers. They were experienced at telling stories to go with expressions like 6×7 (e.g., "There were six parties last week and seven different children attended each party"). But even when students understand arithmetic operations on small numbers, that understanding is often lost when they are faced with learning the long set of rules that goes with doing the same operation on larger numbers.

We began with stories and drawings to go with multiplications like 12×4. Some students interpreted this problem in place-value terms, as, for example, "ten jars with four butterflies in each, and two more jars with four butterflies in each," and they figured that the answer would be $(10 \times 4) + (2 \times 4)$, or forty-eight, butterflies altogether. Others grouped the jars differently and figured differently: $(6 \times 4) + (6 \times 4)$ or $(3 \times 4) + (3 \times 4) + (3 \times 4) + (3 \times 4)$. (These numerical expressions show what students *did*; they did not write down their strategies in this form.) Their solutions were associated with drawings, and the stories and the drawings together served as the standard for deciding whether a particular solution strategy was appropriate. The drawings also made it clear why multiplication was done first and then addition, and why the "4" kept showing up over and over again. The "problem" that students were solving was how to break up the larger number so that they could multiply.

We moved on to multiplications like 28×65, where the problem of breaking down the numbers to operate on them became more complex. Students continued to be able to produce appropriate stories and drawings, however, that kept the *meaning* of multiplication at the center of their work. They were able to appreciate the emphasis on inventing appropriate solutions, in contrast to finding the answer as quickly as possible. The students moved back and forth between operating on numbers and making sense of those operations by referring the numbers to quantities. For example, after the class had discussed representing 28×65 as eight glasses of water, each holding sixty-five drops, plus two jugs of water that would hold ten glassfuls each (20×65), one student came up with this idea:

> You could have three jugs. Two would have 650 drops in them. But if you put 650 drops in the third one, you'd have too much. You'd have to take out two glasses because there are not thirty glasses in the story, but only twenty-eight. And each glass has 65 drops, so you'd have to take out 65 plus 65, or 130, drops. Then in the third jug you'd have 650 – 130, which is 520. You'd have to add the other two, 1300, plus 520. And you'd get 1820.

She drew a picture to go with her solution strategy to illustrate the taking apart and putting together (**fig. 1**), and the drawing helped her and the other students in the class to understand her method.

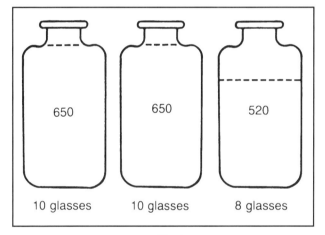

Fig. 1

Practical Problem Solving

During this series of lessons, I also looked for problems that arose in other classroom activities that might be solved by a similar process of taking numbers apart, operating on them, and then putting them back together. The class had been discussing a certain science project and talking about "tons" of this and that, and I asked them if they could explain how heavy a ton really was. My questions drew blank looks, so I posed the problem of finding out how many students would make a ton.

Assuming an average of 50 pounds per person, they first figured that two persons would be 100 pounds, and so ten persons would be 500 pounds. (We might conventionally think of this as a division problem, but as in the long-division algorithm, these students are finding a quotient by multiplication [see Lampert (1986b) for other examples of this kind of problem solving]. One girl continued, "If you double ten persons, you get a thousand pounds and that's twenty persons. If you double that, it's forty persons and it makes one ton." Another explanation came from one of the boys: "If you double two persons you get four people and that equals two hundred pounds. And so for two thousand pounds, you just add a zero to both and it's forty people." This solution troubled Maria, who raised her hand and said she didn't think anyone in the class weighed as little as 50 pounds! The class agreed on 75 pounds as a more appropriate average, and the process of figuring out how many students would be in a ton began again. This suggestion elicited a greater variety of strategies, particularly because the calculation does not "come out even."

Connections with the Conventional Algorithm

All these experiments produced evidence that fourth graders could solve the problem of decomposing large quantities to multiply them and then recompose the partial products appropriately. The next step was to connect this experience with operations on numbers that were not associated with a particular context. As with stories and pictures, I began these lessons with one-digit-by-two-digit multiplications. By breaking the numbers up along place-value lines, the total in 3×86 (which we spoke of as 3 groups of 86) could be figured out in four steps, represented in **figure 2**. We did several multiplications together in this way as a class: I wrote on the chalkboard, the students wrote the procedures on their papers, and we discussed what was happening to the "groups" in each step. This approach gave the students both a spoken and a written language for doing multiplications that could later be connected to the familiar, shorter algorithm.

```
        86  →  80 + 6        (First step)
     ×   3
        18  ←  3 × 6         (Second step)
      +240  ←  3 × 80        (Third step)
       258  ←  18 + 240      (Fourth step)
```

Fig. 2

A major focus of our discussions was on the *order* in which the steps ought to be done. In the foregoing multiplication, for example, students noticed that it did not matter if the second and third steps were reversed. In fact, when we began multiplying larger numbers, like 8×3652, most students began by doing 8×3000, perhaps because three thousand is what comes first when we read the number. They proceeded as shown in **figure 3.** Writing out 3652 as $3000 + 600 + 50 + 2$ as the first step was a reminder of the values those digits represented in the number. By contrast, if the students are taught only to do 8×2, then 8×5, then 8×6, then 8×3, they may learn only a meaningless algorithm.

```
         3 652  →  3 000 + 600 + 50 + 2    (First step)
      ×      8
        24 000  ←  8 × 3 000               (Second step)
         4 800  ←  8 × 600                 (Third step)
           400  ←  8 × 50                  (Fourth step)
      +     16  ←  8 × 2                   (Fifth step)
        29 216  ←  24 000 + 4 800 + 400 + 16   (Sixth step)
```

Fig. 3

Of course, the use of numbers arranged in such a way as to make the mathematical principles involved in multiplication more explicit becomes more cumbersome as the numbers get larger. But one finding of this study that surprised me, given my previous experience with teaching this topic, was that students often preferred to write out these steps, even after they had mastered the shortcut. (See Lampert [1986a] for an explication of errors students made as they made this transition.) They called it the "no carry way" and felt that it was more reliable as a strategy than the usual algorithm.

Implications for Teaching

The outcome of this teaching experiment was that students of diverse abilities were able both to do multiplication of large numbers and to make sense of why they were doing what they were doing in mathematics lessons. Although we spent more time on this topic than is ordinarily allocated in textbooks, they learned not only how to do multiplication but also how to connect actions on quantities with the operations of arithmetic. When we came to related topics later in the year, the students used this connection to make sense of what they were learning. We spent less time, therefore, on practicing the rules because students were able to talk about why the rules worked the way they did.

References

Lampert, Magdelene. "Knowing, Doing, and Teaching Multiplication." *Cognition and Instruction* 3 (1986a): 334–36.

———. "Teaching Multiplication." *Journal of Mathematical Behavior* 5 (December 1986b): 241–80.

Creating a Problem-Solving Atmosphere

Paul Cobb, Erna Yackel, Terry Wood, Grayson Wheatley, and Graceann Merkel

Our ongoing project involves twenty-three second-grade teachers who are teaching all their mathematics, including computation, through small-group problem solving and whole-class discussion. Typically the children first work on problem-centered mathematical activities in pairs or occasionally in groups of three. During this phase of the lesson, the teacher moves from group to group, observing and interacting with the children while they do mathematics. After fifteen or twenty minutes, the teacher asks the children to stop working and begins a whole-class discussion of their solutions to the problems.

These teachers are generally successful in creating what we call a "problem-solving atmosphere" in their classrooms. As a result, the children persist in attempting to solve the problems and do not worry if they are still on the first mathematical activity while other groups have completed three or four. They view mathematical problems as personal challenges and become upset if someone tells them the answer. They believe that mathematics should make sense, and they achieve personal satisfaction when they figure something out for themselves. Furthermore, they feel free to discuss their mathematical understandings both in their small groups and in the whole-class discussions, and they accept that their solutions should be explainable and justifiable (Cobb, Wood, and Yackel 1991; Cobb, Yackel, and Wood 1989).

To understand the teachers' success, we shall focus here on the ways in which they communicate their expectations to the children and thus attempt to place the children under certain obligations for their conduct in the classroom (Voigt 1985). For example, one teacher expectation is that the children should explain how they understand and attempt to solve the mathematical activities. The following incident occurred on the first day of school in one classroom and illustrates how the teacher exploited a potentially damaging incident to communicate her expectations. The discussion centers on the following word problem:

> How many runners are there all together? There are six runners on each team. There are two teams in the race.

Teacher: Peter. What answer did you come up with?

Peter: Fourteen.

Teacher: Fourteen. How did you get that answer?

Peter: Because six plus six is twelve. Six runners on two teams . . . (Peter stops talking, puts his hands to the side of his face, and looks down at the floor.)

Teacher: Would you say that again. I didn't quite get the whole thing. You had . . . . Say it again, please.

Peter: (Softly, still facing the front of the room with his back to the teacher.) It's six runners on each team.

Teacher: Right.

Peter: (Turns to look at the teacher.) I made a mistake. It's wrong. It should be twelve.

Peter's acute embarrassment at having made a mistake in front of his classmates confounded the teacher's goal of having the children talk about *their* mathematics. She made an on-the-spot decision to use this incident to talk about her expectations.

Teacher: (Softly.) Oh, okay. Is it okay to make a mistake?

Andrew: Yes.

Teacher: Is it okay to make a mistake, Peter?

Peter: (Still facing the front of the class.) Yes.

Teacher: You bet it is. As long as you're in my class it is okay to make a mistake. Because I make them all the time, and we learn a lot

---

The research reported in this paper was supported by the National Science Foundation under Grant No. MDR-8740400. All opinions and recommendations expressed are, of course, solely those of the authors and do not necessarily reflect the position of the foundation.

from our mistakes. Peter already figured out, "Oops. I didn't have the right answer the first time" [Peter turns and looks at the teacher and smiles], but he kept working at it and he got it right.

Here, in context, the teacher emphasized that Peter's attempt to solve the problem was appropriate in every way. She demonstrated to the children that she was genuinely interested in their mathematical thinking and that they can learn from errors.

The teacher, also capitalized on situations in which the children acted in accordance with her expectations. For example, during one small-group problem-solving session a number of children had completed several problems, but one pair volunteered that they had spent twenty minutes working on a single problem.

Kara and Julia: Because at first we didn't understand it.
Teacher: How did you feel when you finally got your solution?
Kara and Julia (almost jumping up and down): Good!

By calling this incident to the attention of the entire class and asking the two girls further questions about what had happened, the teacher demonstrated that she expected the children to persist and figure problems out for themselves. She tried to show them that they should take pride in their own accomplishments. In contrast, the teacher never drew attention to a group that had completed a relatively large number of mathematical activities.

In both these examples, the teacher acted as a practical reasoner who continually had to find ways to deal with the unexpected. She displayed her skill and expertise by using problematic classroom situations, such as Peter's response, to further her goals for mathematics instruction. The manner in which the teacher negotiated her expectations with the children illustrates a crucial aspect of what it means to be an effective mathematics teacher.

The teacher's expectations for the children during whole-class discussions included that each child be able to do the following:

- Explain how he or she understood and attempted to solve a mathematical activity that the group has completed
- Listen to, and try to make sense of, explanations given by other children
- Indicate his or her agreement or disagreement with solutions given by other children
- In the event of conflicting solutions, attempt to justify a solution and question alternatives and thus work toward the achievement of a consensus

In the small-group problem solving, the teacher expected that the children would do the following:

- Cooperate to complete the activities
- Agree on an answer and, ideally, on a solution method
- Explain their solutions to one another and listen to one another's explanations
- Persist to figure out problems for themselves

During small-group problem solving in particular, the teacher had to draw on her expertise to help the children find ways to fulfill their obligations. For example, the children's initial attempts to think things through for themselves and to explain their solution methods to their partners sometimes led to conflicts. If one child was trying to figure something out while her partner was simultaneously explaining his solution method, the first child typically complained to the teacher. The teacher then initiated a discussion to help the children take each other's viewpoint into account. She might first ask each child what the difficulty was and then ask how they might resolve it. By giving the children the primary responsibility for their learning and conduct in the classroom, the teacher was generally successful in helping them develop productive working relationships.

Thus far, we have focused on the expectations that the teacher had for the children. The teacher also had to accept obligations for herself if she was to realize her expectations in the classroom. For example, the teacher had to accept the children's explanations and justifications in a nonevaluative way if she wanted them to say what they really thought. The discussions came to an abrupt halt on the very rare occasions in which the teacher attempted to steer the children to a solution that she had in mind. The children failed to volunteer solutions and, when called on, typically made excuses to avoid having to present a solution (e.g., "We didn't do that one" or "We forgot how we did it"). An open, accepting attitude in which the teacher assumes that the children's solutions make sense to them is a prerequisite for the creation of a problem-solving atmosphere during mathematics instruction. This crucial attitude is captured by asking, "What is this child trying to say and how is he or she thinking?" (Labinowicz 1987). We cannot overemphasize the importance of viewing children's solution attempts as expressions of their mathematical thinking that should be treated with respect rather than as examples of faulty thinking that need to be corrected.

In conclusion, we note that our view of teachers as practical reasoners implies that experiences of interacting with children in the classroom are vital to the development of the wisdom and judgment that characterize the expertise of successful teachers. The teachers we are working with are developing this expertise further as they make sense of classroom occurrences that arise as they implement a problem-solving approach. We are not translating theory into practice in the sense of telling teachers what they ought to do. In certain domains, the teachers are far more knowledgeable than we are for the simple reason that they have far richer experiences of interacting with children within the institutionalized constraints of the school. However, we have greater expertise in other domains. Thus, it is not a matter of transporting readymade theories into the classroom, but one of researchers and teachers learning from each other. In this way, theory and practice grow together, with each informing the other.

References

Cobb, Paul, Terry Wood, and Erna Yackel. "A Constructivist Approach to Second Grade Mathematics." In *Radical Constructivism in Mathematics Education,* edited by Ernst von Glasersfeld. Dordrecht, Netherlands: Kluwer Academic Press 1991.

Cobb, Paul, Erna Yackel, and Terry Wood. "Young Children's Emotional Acts While Doing Mathematical Problem Solving." In *Affect and Mathematical Problem Solving: A New Perspective,* edited by Douglas B. McLeod and Verna M. Adams. New York: Springer-Verlag, 1989.

Labinowicz, Ed. "Children's Right to Be Wrong." *Arithmetic Teacher* 35 (December 1987):2, 20.

Voigt, Jorg. "Patterns and Routines in Classroom Interaction." *Recherches en Didactique des Mathematiques* 6 (January 1985):69–118.

Mathematical Problem Solving in and out of School

Frank K. Lester Jr.

An ice cream cone can be bought for 60 cents, and you have a quarter, a dime, and two pennies. How much more money do you need to buy the cone?

Adapted from Resnick (1987, p. 15)

Imagine yourself as a fourth-grade teacher. You give the problem shown above to your students to solve cooperatively. How do you think they will solve it? What kind of solution processes do you hope they will use? Now imagine that three of your students are actually in an ice-cream shop and that they have encountered this very same problem. Will their approach to solving the problem be the same in the ice-cream shop as in the classroom?

Resnick (1987), who has been studying the differences between children's out-of-school modes of thinking and their in-school modes of thinking, suggested that a practical, everyday solution to this problem would probably involve looking through pockets for a quarter, or perhaps some combination of nickels and dimes to make a quarter, to add to the quarter and dime already in hand. However, when she presented the problem in school to a class of fourth graders, some students interpreted it as a straightforward calculation exercise and did something like this:

$$25¢ + 10¢ + 2¢ = 37¢$$

and

$$60¢ - 37¢ = 23¢.$$

Of course, both solutions are correct. But in the "really real world"—as contrasted with the "real world" of textbook problems—it simply would not make sense to spend your time going through the exact calculation and then putting together some combination of coins to get twenty-three cents. In everyday situations that involve mathematics, using the standard procedures typically learned in school often does not make sense.

The purpose of this article is to discuss the differences between mathematical problem solving that typically goes on inside classrooms and the problem solving that is part of everyday situations outside of school.

The Use of Mathematics in Everyday Situations

A substantial and growing body of research has emerged on the use of mathematics in everyday situations. Much of this research has been devoted to the study of how adults use mathematics to solve problems in the course of their activities: notably, dairy workers (Scribner 1984), carpenters (Schliemann 1985), tailors (Lave 1977), grocery shoppers (Lave, Murtaugh, and de la Rocha 1984; Murtaugh 1985), weight watchers (de la Rocha 1985), and even bookies (Acioly and Schliemann 1986). Lave (1988) and Stigler and Baranes (1988) have presented excellent discussions of this body of research.

Only a few studies have investigated the mathematical behavior of children in out-of-school settings. A good example of this research is the work of Carraher and her colleagues in Brazil, who have studied the informal procedures children create to solve problems in natural, out-of-school settings (Carraher, Carraher, and Schliemann 1985, 1987; Carraher and Schliemann 1988). In one study of youngsters, ages 9–15, who worked in markets, they found that those who were capable of solving a computational problem in a natural, everyday setting often failed to solve the same problem when it was presented to them out of its context (Carraher, Carraher, and Schliemann 1985). They suggested that the reason for the superior performance in everyday contexts was that in natural settings the youth used their common sense and relied on mental calculations that were closely tied to the quantities involved in the problems. By contrast, their attempts to solve the more formal, schoollike problems typically were restricted to the rote manipulation of symbols.

Two common threads run through all this research. First, problem solving in the really real world usually is sustained by the fact that the contexts in which problems are embedded make sense to the problem solver. As a result, the individual is able to apply informal knowledge to a problem that has developed from direct experience. The second thread is that in the everyday world, people often use mathematical procedures and thinking processes that are quite different from those learned in school. Furthermore, people's everyday mathematics often reflects a higher level of thinking than is typically expected or accomplished in school. Estimation and approximation are often used during everyday activities. And when exact calculations are needed and made, school-learned algorithms are often *not* used, and the answers are almost always correct.

Solving Problems in School

To a large extent, school mathematics bears little resemblance to the kinds of problems with which people are confronted in out-of-school settings. Perhaps even more important, school tasks often require little or no understanding to get correct answers. A few years ago, in a study of the problem-solving behavior of some third- and fifth-graders (Lester and Garofalo 1982), students were asked to solve the following problem. It is not an everyday problem, but it does require a level of understanding not found in typical school tasks. Using such a problem allowed us to see more clearly the inappropriate procedures the students were using in school.

> Tom and Sue visited a farm and saw chickens and pigs in the barnyard. Tom said, "There are 18 animals." Sue said, "Yes, and there are 52 legs." Can you tell me how many of each kind of animal they saw?

Almost all the third-graders solved the problem by adding 52 and 18 to obtain 70 as the answer. (When asked "Seventy what?" a common reply was "Chickens, pigs, and legs.") Most fifth-graders solved the problem by attempting to divide 52 by 18, usually unsuccessfully. Prior to our study, these students had had experience only with routine textbook problems, primarily single-step problems. The vast majority of them had come to believe that all mathematics problems could be solved by a single application of one of the four arithmetic operations. As a result of this narrow conception, their sole approach to problem solving was to determine the operation to use and then to carry out that operation. They made little or no attempt to make sense of the problems because they had rarely encountered problems for which their method would not be successful. The responses of these students suggest that they were not learning in school that doing mathematics should be an act of sense making (Schoenfeld 1988).

Why Is Out-of-School Performance Often Better Than School Performance?

In classrooms, students tend to focus on the syntax (i.e., the symbols and rules) of mathematics rather than its semantics (i.e., meaning). Research suggests that the focus in school mathematics on formal manipulation of symbols discourages students from bringing their developed intuitions to bear on school learning tasks (Resnick 1989). All too often, memorization and written computation are stressed. The informal, sensible methods children have learned outside of school are ignored or discouraged, and little oral arithmetic or substantive discussion is related to the meaning of the formal procedures that are taught. Thus, it is not surprising that students spend very little time using their intuition or making sense of what they do in school mathematics; they rarely are expected or given the opportunity to do so.

A comparison of in-school and out-of-school contexts indicates that people's superior performance in everyday situations is attributable in large part to the fact that it is much easier to make sense out of everyday situations than schoollike situations. Research suggests that the process of making sense of a situation is very closely linked to five factors that are usually found in out-of-school contexts: (1) everyday problem solving typically takes place in a *familiar setting;* (2) problem solving in everyday life is often *dilemma driven,* in the sense that a problem arises as a part of some ongoing activity in which the individual is forced to make decisions; (3) everyday problem solving is almost always *goal directed,* and the goal is relatively near at hand; (4) children's own *natural language* can be used to solve the everyday problems they confront, thereby making the techniques and processes they use more personalized and more meaningful to them; (5) successful problem solving outside of school is commonly the result of the problem solver's having served some kind of *apprenticeship* that allows the novice the opportunity to observe the skills as well as the thinking involved in expert performance (Lave 1988).

Bringing Sense Making into the Classroom

Teachers can bring more sense making into the classroom in several ways. For example, they should feel free to break away from reliance on the textbook as the sole source of problems. By asking students to create their own story problems (this is an especially good activity for cooperative learning groups) that are then solved by their classmates (also working in groups), sets of problems will result that contain familiar settings and that are stated in the students' own language. Another suggestion is for teachers to encourage their students to solve problems in more than one way and to share their approaches with one another. By sharing approaches, students will learn about methods used by their peers and also that it is acceptable to use the informal methods they have developed on their own.

An especially good way to bring aspects of really real-world problem solving into the classroom is to involve students in mathematics projects. A mathematics project in designed to give students experience working in situations that can be better understood with the use of mathematical methods and reasoning. These situations can arise from the students' own surroundings or can come from a current news item or some other source.

A mathematics project that was completed recently by Tammy Miller's sixth-grade class in Bloomington, Indiana, involved building model bridges using toothpicks (see Pollard [1985] for a complete description of this project). The project was designed to take about ten days with students working in teams (construction companies) to build bridges according to certain rules and specifications. Bridges were built using only toothpicks (lumber), thread (cable), and glue (welding material), and every construction company began with a certain amount of money. The first few days of work on the project were devoted to discussions of basic bridge design and the history of bridge development. During this time, students had a chance to study pictures of different bridge types and to form construction companies (including identifying who would be the architect, the carpenter, the accountant, etc.). These orientation days were followed by planning days during which decisions were made about the schedule to be followed, the kind of bridge to build, and the material to be ordered. The remainder of the project was devoted to actual construction of the bridge. Once all the bridges were built, they were judged according to the quality of the building plans and the strength of the bridge. The project proved to be both challenging and interesting for the students partly because it involved a lot of physical activity and partly because it involved many of the factors that encourage sense making: All students were *familiar* with bridges and were motivated to participate; the problem solving that took place in the context of the project was *dilemma driven* (i.e., many decisions had to be made, often on the spot, in response to unforeseen problems that arose); both the overall project and the individual problems that arose as the project progressed were *goal directed;* and the students' *natural language* was used almost exclusively (new terms were introduced as they arose naturally).

Conclusion

Research informs us that young children begin school having already learned quite a lot of mathematics from their everyday problem solving, that they continue to create and acquire new mathematical procedures on their own throughout the years they spend in school, and that schools often suppress these naturally developed methods. Teachers should capitalize on the knowledge their students bring to school by having it serve as a basis for the development of mathematical knowledge that is more deeply understood and applicable to a wider variety of situations than those they have experienced in their narrowly defined worlds.

References

Acioly, N. M., and Analúcia D. Schliemann. "Intuitive Mathematics and Schooling in Understanding a Lottery Game." Paper presented at the Tenth Psychology of Mathematics Education Conference, London, July 1986.

Carraher, Terezinha N., and Analúcia D. Schliemann. "Research into Practice: Using Money to Teach about the Decimal System." *Arithmetic Teacher* 36 (December 1988):42–43.

Carraher, Terezinha N., David W. Carraher, and Analúcia D. Schliemann. "Mathematics in the Streets and in Schools." *British Journal of Developmental Psychology* 3 (1985):21–29.

———. "Written and Oral Mathematics." *Journal for Research in Mathematics Education* 18 (March 1987):83–97.

de la Rocha, Olivia. "The Reorganization of Arithmetic Practice in the Kitchen." *Anthropology and Education Quarterly* 16 (Fall 1985):193–98.

Lave, Jean. "Cognitive Consequences of Traditional Apprenticeship Training in West Africa." *Anthropology and Education Quarterly* 7 (Fall 1977):177–80.

———. *Cognition in Practice: Mind, Mathematics, and Culture in Everyday Life.* Cambridge, England: Cambridge University Press, 1988.

Lave, Jean, Michael Murtaugh, and Olivia de la Rocha. "Dialectic of Arithmetic in Grocery Shopping." In *Everyday Cognition: Its Development in Social Context,* edited by Barbara Rogoff and Jean Lave, pp. 67–94. Cambridge, Mass.: Harvard University Press, 1984.

Lester, Frank K., and Joe Garofalo. "Metacognitive Aspects of Elementary School Students' Performance on Arithmetic Tasks." Paper presented at the annual meeting of the American Educational Research Association, New York, March 1982.

Murtaugh, Michael. "The Practice of Arithmetic by American Grocery Shoppers." *Anthropology and Education Quarterly* 16 (Fall 1985):186–92.

Pollard, Jeanne. *Building Toothpick Bridges.* Palo Alto, Calif.: Dale Seymour Publications, 1985.

Resnick, Lauren B. "Learning in School and Out." *Educational Researcher* 16 (December 1987):13–20.

———. "Developing Mathematical Knowledge." *American Psychologist* 44 (February 1989):162–69.

Schliemann, Analúcia D. "Schooling versus Practice in Problem Solving: A Study of Mathematics among Carpenters and Carpentry Apprentices." Unpublished manuscript, Universidade Federal de Pernambuco, Recife, Brazil, 1985.

Schoenfeld, Alan H. "Problem Solving in Context(s)." In *The Teaching and Assessing of Mathematical Problem Solving.* Research Agenda for Mathematics Education, vol. 3, edited by Randall I. Charles and Edward A. Silver. Reston, Va.: National Council of Teachers of Mathematics, 1988.

Scribner, Sylvia. "Studying Working Intelligence." In *Everyday Cognition: Its Development in Social Context,* edited by Barbara Rogoff and Jean Lave, pp. 9–40. Cambridge, Mass.: Harvard University Press, 1984.

Stigler, James W., and Ruth Baranes. "Culture and Mathematics Learning." In *Review of Research in Education* 15, edited by Ernst Z. Rothkopf, pp. 253–306. Washington, D.C.: American Educational Research Association, 1988.

PART 2
Children's Thinking in the Content Domains

Section 4

Numbers, Number Sense, and Number Operations

Mon 75-92
Wed 93-103
Fri 105-127

Number Sense–Making

Judith Sowder and Bonnie Schappelle

In *Everybody Counts,* a document from the National Research Council (1989), we are told that the major objective of elementary school mathematics should be to develop number sense. This strong statement, if taken seriously, can change the way many—but not all—teachers teach mathematics in elementary school.

What is number sense? Reys and others (1991, pp. 3–4) describe it well:

> Number sense refers to an intuitive feeling for numbers and their various uses and interpretations; an appreciation for various levels of accuracy when figuring; the ability to detect arithmetical errors, and a common-sense approach to using numbers. . . . Above all, number sense is characterized by a desire to make sense of numerical situations.

More and more, researchers have documented examples of classrooms wherein children acquire good number sense (e.g., Fennema et al. [1993]; Howden [1989]; Lampert [1990]). Common elements about instruction in these classrooms include the following:

1. *Sense-making is emphasized in all aspects of mathematical learning and instruction.* This statement is particularly true of number-related aspects.

2. *The classroom climate is conducive to sense-making.* Open discussion about mathematics occurs both in small groups and with the class as a whole. Brown and Palincsar (1989, p. 395) have described why this type of climate encourages sense-making:

   > Environments that encourage questioning, evaluating, criticizing, and generally worrying knowledge, taking it as an object of thought, are believed to be fruitful breeding grounds for *restructuring.* . . . Change is more likely when one is required to explain, elaborate, or defend one's position to others, as well as to oneself; striving for an explanation often makes a learner integrate and elaborate knowledge in new ways.

3. *Mathematics is viewed as the shared learning of an intellectual practice.* Thus it is more than simply the acquisition of skills and information. Children learn how to make and defend mathematical conjectures, how to reason mathematically, and what it means to solve a problem.

4. *Children learn more mathematics than they do in more traditional classroom settings.* Vygotsky (1978) speaks of the *zone of proximal development* as a place in the learning process where a child is just ready to learn something, and interacting with peers and other people helps the child reach the next level of understanding. When children are operating at the edge of their understanding, they can learn more than when they lack this challenge.

Establishing classrooms in which such rich, connected learning takes place is not an overnight task. Several sites exist in our present curriculum wherein we as teachers can begin to think seriously about sense-making with numbers as a focus of instructional activity. The remainder of this article discusses two of these areas: first, understanding number symbols and number size; and second, alternative ways to think about computation.

Number and Symbol Meaning

The most important single element of number sense is an understanding of numbers. One cannot make sense of numbers without attaching meaning to them. This meaning should be developed by ensuring that the use of numbers in concrete settings precedes the introduction of symbols for the numbers. Examined here are some of the issues involved in making sense of whole numbers and decimal fractions through relative number-size and place-value understanding and of common fractions through part-whole ideas and fraction size.

Number size
Sowder and Markovits (1989) found that little curriculum development had been devoted to "the poten-

tial power of work on number magnitude for increasing understanding of numbers and their systems of symbolization" (p. 106) and found, in investigating the effect of meaning-based instruction on understanding the size of fraction and decimal numbers, that time would be well spent in developing number sense. Knowledge of relative and absolute number size is essential to judging the reasonableness of computation. A twenty-eight-year-old teacher seems "old" to her students but "young" to her fifty-six-year-old colleague. Students who find it difficult to ignore the 6 in estimating

$$3475 + 5872 + 1983 + 6$$

exhibit inadequate awareness of relative magnitude. An activity suggested to give students a sense of the orderliness of numbers and the regularity of the number system (McIntosh, Reys, and Reys 1992) but that also promote a discussion of relative and absolute size is the following: On this number line, name the number marked at point P.

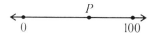

Change the right-most label from 100 to 10 then to 1, and perhaps to 0.1. Through discussing the fact that 0.6 is relatively the same distance from 1 as 60 is from 100, although the distances are very different in absolute terms (0.4 vs. 40), students might see the folly in rounding 0.6 to 1 to estimate the product of 0.57 and 789. The alternative rounding to 0.5 as producing a more reasonable estimate could emerge from this discussion.

Place value

The beauty and seeming simplicity of the base-ten number system can contribute greatly to the students' acquisition of mathematical power as well as their ability to view mathematics as a sense-making endeavor. However, such an appreciation does not develop unnurtured for most students. It results only if we present frequent opportunities over an extended time for students to construct understanding of place value, a concept that is difficult and slow to develop for most students (Ross 1989). The ability to multiply and divide mentally by powers of ten is an important skill. A deep understanding of place value would allow students to see 440 as 44 tens so they would know that 440 divided by 10 equals 44 not merely because of a rule for crossing off zeros but because 44 is the number of tens in 440. With such understanding, students could generalize: for example, when told that a candy factory had 48 638 candy bars to put into boxes holding 100 candy bars each, they could determine the number of boxes required by viewing 48 638 as 486 hundreds and 38 ones, without needing to perform long division. Yet this problem was solved by only 9 percent of typical sixth graders in one research study (Sowder 1992). This sort of place-value understanding used in this example is not sterile like the typical place-value question about the number of tens in 378 where the expected answer is 7, an answer that can be given without actual understanding of place value and one that reveals no useful information in problem solving. Tasks similar to one designed by Bednarz and Janvier (1982) are also useful in developing and assessing place-value understanding: Given digit cards for 0 through 5, make the largest number you can. Third- and fourth-grade students made such errors as using only four of the cards, not using the zero, or choosing only the first digit correctly.

Place value is one basis for the flexible decomposition and recomposition of numbers, a key element in other skills related to number sense—mental computation and estimation. Ross (1989) says, "Pupils need to engage in problem-solving tasks that challenge them to think about useful ways to partition and compose numbers" (p. 50). She sees such activity as having a reciprocal effect—developing understanding of place value and the numeration system. The ability to view 91 as 7 tens and 21 ones enables a child to decompose the dividend in a useful way to calculate $91 \div 7$ as 7 tens and 21 ones divided by 7, or 1 ten and 3 ones, or 13.

Without understanding how place value extends to decimal fractions, students are likely to overgeneralize their whole-number knowledge and believe that $0.31 > 0.4$ because $31 > 4$. Some erroneously apply their knowledge of common fractions to decimals in claiming that $0.683 < 0.68$ because thousandths are smaller than hundredths (Resnick et al. 1989). Conversely, students who do understand place value could generalize this understanding to see 1 as 10 tenths, knowledge not demonstrated by those who count by two tenths as follows: 0.2, 0.4, 0.6, 0.8, 0.10. Ultimately students need to realize that although the decimal point signals that numbers to the right are parts of 1, a consistent pattern exists in terms of place value and that all place values are related to 1, the unit. The number 1, not the decimal point, is the focal point of this system. As students become aware that in the foregoing counting, the 0.2 is two-tenths of 1, then the fact that ten-tenths of 1 is 1 allows them to correct their counting error.

Understanding that any whole number can be viewed as a number of 1's should facilitate understanding of the decimal portion as a part of 1—the number 342 is three hundred forty-two 1's; 342 000 is 342 thousand 1's; 0.342 is 342 thousandths of 1. An understanding of place value enables students to compare decimal fractions meaningfully rather than merely to equalize the number of decimal places to compare 0.48 and 0.347. It also helps them understand that 148.26 is one number, not two numbers separated by a decimal point, so they are unlikely to round the number 148.26 to 150.3, as some students are apt to do (Sowder 1992).

Fractions

Students need experiences that build on their informal fraction knowledge before they are introduced to fraction symbols. It is necessary to move back and forth between symbols and concrete experiences with fractions while this knowledge is solidifying. Two of the critical mathematical ideas for understanding fractions lay a solid foundation for future work with fractions: (1) As the whole is divided into more parts, the parts become smaller. This idea enables students meaningfully to compare unit fractions and fractions with the same numerator in terms of piece size, but they can go on to compare such fractions as 2/3 and 3/4 in terms of the additional piece size needed to make 1 for each fraction. (2) A fraction is a single number with specific value rather than two independent whole numbers (Mack 1993). Kieren (1993) has found that before viewing 3/4 as a single entity students go through a stage of viewing it as 3 one-fourths, and he suggests that activities in which students divide the unit in various ways will enable them to see fractional numbers as amounts. Although 3/4 can be viewed in many ways (e.g., as part of a single whole or of a set, as a quotient, as a measure—a number on a number line, as a ratio, as an operator), development of understanding associated with the symbols within the part-whole model is the usual introduction to fractions. Fraction meaning established in this context can serve as a basis for experience with other models. Students who learn to use benchmarks for fractions (e.g., 7/8 is close to 1, 4/7 is a little more than 1/2) develop a deeper understanding of fraction size and are more able to work with fractions in computational settings. They can estimate the sum of 12/13 and 7/8 as 2 rather than as 19 or 21, as did 55 percent of thirteen-year-olds in one study (Carpenter and Lindquist 1989).

Rethinking Computation

The ubiquity of calculators in today's world is forcing us to rethink both the need for solid number sense and the role of computation in elementary school classrooms. The standard computational algorithms that have been the mainstay of elementary school mathematics were developed because people needed fast, efficient computational schemes. This need no longer exists, and with its demise, a world of opportunities opens up to us to combine developing number sense and teaching computation in useful, meaningful ways.

No one is advocating elimination of all paper-and-pencil calculation. But many do advocate doing away with many of our traditional algorithms because they do not correspond to the way we think about numbers, they are difficult to learn, and they are unnecessary. Perhaps even more important is the fact that focusing instruction on letting children find ways to calculate allows them to focus on number meaning instead of on routines. In an elementary school project in England (Shuard et al. 1991), teachers worked together to develop ways of teaching nonstandard computation. Children were allowed free access to calculators in classrooms from the time they entered school, but they often chose not to use them. When the children were faced with problems involving addition, subtraction, multiplication, and division, they developed sensible methods of operating mentally when the numbers were small, and they invented written methods when the numbers were too large for mental computation. Here are some of the ways eight-year-olds wrote about their subtraction methods (Shuard et al. 1991, p. 26):

> Student 1: 549 − 331. 5 take away 3 = 2 (and you can make it into hundreds) so that is 200, and you add 40→240 − 30 and it comes to 210 + 9 = 219 − 1 = 218.
>
> Student 2: 135 − 72 = 63. First I take 70 away from

100. That leaves me with 30. Then I add the other 30 back. That makes 60. Then I take 2 from 5; that left 3 so the answer is 63.

In another project, first-grade teachers set their textbooks aside and, listening carefully to their students, built on what students already knew. One result was that children came to a deep understanding of numbers and were able to work with them in meaningful ways. One student, asked to find 246 + 178, replied as follows (Carpenter 1989, 90):

> Well, 2 plus 1 is 3, so I know it's 200 and 100, so now it's somewhere in the three hundreds. And then you have to add the tens on, and the tens are 4 and 7 . . . Well, um. If you started at 70, 80, 90, 100. Right? And that's four hundreds. So now you're already in the three hundreds because of the [100 + 200], but now you're in the four hundreds because of that [40 + 70]. But you're still got one more ten. So if you're doing it 300 plus 40 plus 70, you'd have 410. But you are not doing that. So what you need to do then is add 6 more onto 10, which is 16. And then 8 more: 17, 18, 19, 20, 21, 22, 23, 24. So that's 124. I mean 424.

Several things are notable in the work of the children studied in this project. One is that children worked with hundreds, tens, and ones from left to right rather than with individual digits from right to left as they do with standard algorithms. Their understanding of place value was reinforced. Each child found a way to do the computation in a manner personally meaningful. The methods they developed showed deep understanding of properties of numbers. These children were not harmed by their lack of knowledge of standard computational procedures; on the contrary, their methods were far less likely to lead to errors than standard procedures memorized but poorly understood. These nonstandard methods are useful both for mental computation and for paper-and-pencil computing. (See Sowder [1990] for more information on mental computation and its role in developing number sense.)

Computational estimation
Competence in computational estimation requires the interplay of all of the previously described components of number sense. It requires one to be able to coordinate the skills of rounding and of computing mentally (Case and Sowder 1990) and is, therefore, more difficult than many realize. It is a worthwhile topic because of its usefulness both as a situation for developing number sense and as a skill in and of itself. Students must first learn to round in flexible ways; estimating 36 + 49 as 35 + 50 is more appropriate than as 40 + 50. It is therefore important that we let go of the inflexible rounding rules of the past, rules that were most useful in carrying out the standard long-division algorithm, another routine happily left by the wayside. Estimation allows for various methods and invented algorithms, the development of which calls for thinking about the meaning of numbers (Sowder 1992).

Planning for Instruction

If number sense is truly a way of thinking that must permeate all aspects of learning mathematics, then planning calls for vigilantly watching for ways to lead children to make sense of numbers. We do so both through what we emphasize in the curriculum (e.g., exploring meaningful ways to compute rather than memorizing algorithms) and through the learning climate we establish for students (e.g., encouraging students to state reasons for the way they use numbers rather than allow them to give only answers). When planning focuses on developing number sense, students gain mathematical power and the belief that mathematics is supposed to make sense. As Resnick (1986, 191) has told us, "Good mathematics learners expect to be able to make sense of the rules they are taught, and they apply some energy and time to the task of making sense." As more of our instruction promotes sense-making in mathematics, we can expect a greater proportion of our students to qualify as "good mathematics learners."

References

Bednarz, Nadine, and Bernadette Janvier. "The Understanding of Numeration in Primary School." *Educational Studies in Mathematics* 13 (February 1982): 33–57.

Brown, Ann L., and Annemarie S. Palincsar. "Guided, Cooperative Learning and Individual Knowledge Acquisition." In *Knowing, Learning, and Instruction,* edited by Lauren B. Resnick, pp. 393–451. Hillsdale, N.J.: Lawrence Erlbaum Associates, 1989.

Carpenter, Thomas P. "Number Sense and Other Nonsense." In *Establishing Foundations for Research on Number Sense and Related Topics: Report of a Conference,* edited by Judith T. Sowder and Bonnie P.

Schappelle, pp. 89–91. San Diego: San Diego State University Center for Research in Mathematics and Science Education, 1989.

Carpenter, Thomas P., and Mary M. Lindquist. "Summary and Conclusions." In *Results from the Fourth NAEP Mathematics Assessment,* edited by Mary Montgomery Lindquist, pp. 160–69. Reston, Va.: National Council of Teachers of Mathematics, 1989.

Case, Robbie, and Judith T. Sowder. "The Development of Computational Estimation: A Neo-Piagetian Analysis." *Cognition and Instruction* 7 (1990): 79–104.

Fennema, Elizabeth, Megan L. Franke, Thomas P. Carpenter, and Deborah A. Carey. "Using Children's Mathematical Knowledge in Instruction." *American Educational Research Journal* 30 (1993): 555–83.

Howden, Hilde. "Teaching Number Sense." *Arithmetic Teacher* 36 (February 1989): 6–11.

Kieren, Thomas E. "Creating a Space for Learning Fractions." Unpublished paper, 1993.

Lampert, Magdalene. "Connecting Inventions with Conventions." In *Transforming Children's Mathematics Education: International Perspectives,* edited by Leslie P. Steffe and Terry Wood, pp. 253–65. Hillsdale, N.J.: Lawrence Erlbaum Associates, 1990.

Mack, Nancy K. "Critical Ideas, Informal Knowledge, and Understanding Fractions." Unpublished paper, 1993.

McIntosh, Alistair, Barbara J. Reys, and Robert E. Reys. "A Proposed Framework for Examining Basic Number Sense." *For the Learning of Mathematics* 12 (November 1992): 2–8, 44.

National Research Council. *Everybody Counts.* Washington, D.C.: National Academy Press, 1989.

Resnick, Lauren B. "The Development of Mathematical Intuition." In *Perspectives on Intellectual Development: The Minnesota Symposia on Child Psychology,* vol. 19, edited by M. Perlmutter, pp. 159–94. Hillsdale, N.J.: Lawrence Erlbaum Associates, 1986.

Resnick, Lauren B., Perla Nesher, Francois Leonard, Maria Magone, Susan Omanson, and Irit Peled. "Conceptual Bases of Arithmetic Errors: The Case of Decimal Fractions." *Journal for Research in Mathematics Education* 20 (January 1989): 8–27.

Reys. Barbara J., and others. *Developing Number Sense in the Middle Grades.* Addenda Series, Grades 5–8. Reston, Va.: National Council of Teachers of Mathematics, 1991.

Ross, Sharon H. "Parts, Wholes, and Place Value: A Developmental View." *Arithmetic Teacher* 36 (February 1989): 47–51.

Shuard; Hillary, Angela Walsh, Jeffrey Goodwin, and Valerie Worcester. *Primary Initiatives in Mathematics Education: Calculators, Children and Mathematics.* New York: Simon & Schuster, 1991.

Sowder, Judith T. "Mental Computation and Number Sense." *Arithmetic Teacher* 37 (February 1990): 18–20.

———. "Making Sense of Numbers in School Mathematics." In *Analysis of Arithmetic for Mathematics Teaching.* edited by Gaea Leinhardt, Ralph Putnam, and Rosemary A. Hattrup, pp. 1–51. Hillsdale, N.J.: Lawrence Erlbaum Associates, 1992.

Sowder, Judith, and Zvia Markovits. "Effects of Instruction on Number Magnitude." In *Proceedings of the Eleventh Annual Meeting: North American Chapter of the International Group for the Psychology of Mathematics Education,* edited by Carolyn A. Maher, Gerald A. Golding, and Robert B. Davis, pp. 105–10. New Brunswick, N.J.: Rutgers University Center for Mathematics, Science, and Computer Education, 1989.

Vygotsky, L. S. *Mind in Society: The Development of Higher Psychological Processes.* Cambridge, Mass.: Harvard University Press, 1978.

Number Sense and Nonsense

Zvia Markovits, Rina Hershkowitz, and Maxim Bruckheimer

Common sense has many aspects and is developed by a variety of experiences in and out of school. Number sense is one aspect of common sense that we rightly expect schooling to improve. But does it? Given a problem, do students pay any attention to the meaning of the numbers in the data or in the solution they obtain? In study devoted to estimation and reasonableness of results, we found that sixth- and seventh-grade students either do not have a reasonably developed number sense or, if they have it, do not apply it to simple tasks in a mathematical context.

This "number nonsense" occurred in purely numerical tasks as well as in tasks with a contextual setting. The former kinds of tasks are abstract and confined to the classroom, but the latter tasks refer to real situations that should naturally encourage the application of number sense. Consider this classic problem:

> The height of a 10-year-old boy is 1.5 meters. What do you think his height will be when he is twenty?

About a third of the students in our study answered 3 meters! These students apparently invoked a standard algorithm without paying any attention to its relevance or to the meaning of the number obtained. One suspects that their mathematical experience has been confined almost exclusively to applying one standard algorithm at any one time and obtaining one answer. Neither the choice of algorithm nor the meaning of the answer was explicitly questioned. In other words, number sense has not been explicitly and meaningfully developed.

We investigated the performance of sixth and seventh graders on a variety of tasks involving estimation and reasonableness of results, both of which invoke number sense. We also examined the effect of a short unit of activities on these two topics. Although the unit was written specifically for these grades, most of the tasks could be given to students in lower grades with some adjustment where necessary.

Number Sense in Purely Numerical Tasks

Students were asked to locate the decimal point in the answers given in these two exercises:

$$3.5 \times 4.5 = 1575 \qquad 5.5 \times 3.2 = 176$$

Students with number sense would find these exercises easy, even if they did not know the standard algorithm. But we found that students counted the digits after the decimal point, obtaining 1.76 as the second answer. It never crossed their minds to consider whether the answer made sense; apparently they think that algorithms never lie.

Experiences with whole numbers often carry over into work with fractions and decimals and dominate students' thinking to the extent that number sense never appears to play a role in obtaining an answer. For example, consider another exercise we used:

The result of $426.5 \div 0.469$ is . . .
a) less than 426.5
b) equal to 426.5
c) greater than 426.5

Explain your answer.

The correct answer, *(c)*, was given by 40 percent of the 328 students. The explanation for most wrong answers was that "when divided, a number becomes smaller." Slightly better, but similar, results were obtained for multiplication.

To improve number sense, students throughout their schooling should be given many varied tasks that require little, if any, computation. Here are some examples of tasks that increase number sense:

1. Which is greater—

    13 + 11 or 9 + 8?

    46 − 19 or 46 − 17?

    1/2 + 3/4 or 1 1/2?

    0.0358 or 0.0016 + 0.393?

    Many students will do exact calculations before answering, but the teacher can use various strategies to wean them from this practice, for instance, by giving the questions orally and imposing a time limit.

2. The result of 46 × 91 will be in the—

    a) hundreds

    b) thousands

    c) ten thousands.

    Is 46 × 91 more or less than 5000? More or less than 3600?

3. If you could round only one of the numbers in the problem 32 × 83, rounding which factor would produce a product closest to the exact answer?

4. *a)* 127 × □ ends in 3. What can you say about the missing whole number?

    If you also know that the product is less than 4000, what can you then say about it?

    If the product is also greater than 3000, what is the missing number?

    b) The sum of five two-digit numbers is less than 100. For each of the following statements decide whether it is necessarily true, necessarily false, or possibly true. Explain your answers.

    (1) Each number is less than 20.

    (2) One number is greater than 60.

    (3) Four numbers are greater than 20, and one is less than 20.

    (4) If two numbers are less than 20, at least one is greater than 20.

    (5) If all five numbers are different, then their sum is greater than or equal to 60.

5. Without computing, explain why each of the following is incorrect:

    a) 310
    520
    630
    150
    + 470
    2081

    b) 119
    46
    137
    940
    + 300
    602

    c) 27 × 3 = 621

    d) 36 ÷ 0.5 = 18

    e) 0.46
    1.93
    2.46
    0.99
    + 0.87
    672

Number Sense in Context

Another question we used in the study was the following:

A bath has two outlets. The first alone empties the bath in 10 minutes, and the second alone in 4 minutes. Both were opened at the same time. In how many minutes will the bath be empty?

Choose the most appropriate answer from among the following and explain your answer.

a) 40 minutes
b) 14 minutes
c) less than 14 but more than 7
d) 7 minutes
e) 6 minutes
f) 4 minutes
g) about 3 minutes
h) about 1/2 minute

The students had not been taught to solve such problems using algebra, but number sense is all that is necessary: if one outlet empties the bath in four minutes, then the two outlets will empty the bath in less than four minutes but not in as little as half a minute.

Once again we found that students tended to apply an algorithm and operate on the numbers in the question without thinking. They added, multiplied, computed the mean, and so on. Even if we also allow one-half minute as a reasonable option, since it is less than four minutes, only 36 percent of the students who had not received the special instruction gave a reasonable answer. After receiving the brief special instruction 85 percent of the students gave a reasonable answer to a question in a similar context.

In this task, the algorithm—even if known—is irrelevant, but there is nothing inherently ridiculous in the answers obtained by incorrect algorithms. To conclude that an answer is incorrect, the student must relate it to the data in the question. This ability demands a more sophisticated number sense than that required by the height problem at the beginning of this article—three meters is ridiculous for a man's height.

The general problem of learning to judge the reasonableness of a result in a given context involves number sense. Although students may be encouraged to check their answers, they do so far too often only in purely mathematical situations, where a standard algorithm has been applied, and they check only whether that algorithm has been performed correctly. The numerical work itself has no real-life context. This situation leads students to divorce the mathematical answer completely from its implication in the context; the answer, they think, is mathematically correct even though it makes no sense in the context. This phenomenon is again demonstrated by the following task from our study:

> A bookstore delivered 188 books to 6 libraries. How many books do you think each library received?

This question cannot be answered algorithmically; nevertheless, many students did use the division algorithm. Fractional answers (31 $\frac{1}{3}$ books, 31.3 books, etc.) were given by 34 percent of the students before doing our instructional activities, but only 3 percent gave such answers after the instruction. When we asked whether a man can be three meters high or whether a library can receive thirty-one and one-third books, the answer was definitely "no!" But if these answers are given by the algorithms, then students assume that they are correct!

We also found that students know little of various everyday measurements that can serve as reference data for number sense in contextual problems, for example, the speed of a car, the height of an eight-story building, or the number of liters of water in a bucket. When we asked for the speed of a plane flying from London to New York, a reasonable answer (we allowed anything from 500 to 1500 km/h) was given by 29 percent of the students who had no instruction. We also found that if we asked for the measures of various objects of the same kind (e.g., the width of a car, a truck, a main road, etc.), then often the answers were internally inconsistent (e.g., the width of car would be given as greater than that of a road). Therefore, we recommend that students first be asked to order the given objects qualitatively and only then be asked to assign individual measures to them. For example, order the following from the heaviest to the lightest and estimate the weight of each: car, bicycle, truck, dog, chair, elephant, kitten. Correct responses approximately doubled after our instruction, indicating that similar instruction needs to be extended or improved in the classroom. We feel that this sort of number sense is not easy to acquire and probably needs to be developed gradually over a number of years of schooling.

Conclusion

Number sense is not a single curricular topic but has many aspects. The development of number sense is not something that occurs naturally for most students in the regular curriculum, as we have seen in this study. Evidence clearly suggests, however, that number sense can be developed with appropriate activities. The short instruction we gave to sixth and seventh graders improved their number sense, in some aspects very significantly. They also learned to look at all the answers they obtained in a different way: the numbers in the answer had a meaning that needed to be examined critically.

We suggest that activities of the types described here should be integrated into the curriculum, even before sixth and seventh grades. We should begin in the first grade with appropriate tasks that call on number sense and give the children a less mechanical view of mathematics.

Children's Multiplication

Kurt Killion and Leslie P. Steffe

The purpose of this article is to discuss how children give meaning to situations which, from an adult's perspective, could be solved by multiplying or dividing (Steffe and von Glasersfeld 1985). We have found in our research that a child's understanding of these situations may be significantly different not only from those of the teacher but also from those of other children in the same classroom. Thus our teaching decisions must be based on knowledge of the children's understanding of mathematics.

To ascertain the mathematical knowledge of individual children, the teacher must engage them in situations that reveal their ways of operating. Through interactive communication, the teacher must try to make sense of children's activity. To illustrate this process, we present examples of interactive communication with two children and explain how we interpret the children's mathematical activity. From an adult perspective, children often made mistakes. As researchers, we treat these mistakes as valuable indicators of the nature of children's understanding.

Children's understanding of multiplication depends on their ability to work with what we call "composite wholes." In its simplest form, a composite whole is a collection of items, such as cups on a table, houses on a hill, or candies in a bag. Zachary, at the beginning of third grade, focused on the individual items in a collection and could not view the collection as one thing. For example, although he had a concept of "three," he viewed three items as 3 ones rather than 1 three. For this reason, he could not view three items as a countable unit as he could units of one.

Because Zachary used counting to make sense of mathematical situations, the following task was presented to him in a counting context. We knew he was capable of keeping track of counting by ones using his fingers. For example, he could easily determine how many times he would count by ones from fifteen to twenty-one by first putting up a finger for each of the number words he said from sixteen through twenty-one and then reading a finger pattern for six when he was finished. We wanted to know if he could take advantage of his counting scheme for units of one and modify it to count composite wholes consisting of three items:

T (*Teacher*): If you count to twelve by threes, how many threes would you count?

Z (*Zachary*): (Squeezes his thumb, index finger, and middle finger together and places them on the table) Three. (Squeezes his ring finger and little finger together and places them on the table) Six. (Proceeds to his other hand and places his thumb and index finger on the table) Eight-nine! (Places his middle finger and ring finger on the table) Twelve. (Looks at his hands) Nine times!

Zachary then seemed unsure, started over again—this time putting down only two fingers for each of the number words—and said "eight" for his answer. Because he counted to twelve and answered "nine times" and then "eight," we can assume that he intended to keep track of the composite wholes he was making—that he did understand the task. Zachary was attempting to track his production of composite wholes in the same way he did his singleton units, by placing fingers down and reading a finger pattern when done. We could see on the videotape that he said "one" and "two" to himself before saying "three" out loud, "four" and "five" to himself before saying "six" out loud, and so on. Zachary's concept of three was a numerical pattern he could use to segment the number words from one to twelve into trios, but he still could not look back and isolate each trio as one thing to be counted.

Our explanation of why Zachary used two fingers to record the trios of number words is that after he initially placed three fingers on the table, he had only

---

The research reported here was supported by the National Science Foundation, grant no. MDR-8550463. The opinions and suggestions do not necessarily reflect those of the National Science Foundation.

two fingers remaining on that hand. He was consumed with producing the trios, so he paid little attention to how he was recording them. However, he did notice his inconsistency in recording because he started over this time using two fingers throughout. Zachary's method of recording does furnish valuable insight into the nature of the composite wholes he made because they were not recorded as single items.

The very next day, Zachary modified his recording method when he was asked to find how many twos he would count if he counted by twos up to twelve. This time he used one finger to record each pair of number words. So he was beginning to focus on the unitary nature of a pair of items. Zachary soon progressed to being able to count patterns of three as he made them. Because Zachary was able to count composite wholes of two and three as one thing, we say he could create "abstract composite units" of two and three.

Six months later, we gave Zachary the following task to see if he could count patterns of ten as he produced them. More specifically, we wanted to know if he could interpret the eight in eighty-seven as ten, eight times:

T: (Collects a group of blocks and puts them by Zachary) There are eighty-seven blocks in this group of blocks. Can you make stacks of ten? (Interrupts after Zachary makes one stack.) How many stacks of ten could you make like that?

Z: (Silent for five seconds, then shrugs shoulders)

T: We don't know for sure, do we? Can you find a way to find out?

Z: I could put ten up again and find out how many columns.

T: How many could you stack up?

Z: (Thinks fifteen seconds) I don't know.

T: But we have one way to find out, right?

Z: Yeah, we could stack them up.

T: Well, just pretend you stack them up.

Z: (Moves index finger up in air next to existing stack three times while subvocally uttering number words ten, twenty, thirty. Looking puzzled, starts over but again appears to lose track. Again starts over, this time coordinating putting up fingers on left hand with each counting-by-ten act. As he silently says, "ten, twenty, thirty, forty," makes a vertical sweeping motion with right index finger for each number word, each motion indicating making a stack of ten. Then stops the sweeping motions and tracks the production of number words fifty, sixty, seventy, eighty by continuing to put up fingers sequentially, completing a finger pattern for eight.) "Eight rows!"

To solve the problem, Zachary pretended to make stacks of ten. Each of the imagined stacks could be considered as one thing and counted. Zachary did make abstract units of ten, but he could not immediately see eighty-seven as being composed of tens. The tens did not exist for him prior to his production of them in counting.

Some children begin third grade with a more advanced way of dealing with composite wholes than Zachary had developed by the end of third grade. For these children, "eight times ten" refers to the possible action of producing the unit "ten" eight times without having actually to produce it in counting. At the other end of the spectrum are those children who do not make abstract composite units until the end of fourth grade. Keisha is an example of such a child. The following interactive communication occured at the end of fourth grade after we had worked with her for two years. At this point, she finally could view a composite whole as a countable item:

T: (Asks Keisha to close her eyes and places twelve tiles, glued together in pairs, under a paper plate. Then tells Keisha to open her eyes and shows her a single tile.) There are twelve of these tiles under here. (Then shows Keisha a pair of tiles glued together on a strip of paper) How many twos do you think are under here?

K (*Keisha*): (Moves her elbow across the table in six long sweeps) Twelve!

T: Twelve little ones, right? I want to know how many groups of two.

K: Five? (very quickly)

T: How do you figure that?

K: (Places thumb and index finger together on the table) One. (Places middle and ring finger down on the table) Two. (Places little finger and thumb of other hand down on the table) Three. (Places next two fingers of second hand down on the table) Four. (Places last two fingers of other hand down on the table) Five. Six! (Does not place any new fingers down)

Her success, at this point, is limited to counting units of two that she makes in her visual field. But this modification of counting to include units of two will open a path for her to begin restructuring her operations with composite wholes. Only with this modification can we expect Keisha to make any progress toward meaningful experiences with multiplying. This occasion was one of the few times Keisha was visibly excited about her mathematical activity.

We need to emphasize that both Keisha and Zachary were participants in a typical classroom and were by no means isolated examples. Furthermore, the tasks described in this article do not represent the only tasks we used with the children. We had them manipulate physical objects, and we used arrays, cross products, and measurement situations. In our examples, Zachary and Keisha were engaged in more abstract tasks than the manipulation of objects. The advancements children made in being able to operate without dependence on objects in their visual field were facilitated by having them work in situations where objects were screened from view (Keisha) or where they were asked to work in their imagination (Zachary).

Our interactions with Zachary, Keisha, and the other children in our study support the following suggestions for classroom practice:

1. As teachers, we must try to understand our students' ways of doing mathematics. With such an understanding, teachers can make more informed decisions about appropriate activities. This understanding can be gained by watching the children interact as well as by interacting with the children.

2. In the communications with Zachary and Keisha, we can see modifications of their previous ways of operating. Both made changes to more advanced ways of operating because they were involved in solving a task for which their prior operations were not adequate for a solution. This observation implies that emphasis should be placed on the children's active mathematical learning in situations that might encourage modification of their present ways of operating.

3. Children do have fundamentally different ways of organizing their mathematical experiences. This conclusion suggests that we must give up the idea of one curriculum for all and design learning environments that are sensitive to children's ways and means of operating.

4. We should not try to teach children mathematical concepts and operations they are not capable of understanding. Both Keisha and Zachary were in no position to make sense of the multiplication or division algorithms they were expected to perform in the pages being covered in the textbook. We have found an emphasis on computational skills to be counterproductive in fostering growth in reasoning with composite wholes.

5. Even though this approach requires extra patience, teachers should encourage children to make and reason with composite wholes, for that ability is the basis of the quantitative reasoning we expect of them throughout their school experience. The ability to reason with units of more than one should be an elaboration of their ability to reason with units of one, and we should encourage children to count when making composite wholes.

6. Finally, we do not advocate that teachers teach children only in one-on-one situations. We do, however, advocate the interactive communication we have described. Learning children's understanding of mathematics and how to foster its growth is an exciting challenge that is well worth the time and effort expended. But teachers must take the time and be given support in such an effort.

Reference

Steffe, Leslie P., and Ernst von Glasersfeld. *Child Generated Multiplying and Dividing Algorithms: A Teaching Experiment* (Grant no. MDR-8550463). Washington, D.C.: National Science Foundation, 1985.

Multiplication and Division: Sense Making and Meaning

Vicky L. Kouba and Kathy Franklin

Young children often perform poorly on multiplication and division problems (Anghileri and Johnson 1988; Kouba et al. 1988), and their persistent use of immature strategies on these problems is inadvertently reinforced by several factors. Recent research suggests ways that teachers can counteract these debilitating factors.

Expand the Range of Numbers from Whole to Fraction

Multiplication and division problems in grades K–4 are often limited to whole numbers. This situation does not reflect reality and leads children to conclude falsely that multiplication always "makes larger" and division always "makes smaller" (Greer 1988). Children need hands-on experiences with fractions, such as sharing three candy bars among six children or sharing two and one-half pizzas among nine children.

Expand the Types of Situations

Two types of "grouping" division problems are possible: (1) measurement division, in which the unknown is the number of groups, and (2) partitive division, in which the unknown is the number of elements in each group.

Young children are competent and creative at modeling both of these division types. Measurement-division problems are represented differently from partitive-division problems (Kouba 1989). To solve the measurement-division problem of 88 crayons grouped 8 crayons to a box, many young children will count out 88 crayons and then make groups of 8 until the pool of 88 crayons is used up. To solve the partitive-division problem of 88 crayons placed equally into 11 boxes, about half the children will use a systematic-trial-and-error strategy. They count out the 88 crayons, guess at the number that would go into a box, and see if grouping by that guess results in 11 groups. If not, they try a reasonable second guess, and so on. The other half of the children may try a dealing-out strategy. They deal out the 88 crayons one by one to eleven boxes and count the number in one box to get the answer.

Some children may use a building-up approach in conjunction with either the systematic-trial-and-error or the dealing-out strategy (Kouba 1989). They first decide on a grouping to test or deal out and perform that action while keeping a running count of the total objects placed in groups or dealt. They stop when the count reaches 88. This division strategy is quite close in structure to strategies used when making sense of multiplication and can be used to advantage when linking division to multiplication. Children should be encouraged to use strategies that make sense to them but also to share and try to understand their classmates' approaches.

Scalar relationships from another type of multiplicative situation. One factor in scalar problems is the number in a set—Sheila has 12 pencils—whereas another factor is the number of sets—Corey has 3 times as many pencils as Sheila. Most children would solve the problem "How many pencils does Corey have?" by adding 3 twelves. Children need experience with tactile situations and guidance in understanding and modeling scalar relationships (Hendrickson 1986). Children who have difficulty with questions like "How tall is a giant who is 5 times as tall as you?" can lie down and physically trace their height five times along a hallway. This problem can be tackled in the second and third grades.

Other multiplicative situations involve factors with interchangeable roles. The most common examples are area and cross-product, respectively:

This article is based on "Multiplication and Division: Sense Making and Meaning" by Vicky L. Kouba and Kathy Franklin, which appears as chapter 5 in *Research Ideas for the Classroom: Early Childhood Mathematics,* a publication from the National Council of Teachers of Mathematics's Research Interpretation Project, edited by Robert J. Jensen and published by Macmillan Publishing Company. Interested readers should consult that chapter and its extensive bibliography.

- Alvinia's desk is 5 decimeters wide and 6 decimeters long. What is its area?
- A shop sells sandwiches made with one slice of meat and one slice of cheese. It has 3 kinds of meat (harm, turkey, and beef) and 2 kinds of cheese (Swiss and provolone). How many different kinds of sandwiches are possible to buy?

These situations can be represented by area models and arrays (see **fig. 1**). Research indicates (Kaput 1989) that children function best when they can use various representations for all multiplication and division situations and can explain the relationships among those representations. Children's modeling and explaining of multiplication and division situations can be viewed as performance assessments. Their explanations can be presented to small groups or to the class.

Reward for Critical Thinking

Current practices frequently reward computations and answers, not exploring, analyzing, interpreting, representing, hypothesizing, and verifying. If multiplication and division are taught as tested facts and algorithms to be unquestioningly memorized and used rapidly to produce numerical answers (Sowder 1988), what incentive do children have for engaging in critical thinking? Research offers some suggestions for change.

Build on children's knowledge

To get children to think critically, teachers should value, acknowledge, and build instruction on children's thinking and knowledge base. Children can and do use different, although limited, levels of representations to make sense of multiplication and division situations before they have been formally introduced to these two operations, before they have memorized any multiplication facts, and before they know the words or formal symbols for multiplication (Baroody 1987; Steffe 1988; Kouba 1989).

Many children first make sense of multiplication and division situations by using representations related to counting, addition, and subtraction. They begin by physically modeling what is occurring (Baroody 1987; Anghileri and Johnson 1988; Kouba 1989). For example, to resolve the situation "If 8 plates hold 4 cookies each, how many cookies are on all the plates?" children set out 8 plates and put 4 cookies or objects on each plate. Then they count the total number of cookies. Many children progress to being able to make 8 groups of 4 without having to use separate objects for the 8 plates. Next, they are able to make one group of 4 and count it repeatedly 8 times. When so doing, they track how many groups they have counted by using their fingers or some other memory device, such as nodding the head, seeing or writing tally marks, or counting rhythmically.

The physical modeling produced at these progressively more sophisticated levels of representation forms a concrete basis for beginning to explore the relationships between multiplication and division. The final arrangement of the physical materials is usually the same for multiplication and division situations involving equal groupings. The teacher may ask why that result is so or may ask the children to explore a related division situation immediately after working with a multiplication situation.

More advanced levels of representation include counting by fours; counting on when they cannot recall the next multiple, for instance, 4, 8, . . . , 9, 10, 11, 12,

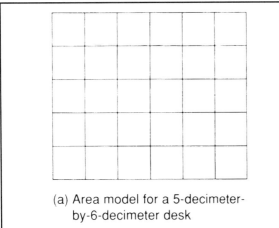

(a) Area model for a 5-decimeter-by-6-decimeter desk

(b) An array model for making a sandwich

Fig. 1. Situations and pictorial representations for area and array

and so on; adding fours; and using such derived facts as "Four groups of 4 are 16 and 16 plus 16 is 32." Although not all children advance through all stages, an examination of how a child solves a specific situation can give the teacher some idea of that child's level of sophistication in problem solving and indicate whether he or she may be ready for a more advanced level. Advancement can be achieved by having the child compare his or her approach with someone's at the next level; by a well-planned question or sequence of questions or discovery activities; or by suggesting a useful strategy, such as skip-counting.

Use student-developed problems and algorithms

Children benefit from making up their own oral or written stories that can be modeled by multiplication and division. For example, on a test or quiz, ask the students to make up a word problem that would match the number sentence $24 \div 6 = \square$. It is commonly answered, "If 6 people share 24 candies fairly, how many candies will each person get?"

Children should be actively involved in devising their own algorithms for solving multiplication and division problems. Research indicates that no single algorithm is the "right" algorithm to teach. As with addition and subtraction, having children create their own multidigit computational algorithms increases their understanding and flexibility in the use of operations (Lampert 1989).

Children should be able to explain what they are doing and demonstrate the validity of their invented algorithms by manipulating physical objects or by creating pictorial arrays. These explanations can be part of the teacher's routine assessment of each child's understanding. Finally, children should be able to link symbolic representations to physical or pictorial representations. An example of the connections between a problem and its pictorial and symbolic representations is shown in **figure 2**.

Using developmental or informal algorithms does take time. However, spending more time on developmental algorithms pays off in the long run by saving time on the practice of rules and by increasing children's understanding and motivation (Lampert 1989).

Conclusion

Overall, the research results imply that at grades K–4, mathematics instruction should emphasize children's development of a sound and varied conceptual basis for multiplication and division situations rather than on memorization of multiplication facts or use of those facts to solve symbolic problems in narrowly prescribed ways. As teachers create and sustain an environment for this development of understanding, they need to *(a)* give children a rich communicative experience with various multiplication and division situations, *(b)* evaluate and reward more than just producing an answer quickly in a prescribed way, and *(c)* help children build from their own experiences and understandings many ways to represent and model multiplication and division situations.

Action Research Ideas

1. The authors indicate that young children solve partitive-division problems using different strategies.

   Pose a partitive-division problem to a class of young children. Note how many use a systematic-trial-and-error strategy, how many use a dealing-

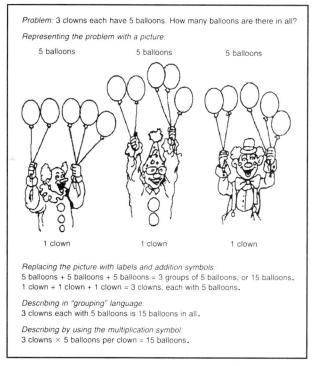

Fig. 2. Moving from pictorial to symbolic representations

out strategy, and how many are unable to solve the problem. Repeat this activity at spaced intervals throughout the year.

   a) Record which strategy each student uses and which students have no strategy.

   b) According to the authors, half of the students will use each strategy. Is that theory true in the class being examined?

   c) Does a different pattern of strategy use emerge over time? For what reasons?

2. Pose a measurement-division and a partitive-division problem on the same day. Use similar numbers in both problems.

   a) Determine which children correctly solve each type of problem.

   b) In this class, does one type seem more difficult than the other?

   c) Do any children use a measurement strategy for a partitive-division problem? If so, try to understand why these children use a strategy that does not seem to match the problem type.

   d) Repeat this activity at spaced intervals throughout the year. Determine what patterns of success emerge.

3. Present the students with a physical or pictorial model, such as that shown in figure 2, using numbers appropriate for their developmental level.

   a) Ask the students to write two or three different problems that could be solved using the model, and ask them to represent each problem by a number sentence. Note: Three types of problems are possible: multiplication, measurement division, and partitive division. Do not explain these types to the students, but wait to see what types emerge from their activity. It is not necessary for students to use such terms as *measurement division or partitive division,* but students should learn to recognize two different types, that is, whether they know the number of groups and want to find the number in each group or they know the number in each group and want to find the number of groups.

   b) Have the students discuss which problems are alike and why.

   c) Note which students seem capable of constructing two or three problems for each model and which students seem capable of articulating a relationship between multiplication and division.

   d) Repeat this activity at spaced intervals throughout the year. Note students' progress toward understanding the relationship between multiplication and division.

References

Anghileri, Julie, and David C. Johnson. "Arithmetic Operations on Whole Numbers: Multiplication and Division." In *Teaching Mathematics in Grades K–8,* edited by Thomas R. Post. Boston: Allyn & Bacon. 1988.

Baroody, Arthur J. *Children's Mathematical Thinking.* New York: Teachers College Press, 1987.

Greer, Brian. "Non-Conservation of Multiplication and Division: Analysis of a Symptom." *Journal of Mathematical Behavior* 7 (December 1988):281–98.

Hendrickson. A. Dean. "Verbal Multiplication and Division Problems: Some Difficulties and Some Solutions." *Arithmetic Teacher* 33 (April 1986):26–33.

Kaput. James J. "Supporting Concrete Visual Thinking in Multiplicative Reasoning: Difficulties and Opportunities." *Focus on Learning Problems in Mathematics* 11 (winter 1989):35–47.

Kouba, Vicky L. "Children's Solution Strategies for Equivalent Set Multiplication and Division Word Problems." *Journal for Research in Mathematics Education* 20 (March 1989):147–58.

Kouba, Vicky L., Catherine Brown, Thomas P. Carpenter, Mary M. Lindquist, Edward Silver, and Jane O. Swafford. "Results of the Fourth NAEP Assessment of Mathematics: Number, Operations, and Word Problems." *Arithmetic Teacher* 35 (April 1988):14–19.

Lampert, Magdalene. "Research into Practice: Arithmetic as Problem Solving." *Arithmetic Teacher* 36 (March 1989):34–36.

Sowder, Larry. "Children's Solutions of Story Problems." *Journal of Mathematical Behavior* 7 (December 1988):227–38.

Steffe, Leslie P. "Children's Construction of Number Sequences and Multiplying Schemes." In *Number Concepts and Operations in the Middle Grades,* edited by James Hiebert and Merlyn Behr. Reston, Va.: National Council of Teachers of Mathematics and Lawrence Erlbaum Associates, 1988.

Misconceptions about Multiplication and Division

Anna O. Graeber

Multiply your options—divide and conquer—these and many other everyday expressions imply that multiplication always makes larger and division always makes smaller. In fact, these two ideas are so self-evident and so pervasive in many people's thinking that they are among the most notorious *mis*conceptions about mathematics. Misconceptions or naive conceptions are commonly held ideas or beliefs that are contrary to what is formally acknowledged to be correct. Mathematics educators study misconceptions because if we can understand how students are apt to see mathematical ideas, we may be better prepared to offer instructional experiences that help them develop accepted conceptions. The two misconceptions just described have been the subject of many research investigations (see, e.g., Bell, Fischbein, and Greer [1984]: Fischbein, Deri, Nello, and Marino [1985]: Greer [1987]).

The two misconceptions "multiplication makes bigger" and "division makes smaller" are often noticed only when students attempt to solve multiplication or division word problems involving rational numbers less than one. Faced with a word problem, students realize from the contextual clues in the problem that the answer should be greater than or less than one of the numbers in the problem. Students multiply if they are looking for a larger number and divide if they are looking for a smaller number. For example, consider the problem "If a car travels 30 miles on 1 gallon of gasoline, how far will it travel on 1/2 gallon?" Students influenced by the misconception will reason that since they have less than one gallon, the car will travel *less* than 30 miles, so the answer must be 30 *divided* by 1/2 rather than 30 multiplied by 1/2. Similar reasoning may be applied to a problem such as "If cookies are to be packaged 0.65 pounds to a box, how many boxes can be filled with 5 pounds of cookies?" Here students offer the alternative argument: Since they are putting less than 1 pound in each box, they will have *more* than 5 boxes, so the answer must be 5 *multiplied* by 0.65 rather than 5 divided by 0.65.

Why do students believe that "multiplication makes bigger"?
First, everyday language suggests it is so. Consider such expressions as "I'm going to multiply my options" or "Rabbits multiply quickly." Another reason for the misconception is the permanence of first impressions. Except for problems involving 0 or 1, students' first contact with multiplication involves problems with whole numbers wherein the product is greater than either factor. Further, multiplication is often defined as repeated addition. The repeated-addition definition, although a useful link between multiplication and addition, is limiting if it is the students' *only* concept of multiplication. With the repeated-addition model, multiplication with mixed numbers, common fractions, or decimal fractions is not easily interpreted. Why? First, because addition with fractions is difficult. In terms of the repeated-addition model, $3 \times 4/5 = 4/5 + 4/5 + 4/5$. But $4/5 + 4/5 + 4/5$ is not trivial. Second, what does repeated addition mean if both factors are common or decimal fractions? Consider $2/3 \times 1\ 3/4$. What does it mean to a student to add 1 3/4 to itself 2/3ds times? Or in 0.5×0.5, what does it mean to add 0.5 to itself 1/2 times? Students' hesitancy to accept 0.25 as the product of 0.5×0.5 is logical in the light of the fact that until faced with decimals and fractions, multiplication (by factors other than 0 or 1) has always "made bigger."

Helping students make sense of "multiplication making smaller."
Upper elementary and middle school teachers are expected to extend multiplication to the rational numbers in a meaningful way. The area model of multiplication is a useful device to use with rational numbers, but students must first understand this interpretation of multiplication with whole numbers. Then the area model can be used to make sense of expressions like $9 \times 2/3$, $1/2 \times 1/10$, or 0.5×0.5 (see **fig. 1**). However, to increase the chance that students will use the area model as an aid in seeing that the prod-

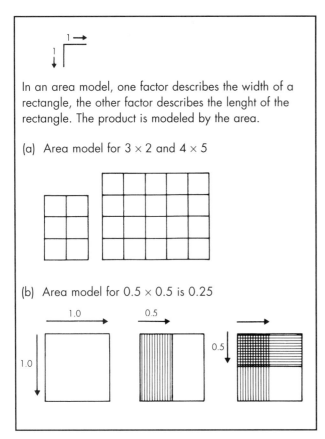

Fig. 1

Finding the length of jumps made by grasshoppers of various sizes*

Grasshopers sometimes jump almost twenty times their body length. In the picture below, the body length of some grasshoppers is given in centimeters. Find out how far each grasshopper would jump if it were to jump twenty times its body length.

A 4-centimeter-long grasshopper can jump 4 × 20, or □, cm.

A 3-centimeter-long grasshopper can jump 3 × 20, or □, cm.

A 2-centimeter-long grasshopper can jump 2 × 20, or □, cm.

A 1-centimeter-long grasshopper can jump 1 × 20, or □, cm.

How far can a 3.5-cm-long grasshopper jump? _____

How far can a 2.4-cm-long grasshopper jump? _____

How far can a 0.6-cm-long grasshopper jump? _____

*Using whole-number factors to estimate the product of a whole number and a decimal fraction can help you check the resonableness of an answer.

Fig. 2

uct of two numbers less than one is a number smaller than either of the two factors, first be sure that the area model is a familiar way for students to interpret multiplication of whole numbers.

Other activities can also help to lay the foundation for accepting that multiplication does not always "make bigger." For example, explorations of products in such patterns as 5 × 20 = □, 4 × 20 = □, ... 1 × 20 = □, 0 × 20 = □ may establish an estimate of the product of 1/2 × 20 as being between 0 × 20 and 1 × 20, that is, as less than 20. If this work is done with real-world problems involving size changes or expansions and contractions of objects, the reasonableness of the results of the calculations may be even more evident (see **fig. 2**) Once students have developed some number sense with decimals, fractions, and mixed numbers, students can connect their knowledge of the relative size of the decimal, fraction, or mixed number and the information they have gained from the set of products, 5 × 20, 4 × 20, and so on, to estimate boundaries for such products as 3.5 × 20, 2.4 × 20, 1.7 × 20, and eventually 0.6 × 20. Products involving decimals can also be found using a calculator (calculators such as the TI Explorer™ are now available that will allow students to enter fractions and mixed numbers). The class discussion should focus on the pattern in the size of the factors and in the size of the product. Students could also be encouraged to write about the pattern. For example, they could be charged with describing it for another student, noting anything they consider to be strange or unexpected and telling why they consider this result strange.

Why do students believe that "division makes smaller"?

Students' early experiences with division are also limited to whole-number divisors and whole-number quotients. This limitation leads students to place restrictions on division that are not necessarily true with rational-number divisors and quotients. For example, students often believe that the divisor must be less than the divided (see Graeber and Baker [1992]); many students will argue that the quotient for $0.25 \overline{)2}$ could certainly *never* be 8 because "division always makes smaller." Furthermore, everyday language reinforces the notion that division means cutting into parts, that is, making smaller.

Isolated numerical examples such as $0.25 \overline{)2}$ offer no context that a student can use to make sense of the expression. Many students have difficulty bringing meaning to such examples as  because they have no interpretation of division other than that it is the opposite of multiplication. If they do attempt to give meaning to such an example, they will likely attempt to use the partitive, or sharing, interpretation of division rather than the measurement interpretation (see **fig. 3**). Unfortunately, in these types of problems the measurement interpretation would be more helpful. On the one hand, the partitive interpretation asks the question "How many are in 1 group if there are 2 in 0.25 of a group?" This situation is not particularly easy to imagine or draw. On the other hand, the measurement interpretation asks the question "How many 0.25s are in 2?" Conceptually this situation is a bit easier to decipher. It can be modeled with fraction pieces or with a drawing. It can also be restated as a related question about money—"How many quarters are in two dollars?"

Contrast between the partitive and measurement models of division

Two people share six cookies. How many cookies does each get if they share fairly?

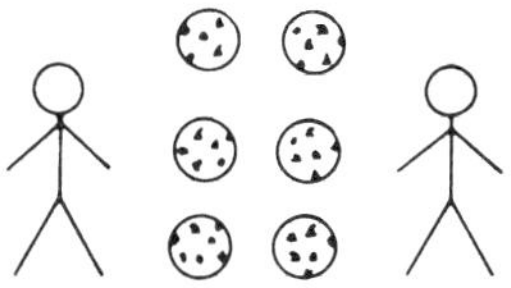

Partitive model for 6 ÷ 2

In partitive situations, a total is known and the number of sets to be made is known. The question asks how large each of these sets will be, or how many individual items will be in each of these sets.

In this example, the total is six and we know we are to make two equal sets. We are trying to find out how many cookies will be in each of the two sets.

Two cookies are to be given out to each person. There are six cookies. How many people will get cookies?

Measurement model for 6 ÷ 2

In measurement situations a total is known and the size of each set is given. The question asks how many such sets can be made.

In this example, the total is six, the set size is two, and we are trying to find how many sets of two cookies each can be made.

Fig. 3

Most elementary-level textbooks introduce division as repeated subtraction: 12 ÷ 3 means. "How many times can 3 be subtracted from 12?" This interpretation leads directly to the measurement interpretation of division: "How many 3s are in 12?" However, once the partitive interpretation of division is introduced, use of measurement word problems in textbooks frequently declines, reinforcing the partitive model. Although this partitive model is certainly important, it is not the model that is most helpful in giving meaning to division by mixed numbers or fractions.

Helping students make sense of "division making bigger"

Before investigating formal algorithms for division by decimals or fractions, students can use the measurement interpretation of division with fraction pieces, drawings, or their knowledge of the monetary system to solve simple division problems including divisors less than one. Familiarity with the measurement model can help students realize that the operation being performed is division. Examples of problems that can be explored with fraction pieces are shown in **figure 4**.

Examples of division-by-a-fraction problems (division that makes bigger) that can be modeled with fraction pieces

How many bags can I fill with 1/4 pound of peanuts if I buy 2 1/2 pounds of peanuts in the bulk-foods section of the store? (2 1/2 ÷ 1/4)

How many nickels are in a quarter? (1/4 ÷ 1/20)

We will take some first graders with us on Billy Goat Trail, and we estimate that we can cover only 3/4 mile each hour. If the trail is 2 miles long, how long will it take to complete the hike? (2 ÷ 3/4)

Fig. 4

Patterns resulting from the division of a constant number by a decreasing divisor can also be explored, for example, 24 ÷ 24 = □, 24 ÷ 12 = □, 24 ÷ 8 = □, 24 ÷ 6 = □, ..., 24 ÷ 1 = □, 24 ÷ 1/2 = □. Again these examples should be drawn from a context. If I have 24 pounds of candy and I package it 24 pounds (12, 8, 6, 4, 3, 2, 1, 1/2, or 1/4 pound) to a bag, how many bags of candy will I get? A discussion of the results; a look at the divisors that "make smaller, don't change, make larger"; and a discussion of the reasons for the common belief that division makes smaller are important parts of such experiences.

Conclusions

- As teachers we should not reinforce students' tendencies to think that when a problem calls for a bigger answer, it means add or multiply; or that when a problem is thought to need a smaller answer, it means subtract or divide (see Sowder [1989]). Although this strategy "works" in lower grades, it does not help students understand the various interpretations of the operations and can lead learners into trouble when confronting such problems as the following:

    A pound of cheese costs $5.59. If one package is marked to show that it holds 0.33 pounds of this cheese, how much will this package of cheese cost?

    I wish to tile the bottom 4 feet of my bathroom walls. If the tiles are squares 1/3 foot × 1/3 foot, how many rows of tiles will I need to fix the walls?

- Develop and make use of the area model of multiplication and the measurement model of division. Include some word problems that are examples of the partitive model of division and some that are examples of the measurement model of division.

- Introduce students to counterintuitive notions as soon as possible. Word problems leading to such calculations as 4 × 1/2, 1/2 × 1/2, 2 ÷ 1/4, or 1 1/2 ÷ 1/4 can be solved with understanding as early as fourth or fifth grade. The use of manipulatives or drawings, not standard algorithms, will facilitate understanding. A firm understanding of the meaning of fractions and much discussion about the apparent "strange" results are needed. Why are the answers unexpected? Why is it reasonable to think that multiplication makes bigger and division makes smaller? When does multiplication make bigger and division make smaller? When does multiplication make smaller and division make larger?

- Neither elementary school students nor adults easily "get rid of" misconceptions. Some researchers even argue that adults never really overcome their misconceptions. They argue that, at best, we come to recognize that we have these misconceptions and learn to be on guard against the faulty thinking that some intuitive notions support. Thus it is important for the teacher to help students realize that checking answers is more than checking computation. "Checking one's work" also involves looking at the reasonableness of an answer in light of the context from which the calculation was derived. Prior to attempting a problem, intuition often plays an important role in estimating the size of the answer. But "checking work" can help all of us catch those times when our intuition about the size of the answer leads us to make an incorrect assumption in planning how to solve the problem.

Bibliography

Bell, Alan, Efraim Fischbein, and Brian Greer. "Choice of Operation in Verbal Arithmetic Problems: The Effects of Number Size, Problem Structure and Content." *Educational Studies in Mathematics* 15 (February 1984): 129–47.

Fischbein, Efraim, Maria Deri, Maria Nello, and Maria S. Marino. "The Role of Implicit Models in Solving Problems in Multiplication and Division." *Journal for Research in Mathematics Education* 16 (January 1985): 3–17.

Graeber, Anna, and Kay Baker. "Little into Big Is the Way It Always Is." *Arithmetic Teacher* 39 (April 1992): 19–21.

Graeber, Anna, and Elaine Tanenhaus. "Multiplication and Division: From Whole Numbers to Rational Numbers." In *Interpreting Research for the Mathematics Classroom: Middle Grades,* edited by Douglas Owens. New York: Macmillan, 1993.

Greer, Brian. "Nonconservation of Multiplication and Division Involving Decimals." *Journal for Research in Mathematics Education* 18 (January 1987): 37–45.

Kouba, Vicky, and Kathy Franklin. "Multiplication and Division." In *Interpreting Research for the Mathematics Classroom: Early Childhood,* edited by Robert Jensen. New York: Macmillan, 1993.

Sowder, Larry. "Story Problems and Students' Strategies." *Arithmetic Teacher* 36 (May 1989): 25–26.

Developing Understanding of Computational Estimation

Judith Sowder

Many recent documents call for a renewed emphasis on computational estimation in the classroom. However, researchers are only now giving attention to how students learn computational estimation. The research described here focused on how children develop the ability to estimate computations. We begin with a brief description of the theory on which the research was based, describe the research study itself, and finally discuss the study's implications and extensions into the classroom.

Theory

Robbie Case is a developmental psychologist interested in applying developmental theory to curriculum design. He describes himself as "neo-Piagetian" because his theory of intellectual development comes from the Piagetian tradition but has been extended and refined to include theories from newer branches of psychology, particularly information processing. Information processing focuses on the *process* of intellectual development rather than its *structure,* with special attention paid to the role of memory in the learning process. Information processing psychologists believe that we can store only about seven items in short-term memory at any one time. We can extend the amount of information stored in short-term memory by "chunking" items together. Suppose, for example, that you are asked to remember a string of digits beginning with one, four, nine, two, and so on. These first four items could be stored as the well-known date 1492. Now you need to store only one item. Chunking is also used in learning procedures. The several steps of a procedure become linked in a chunk and can be recalled from long-term memory into short-term memory as one item rather than many.

Case's theory (1985) proposes that children's cognitive growth proceeds through two stages during the school years. The *dimensional* stage (approximately ages five through ten) is characterized by the number of units, or dimensions, children can focus on at any one time: one, two, or multiple units during the three substages. For example, mentally computing 5 + 4 (if it is not a learned fact) is a unidimensional task because it has only one quantitative dimension, ones. A child could count up to five, then use fingers to count four more ones. Mentally computing 25 + 34 is a bidimensional task because the numbers have two quantitative dimensions, tens and ones. The child must perform two separate additions, remember both, and then relate them in an appropriate fashion.

After children can focus on and integrate three or more units, they pass to the *vectorial* stage (approximately ages eleven through eighteen). The substages at this stage are characterized by growth in the ability to coordinate two or more complex and qualitatively different components of a task. Computational estimation is an example of a task that can be thought of as having two complex and multidimensional components, which Case calls vectors, that must be coordinated. One of the components is arithmetic computation, which in computational estimation is usually done mentally. Approximation, the other component, is an estimation process involving a conversion from exact to approximate numbers. The number of dimensions of each of these components is influenced by the number of things that must be remembered before another step can be executed. Case contends (Case and Sowder 1990) that real understanding of computational estimation involving execution and coordination of both components will not occur until a child enters the vectorial stage (approximately ages eleven through twelve). However, tasks involving only one of these components can be solved during earlier years. An example is a simplified computational estimation (protoestimation) task the solution to which depends primarily on one component. An example of

---

The study described here was carried out under Grant No. MDR-8550614 from the National Science Foundation. Any opinions, findings, conclusions, or recommendations expressed here are those of the author and do not necessarily represent the views of the foundation.

a protoestimation task is to estimate the sum of $14.08, $7.06, and $16.07. The approximation skill required is negligible, and most students would not consciously employ the rounding procedure on this problem, but would instead intuitively drop the cents and work only with the dollars. We presume then that students who have not yet reached the vectorial stage could successfully solve such protoestimation problems. The temptation here is to give the exact answer rather than an estimate because calculating the cents is not difficult after mentally computing 14 + 7 + 16.

Research . . .

Mental computation and approximation (or "nearness") tasks, protoestimation problems, and true computational estimation problems were evaluated in terms of the number of dimensions of each task. According to Case's theory, this analysis will yield the number of items that must be stored in short-term memory while doing the problem. Problems thought to be solvable at each substage of the dimensional and vectorial stages were given to twelve children at the age level corresponding to the substage and to children at an age level corresponding to the next higher substage. None of the children had received any formal instruction on the types of tasks they were given. In order to assess the children's understanding and success at the tasks, each child was interviewed individually and asked many questions about how he or she carried out the tasks.

Examples of the mental computation, nearness, and protoestimation tasks solvable at each of the three substages of the dimensional stage are shown in **table 1**. In each situation, these tasks were not solvable at the preceding substage. By solvable, we mean that at least 50 percent of the students interviewed were able to carry out the task successfully and seemed to understand the task rather than simply to be able to apply rules. In most cases, many more than 50 percent of the students were successful at a task. This means, of course, that in the preceding substage, fewer than 50 percent of the students could understand the task and carry it out successfully.

Examples of the computational estimation tasks solvable at different levels of the more advanced vectorial stage are shown in **table 2**. Again, children at the previous substage were unable to complete the tasks successfully.

In almost all instances, the problems that Case thought appropriate for a particular grade level proved to be so. Most of his errors involved predicting a task to be easier than it actually was for several of the students tested. Thus the data lend support to his hypothesis that computational estimation is a complex task that cannot be fully grasped until children reach the vectorial stage.

. . . Into Practice

Instructional materials for teaching computational estimation are becoming available. These materials

Table 1

Examples of Tasks Successfully Completed by Children in the Dimensional Stage

Substage (Grade Level)*	Mental Computation	Nearness Judgment	Protoestimation
First substage (grade K)	4 + 5	Which is closer to the number 6—5 or 9?	Which box of apples has just about the right number of apples for five people?
Second substage (grade 2)	23 +45	Which is closer to 32—28 or 62?	John has $1.04. Jim has $4.05. If they put their money together, would they have pretty close to $5.00 or pretty close to $9.00?
Third substage (grade 4)	27 +35	Which is closer to $25.35—$20 or $30?	(Sample tried here was not solvable at this level.)

*Grade levels refer to grades tested in research study.

Table 2

Examples of Tasks Successfully Completed by Children in the Vectorial Stage

First substage (grade 7)	About how much is $4.03 + $8.95 + $6.12 + $4.82 + $1.31?
	About how much is 5 × $7.95?
Second substage (grade 9)	About how much is $2\frac{3}{4} + 2\frac{7}{8} + 3\frac{1}{4} + 3 + 2\frac{1}{2} + 3\frac{1}{8}$?
	About how much is 88 + 649 + 897?
Third substage (grades 11 and 12)	About how much is 188 + 249 + 296 + 6 (Students should use same level of significance for each approximation.)
	About how much is 4 × $8.27? (Students were expected to round and multiply, then compensate without being asked to do so.)

frequently advocate several different strategies and give instructions for their use. Explicit teaching of such strategies can be very helpful to individuals who have not discovered them on their own *and who are developmentally ready to understand these strategies.* The research described here supports the hypothesis that true computational estimation is difficult before children have reached the stage (at about ages eleven and twelve) where they can integrate two qualitatively different processes to obtain an answer. The precursors of computational estimation—mental computation and approximation of numbers—*can* be taught to children in earlier grades, however. An instructional program developed to teach these skills has many benefits: (1) the tasks are more readily understood because they are developmentally appropriate, (2) these skills form a necessary foundation for later success in computational estimation, (3) these skills are valuable in and of themselves, and (4) development of these skill areas leads to a deeper insight into the number system, thus increasing "number sense."

We suspect that the performance on the tasks described in tables 1 and 2 of students who are given early opportunities to learn mental computation and approximation skills would be very different from the performance of the students in this study. Furthermore, the number understanding attained by students given these early opportunities will make some of the estimation tasks used in this study with children at the early vectorial stage more accessible to younger children because these tasks will, for them, become more like protoestimation tasks. That is, students may be able to chunk information so that components will not require so much memory space and therefore less coordination between components will be necessary. According to the theory, however, tasks that demand greater coordination between approximation and mental computation, such as those dealing with numbers of greatly different sizes, will continue to elude students who have not yet reached the vectorial stage, even though the students may be skilled at both mental computation and approximation. The important thing for teachers to remember is that if children are taught skills before they are ready to learn them, they will not learn those skills with real understanding. Computational estimation is a higher-order thinking skill, requiring many decisions by the estimator. A rule-based approach may handicap the learner more than a slower approach that develops understanding and the ability to make an on-the-spot choice (or invention) of an appropriate algorithm for obtaining an estimate.

References

Case, Robbie. *Intellectual Development*. Orlando, Fla.: Academic Press, 1985.

Case, Robbie, and Judith T. Sowder. "The Development of Computational Estimation: A Neo-Piagetian Analysis." *Cognition and Instruction* 7 (1990): 79–104.

Section 5

Place Value

Place Value for Tens and Ones

Joseph N. Payne

Two major ways to view the numbers to 100 are counting by ones and grouping by tens and ones.

Counting by Ones

Counting by ones uses a successor or a "next" relation to connect a given number to the next one. Twenty-nine, for example, triggers the "next" name, thirty. A great majority of children enter kindergarten with the ability to say the names from one to ten, and many can count as high as thirty. Most can not only say the names to ten in a rote fashion but also use the string of words to tell the numerosity of sets to ten.

If the string of names is known by children, most errors are coordination errors, errors in matching objects one-to-one with the names. For the set of twelve dots (**fig. 1**), one dot may be counted twice, giving an error of thirteen, or one dot may be omitted, giving eleven. Frequently, children will point to the very last dot and say two number names, as if it is difficult to stop the counting string. These errors reflect faulty and incomplete procedural rules (Gelman 1978; Fuson 1982). When children count by ones, errors occur most often with decade names and with the decade name that follows a number ending in nine. Similar errors have been noted with Korean children (Hong 1987).

Kolnowski (1984) found that the arrangement of objects in linear, circular, or random patterns affected success in counting, with linear arrangements easiest for children. Surprisingly, she found first graders with worse performance than kindergarten children because first graders would look and count instead of touch and count, especially so if she asked them "How many?" instead of "Count and tell how many." When she asked children to mark as they counted dots on paper or to move objects as they counted them, the results were much better. This approach helped them use their procedural rules correctly by supplying a mechanical way to keep track of what had and had not been counted.

Tens and Ones

The second major way to view numbers is to use tens and ones. Rathmell's (1972) analysis of the numeration tasks included three major components; the grouping of objects by tens and ones, the oral name, and the written symbol. He used the diagram in **figure 2** to illustrate the triad of relationships.

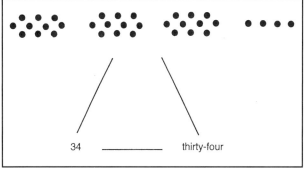

Fig. 2

Rathmell taught a three-week place-value unit to first graders in November and was able to get remarkable mastery, 90 percent or better, on the model-numeral tasks. Success on tasks with oral names did not show as much improvement, even though before instruction children did much better on oral-name

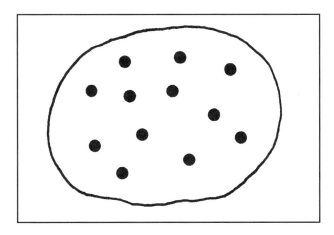

Fig. 1

tasks than on model-numeral tasks. In a similar study in kindergarten using only ten lessons, Barr (1978) found that counting by tens and ones helped, but performance was much lower in Barr's study than in Rathmell's. Both studies supported relating counting by ones to counting by tens and ones.

In telling how many dots are in the picture, two major types of errors are found: counting ones as tens, getting seventy, or counting tens as ones, getting seven. Labinowicz (1985) observed difficulties because the child must see ten in two ways, as the number after nine and as a group. He noted that the child confuses whether to count by tens or to count the ten.

Summary

Children's early understanding of numbers to twenty seems to come from counting by ones using the successor or "next" relation. Counting errors occur because the procedural rules used are faulty or incomplete. Children can be helped by modifying the way counting tasks are presented.

Place-value instruction cannot be hurried. With adequate instructional time and with lessons that include *(a)* groups of tens and ones, *(b)* written two-digit symbols, and *(c)* oral names, place value can be learned by first graders with a relatively high level of performance. Special help is needed in the development of oral names, since they appear to develop more slowly.

Research into Practice

1. *Count.* To count, say, "Cross out as you count" (**fig. 3**). When counting objects, have children move the objects as they count. Alternate the language of "How many?" with "Count to tell how many."

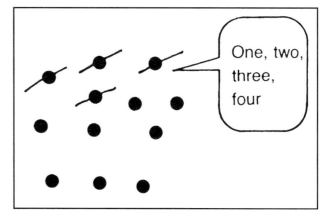

Fig. 3

2. *Group by tens.* Use groups of tens, and say the number using tens. Use children's fingers (**fig. 4**). Or use cups with ten counters in each (**fig. 5**).

3. *Count objects by tens.* Help children relate the decade names with names they know: twenty (twins), thirty (third), forty (four), fifty (fifth), sixty (six), and so on. Then count sets of ten in two ways:

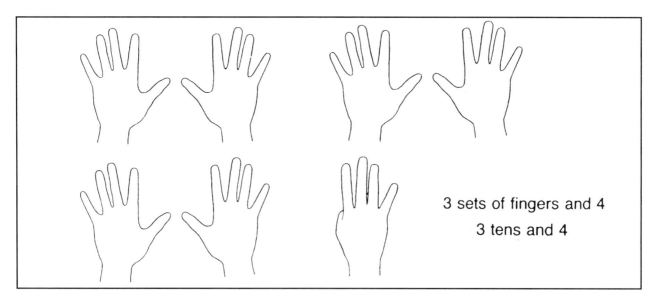

Fig. 4

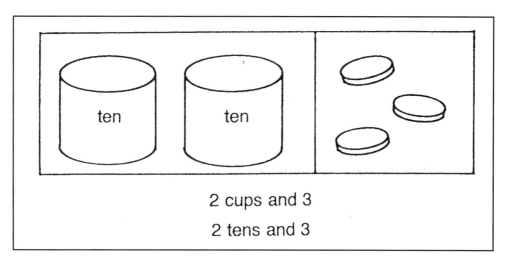

Fig. 5

one ten, two tens, three tens, four tens, ..., nine tens

ten, twenty, thirty, forty, ..., ninety

Show that "-ty" is a special way to say ten. Distinguish clearly such related sounds as "fifty" and "fifteen." Verify that counting by tens gives the same answer as counting by ones. Many examples are required.

4. Relate counting by tens to numerals.
   ten, twenty, thirty, forty, ..., ninety
   10, 20, 30, 40, ..., 90

5. Make a tens-ones chart to relate models and numerals (**fig. 6**). Read the number in two ways, using tens and ones (3 tens, 4 ones) and the usual way (thirty-four). Count by ones to see that the result is the same. Steps 4 and 5 can be reversed without loss of continuity

6. Assess the acquisition of model-, numeral-, and oral-name connections.

(a) Numeral–oral name

Can you read this number, 34? (thirty-four)

Can you write this number when I say it, thirty-four? (34)

(b) Numeral–model name

Can you write the number (or pick cards with the number) to show how many blocks I am holding? (34)

Using blocks, can you show me this many, thirty-four? (The child shows the number with bundles of ten or blocks.)

(c) Model–oral name

Can you look at the blocks and tell me how many? (thirty-four)

Can you show me blocks when I say this number, thirty-four? (The child shows blocks

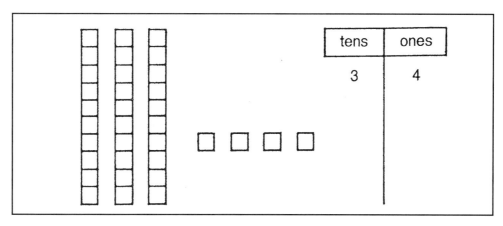

Fig. 6

Place Value for Tens and Ones

or bundles of ten.) Target instruction on tasks that need more development.

7. *Demonstrate place value for digits.* Ask what each digit represents. How many tens does the 3 show? (3) A common wrong answer is thirty.

8. *Extend the ideas to renaming.* Use models to show that 3 tens and 15 ones equals 4 tens and 5 ones, for example, and the reverse, as preparation for the addition and subtraction algorithms.

Bibliography

Barr, David C. "A Comparison of Three Methods of Introducing Two-Digit Numeration. *Journal for Research in Mathematics Education* 9 (January 1978): 33–43.

Fuson, Karen C., and James Hall. "The Acquisition of Early Number Words Meanings." In *The Development of Mathematical Thinking*, edited by Herbert Ginsburg. New York: Academic Press, 1982.

Gelman, Rochel, and Charles R. Gallistel. *The Child's Understanding of Number*. Cambridge, Mass.: Harvard University Press, 1978.

Hong, Haekyang. "Effects of Counting by Tens on Counting Ability of Kindergarten Children." Unpublished paper, School of Education, University of Michigan, 1987.

Kolnowski, Linda W. "The Effects of the Arrangement of Objects and Oral Directions on Success with Counting Tasks." Unpublished paper, Detroit Public Schools, 1984.

Labinowicz, Ed. *Learning from Children*. Menlo Park, Calif.: Addison-Wesley Publishing Co., 1985.

Payne, Joseph N., and Edward C. Rathmell. "Number and Numeration." In *Mathematics Learning in Early Childhood*, Thirty-seventh Yearbook of the National Council of Teachers of Mathematics (NCTM). Reston, Va.: NCTM, 1975.

Rathmell, Edward C. "The Effects of Multibase Grouping and Early or Late Introductions of Base Representations on the Mastery Learning of Base and Place-Value Numeration in Grade One." Ph.D. diss., University of Michigan, 1972.

Van de Walle, John, and Charles S. Thompson. "Partitioning Sets for Number Concepts, Place Value, and Long Division." *Arithmetic Teacher* 32 (January 1985): 6–11.

Place Value and Addition and Subtraction

Diana Wearne and James Hiebert

Marcy and Angela were each solving the following problem in their second-grade classroom:

> The school cafeteria has 347 ice-cream bars in one box and 48 in another box. How many ice-cream bars does the cafeteria have in the two boxes?

Both girls represented the story situation with the same number sentence and got the same answer: 395 ice-cream bars. Their written work looked the same.

$$\begin{array}{r} ① \\ 347 \\ +\ 48 \\ \hline 395 \end{array}$$

Marcy was asked to explain her solution. She said, "Right is right. You always line up the numbers on the right and then you add the numbers starting from the right: 7 and 8 is 15, so I wrote down the 5 and carried the 1: 4 plus 4 is 8 and 1 more is 9, so I wrote down the 9: 3 plus nothing is 3, so I wrote down the 3." When asked why she had aligned the digits as she had in her computation. Marcy said, "That is the way my teacher said to do them. Right is right."

Angela explained her work as follows: "7 and 8 is 15, so I had enough ones to make another 10: 4 and 4 is 8 tens, and one more makes 9 tens: I have nothing to add to the hundreds, so it is 3 hundreds." When asked why she had aligned the digits as she had in her computation. Angela said, "I put the 4s together because they are both tens and the 7 and the 8 together because they are both ones, and it's easier when they are together."

Two years later the girls are in the fourth grade and are adding decimal numbers. The girls solve this problem:

> Jeremy had 3.5 pounds of oranges in one bag and 0.62 pounds of oranges in another bag. How many pounds of oranges does Jeremy have?

Their written work looked like this:

Marcy's work

$$\begin{array}{r} 3.5 \\ +.6\ 2 \\ \hline .9\ 7 \end{array}$$

Angela's work

$$\begin{array}{r} ① \\ 3.5 \\ +\ .62 \\ \hline 4.12 \end{array}$$

Marcy explained her work: "First you line up the numbers and then you add: 5 plus 2 is 7, and 3 plus 6 is 9. Then you bring down the decimal point. I did the one there where there were two numbers behind the point."

Angela's explanation was as follows: "You don't have anything to add to the 2 hundredths, so you just write that down. Then 5 tenths and 6 tenths is 11 tenths, so you have enough tenths to make a one, so you have 4 ones. The answer is 4 and 12 hundredths." When asked why she aligned the digits as she did. Angela said, "You put the 6 and the 5 together because they are both tenths."

How Do the Responses Differ?

What is the difference between the two girls' work? The differences go beyond the fact that Marcy got the decimal problem wrong and Angela got it right. The differences were already apparent in the second grade, in the way that Marcy and Angela explained or justified their procedures. Marcy, on the one hand, depended on the teacher's instructions to know what to do and did not seem to understand very much about the procedure. Angela, on the other hand depended on her own knowledge about how quantities can be combined. These differences are often overlooked and may not even be detected as long as the answer is correct.

Acknowledgment: Some of the work described in the article was supported by grant no. MDR 8855627 from the National Science Foundation. The opinions expressed are these of the authors and not necessarily those of the Foundation.

In the early grades, there may not *appear* to be much difference between Marcy and Angela in their mathematical development—both girls are getting the correct answer—but the girls' responses indicate differences in at least two fundamental ways. First, Angela and Marcy are developing very different ideas about mathematics. For Marcy, mathematics is beginning to look like a series of rules that are hard to remember and apply when slightly different-looking problems are introduced. For Angela, mathematics is solving problems in a way that makes sense. Procedures produce correct answers because they are based on sensible ways of working with quantities.

A second difference between Marcy and Angela is that Marcy's explanation of adding whole numbers does not put her in a good position to add decimal numbers. She can only learn to add decimals by learning a different set of rules. In contrast, Angela is in a good position to add decimals. The idea of combining quantities measured with the same unit carries through from whole numbers to decimals to fractions.

In the story of Marcy and Angela, Marcy did not solve the decimal problem correctly. Of course, her teacher could teach the line-up-the-decimal-points rule and have Marcy practice until she is able to solve this type of problem. But notice that Marcy's learning the rule would not change the two important differences between her and Angela. The differences appear in how they explain or justify their work, not in whether they obtain a correct solution to a problem.

Why are these differences so important if both students get the right answer? It is likely that Marcy will not continue to obtain correct answers, especially on problems that are even slightly different from those that she has practiced in school. Mathematics is simply too complex and extensive to allow a student to learn separate rules to solve each new problem. If procedures are understood, their owners can often modify them to solve new problems. Angela did not need to learn a separate procedure to solve the decimal addition problem. She used what she knew from whole-number addition and just changed a few of the mechanics.

The Effect of Different Goals of Instruction

How did these differences between Marcy and Angela develop? Marcy and Angela probably had different experiences in school mathematics. A glimpse at Marcy's second-grade mathematics class reveals that they spent considerable time practicing computation. Marcy practiced as many as thirty problems each day, usually without a story context. The goal of instruction was efficient, correct computation. Angela's class, in contrast, spent more time developing place-value ideas, using these ideas to develop procedures for combining numbers, and then sharing procedures with other members of the class. Consequently, Angela practiced far fewer problems each day than Marcy. The increased emphasis in Angela's class on meaningful problem solving and discussion and the decreased emphasis on efficient paper-and-pencil computation are consistent with the recommendations in the *Curriculum and Evaluation Standards for School Mathematics* (NCTM 1989) and the *Professional Standards for Teaching Mathematics* (NCTM 1991).

The Importance of Understanding Place Value

Understanding place value involves building connections between key ideas of place value—such as quantifying sets of objects by grouping by ten and treating the groups as units—and using the structure of the written notation to capture this information about groupings. Different forms of representation for quantities, such as physical materials and written symbols, highlight different aspects of the grouping structure. Building connections between these representations yields a more coherent understanding of place value. A first-grade activity to help students develop place-value meanings is the following:

> Pete's Produce Market is selling apples in bags of 10. If the market has 74 apples, how many bags of 10 can Pete fill? How many apples will be left over?

The students put out seventy-four objects and then group the objects by ten to determine the number of full bags and the number of apples left over. The students then see that the 7 in the 74 represents the number of groups of ten that can be formed from seventy-four; that the 7 represents seven groups of ten, or seventy; and that the 4 represents how many apples are left over after all the groups of ten have been made. For a related problem, tell the students how many full bags and extra apples the market has and have the students write the associated numeral. Both exercises assist students in developing meaning for the written symbol.

Developing procedures for the addition and subtraction of multidigit numbers, both whole numbers and decimal numbers, should evolve from the students' understanding of place value. The underlying rationale is that students are more likely to understand the procedures if they are given opportunities to construct them and to share them with others (Hiebert and Carpenter 1992). The procedures arise from the students' understanding of place value and from the physical materials used to represent the quantities. For example, consider the following problem:

> Wendy was having breakfast at a fastfood restaurant. She had scrambled eggs, with 145 calories, and orange juice, with 187 calories. How many calories are in her breakfast?

Students could solve the problem by modeling the two quantities, seeing the regroupings that are necessary to obtain the solution, and making a connection between the models and the symbolic representation of the problem. Or the students could use their understandings of place value to guide their solution—understandings that have developed through the use of physical materials. The understandings would assist the students in recognizing that when they combine the 5 and the 7 in the foregoing problem, they have enough ones to make another ten.

The standard algorithms may not be the first procedures suggested by students. For example, in the breakfast problem, 145 + 187, students may wish to add the numbers from left to right rather than from right to left. A student might say, "100 and 100 is 200; 40 and 80 is 120, so that's 300 and 20 more; and 5 and 7 is 12, so that's 320, 330, 332." Another student might say, "1 and 1 is 2, so there are 2 hundreds; 4 and 8 is 12, so there are 12 tens; and I have enough tens to make another hundred with 2 tens left over; 5 and 7 is 12, so I have enough ones to make another ten. The answer is 332."

Although these methods are different from the standard algorithm, they are only a little less efficient and, most important, show a great deal of number sense that is built on an understanding of place value. Our experience has shown that most students who are encouraged to develop their own procedures for adding multidigit numbers begin with a method something like the methods described previously and then gradually became more efficient, sometimes shifting to a right-to-left procedure. Their success comes from using a method that they understand and that they can explain to their peers. More efficient methods evolve as students watch and listen to their classmates explain their procedures. This scenario is as appropriate in the later grades for working with decimals as it is in the early grades for whole numbers.

Conclusion

A teacher must understand why a student is using a particular procedure. If the student is only following the teacher's rule or is yielding to peer pressure, then he or she is likely to begin performing like Marcy. If the student uses a procedure because it makes sense to her or him, then the student is likely to perform like Angela.

The stories of Marcy and Angela were compiled from research studies the authors have conducted over ten years. The effects of Marcy's and Angela's behavior are based on our interpretations of the results of these studies. Interested readers can consult the articles in the Bibliography for further descriptions of students like Marcy and Angela.

Bibliography

Carpenter, Thomas P., Elizabeth Fennema, Penelope L. Peterson, Chi-Pang Chiang, and Megan Loef. "Using Knowledge of Children's Mathematics Thinking in Classroom Teaching: An Experimental Study." *American Educational Research Journal* 26 (winter 1989): 499–531.

Heibert, James, and Thomas P. Carpenter. "Learning and Teaching with Understanding." In *Handbook of Research on Mathematics Teaching and Learning*, edited by Douglas A. Grouws. New York: Macmillan Publishing Co., 1992.

Heibert, James, and Diana Wearne. "Links between Teaching and Learning Place Value with Understanding in First Grade." *Journal for Research in Mathematics Education* 23 (March 1992): 98–122.

Heibert, James, Diana Wearne, and Susan Taber. "Fourth Graders' Gradual Construction of Decimal Fractions during Instruction Using Different Physical Representations." *Elementary School Journal* 91 (March 1991): 322–41.

National Council of Teachers of Mathematics (NCTM). *Curriculum and Evaluation Standards for School Mathematics*. Reston, Va.: NCTM, 1989.

———. *Professional Standards for Teaching Mathematics*. Reston, Va.: NCTM, 1991.

Wearne, Diana. "Acquiring Meaning for Decimal Fraction Symbols: A One-Year Follow-Up." *Educational Studies in Mathematics* 21 (1990): 545–64.

Wearne, Diana, and James Hiebert. "Cognitive Changes during Conceptually Based Instruction on Decimal Fractions." *Journal of Educational Psychology* 81 (December 1989): 507–13.

Yackel, Erna, Paul Cobb, and Terry Wood. "Small-Group Interactions as a Source of Learning Opportunities in Second-Grade Mathematics." *Journal for Research in Mathematics Education* 22 (November 1991): 390–408.

Place Value as the Key to Teaching Decimal Operations

Judith Sowder

Some years ago I examined several middle school students' understanding of numbers (Threadgill-Sowder 1984). The answers that students gave me during that study showed me that their understanding, developed largely through experiences in the elementary grades, was fuzzy and led me to undertake a decade of research on children's number sense in the elementary and middle school grades. I will set the stage for this article by sharing two of the questions I gave the students during that study and some of the responses I received.

Question 1

The sum of 148.72 and 51.351 is approximately how much?

One student said, "Two hundred point one zero zero. Because the sum of 72 and 35 is about 100, and then 148 and 51 is about 200." (Note: I have used words for numerals where there is confusion about how the students read the numbers.) Another said, "One hundred fifty point four seven zero, because one hundred forty-eight point seven two rounds to one hundred point seven and fifty-one point three five one rounds to fifty point four zero zero. Add those." Fewer than half the students gave 200 as an estimate of this sum. The others saw a decimal number as two numbers separated by a point and considered rounding rules to be inflexible.

Question 2

789 × 0.52 is approximately how much?

One response was "789. I rounded point five two up to 1 and multiplied." A second student said, "Zero. This (789) is a whole number, and this (0.52) is not. It (0.52) is a number, but it is very small. You round 789 to 800, times zero is zero." Only 19 percent of the students rounded 0.52 to 0.5 or 1/2 or 50 percent. Several of them said that answering this question without paper and pencil was impossible and refused to continue. The majority of students had little idea of the size of a decimal fraction and applied standard rounding rules that were inappropriate for this estimation.

Others who have studied elementary school children's understanding of decimal numbers have found that when students are confronted with decimal numbers, many are confused about what the symbols mean. In a study of fourth through seventh graders by Sackur-Grisvard and Léonard (1985), children devised two "rules" to help them compare decimal numbers (p. 161). These rules worked just often enough that students did not recognize that they were in error. (I suspect that many teachers will recognize them.)

> Rule 1: Select as smaller the number whose decimal portion, as a whole number, is the smaller (e.g., 12.4 is smaller than 12.17, because 4 is smaller than 17.
>
> Rule 2: Select as smaller the number whose decimal portion has more digits (e.g., 12.94 and 12.24 are smaller than 12.7, because they each have two digits and 12.7 has only one).

The first rule has its origins in the separation of a decimal number into two numbers; that is, children treat the portions separately, and in this case treat the portion to the right of the decimal point as though it is itself a whole number separate from the number to the left of the decimal point. The second rule is slightly more sophisticated; it is based on the thinking that tenths are larger than hundredths.

More recent research on decimal-number understanding confirms that many students have a weak understanding of decimal numbers. For a summary of this work, see Hiebert (1992). The children in these studies were primarily from classes where the introduction to decimal numbers was brief so that sufficient time would remain for the more difficult work of learning the algorithms for operating on decimal numbers. But time spent on developing students' understanding of the decimal notation is not time wasted. Teachers with whom I have worked claim that much less instructional time is needed later for operating on decimal numbers if students first understand decimal notation and its roots in the decimal-place-value system we use. In the remainder of this article I will discuss decimal notation and how we can help students construct meaning for decimal number.

Giving Meaning to Decimal Symbols

The system of decimal numbers is an extension of the system of whole numbers and, as such, contains the set of whole numbers. For the sake of convenience, this article refers to decimal numbers as those numbers whose numerals contain a decimal point.

Decimal numbers, like whole numbers, are symbolized within a place-value system. Place-value instruction is traditionally limited to the placement of digits. Thus, children are taught that the 7 in 7200 is in the thousands place, the 2 is in the hundreds place, a 0 is in the tens place, and a 0 is in the ones place. But when asked how many $100 bills could be obtained from a bank account with $7200 in it, or how many boxes of 10 golf balls could be packed into a container holding 7200 balls, children almost always do long division, dividing by 100 or 10. They do not read the numbers as 7200 ones or 720 tens, or 72 hundreds, and certainly not as 7.2 thousands. But why not? These names all stand for the same number, and the ability to rename in this way provides a great deal of flexibility and insight when working with the number. (Interestingly, we later expect students to understand such newspaper figures as $3.2 billion.)

Therefore, before we begin instruction on decimal numbers, we need to provide more instruction on place value as it is used for whole numbers, by asking such questions as the bank question and the golf-ball question, and we need to practice reading numbers in different ways. Problems that require working with powers and multiples of 10, both mentally and on paper, give students a flexibility useful with whole numbers, and this flexibility makes it easier to extend instruction to decimal numbers.

The naming of decimal numbers needs special attention. The place-value name for 0.642 is six hundred forty-two *thousandths*. Compare this form with 642, where we simply say six hundred forty-two, not 642 *ones*. This source of confusion is compounded by the use of the *dths* (thousa*ndths,* hundre*dths*) or *nths* (te*nths*) with decimal numbers and the use of *d* (thousa*nd,* hundre*d*) or *n* (te*n*) with whole numbers. The additional digits in the whole number with a similar name is another source of confusion. Whereas 0.642 is read 642 *thousandths,* 642 000 is read 642 *thousand,* meaning 642 thousand *ones.*

In a number containing a decimal point, the units place, not the decimal point, is the focal point of the number, as shown in **figure 1.** The decimal point identifies where the units, or ones, place is located; it is the first place to the left of the decimal point. The decimal point also tells us that to the right the unit one is broken up into tenths, hundredths, thousandths, and so on. So really, 0.642 is 642 thousandths of 1. Put another way, 0.6 is six-tenths of 1, whereas 6 is 6 ones and 60 is 6 tens, or 60 ones. But just as 0.6 is six-tenths of 1, 6 is six-tenths of 10, 60 is six-tenths of 100, and so on. These relationships can be more clearly seen in the base-ten-blocks representations shown in **figure 2.**

Similarly, starting with the smaller numbers, 0.006 is six-tenths of 0.01, whereas 0.06 is six-tenths of 0.1.

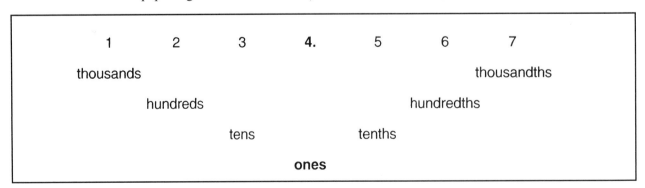

Fig. 1. Ones as the focal point of the decimal system

Moving in the opposite direction, 6000 is 60 hundreds, 600 is 60 tens, 60 is 60 ones, 6 is 60 tenths, 0.6 is 60 hundredths, 0.06 is 60 thousandths, and so on. Although this convention might seem confusing at first, it becomes less so with practice. These issues are discussed more fully in Sowder (1995).

Students who try to make sense of mathematics must become very confused when they are told to "add zeros so the numbers are the same size" when comparing numbers such as 0.45 and 0.6. This strategy does not develop any sense of number size for decimal numbers. Instead of annexing zeros, it would be more appropriate to expect students to recognize that another name for 6 tenths is 60 hundreths, which is more than 45 hundredths, or that 45 hundredths has only 4 tenths and what is left is less than another tenth, so it must be less than 6 tenths. Students do come to think this way when comparing decimal numbers if they have had sufficient opportunities to explore and think about place value, using manipulatives as representations for numbers. Heibert (1992) discusses research showing that if students do not have a sound understanding of place value when they learn to add and subtract decimal numbers, they make many errors that are very difficult to overcome because they are reluctant to relearn how to operate on decimal numbers in a meaningful way.

An Instructional Unit on Decimal Numbers

The unit summarized here was developed for a research study (Markovits and Sowder 1994) and resulted in students' performing much better on later decimal topics in their textbook. This unit has also been used by teachers who asked me for a way to teach decimals meaningfully. These teachers later

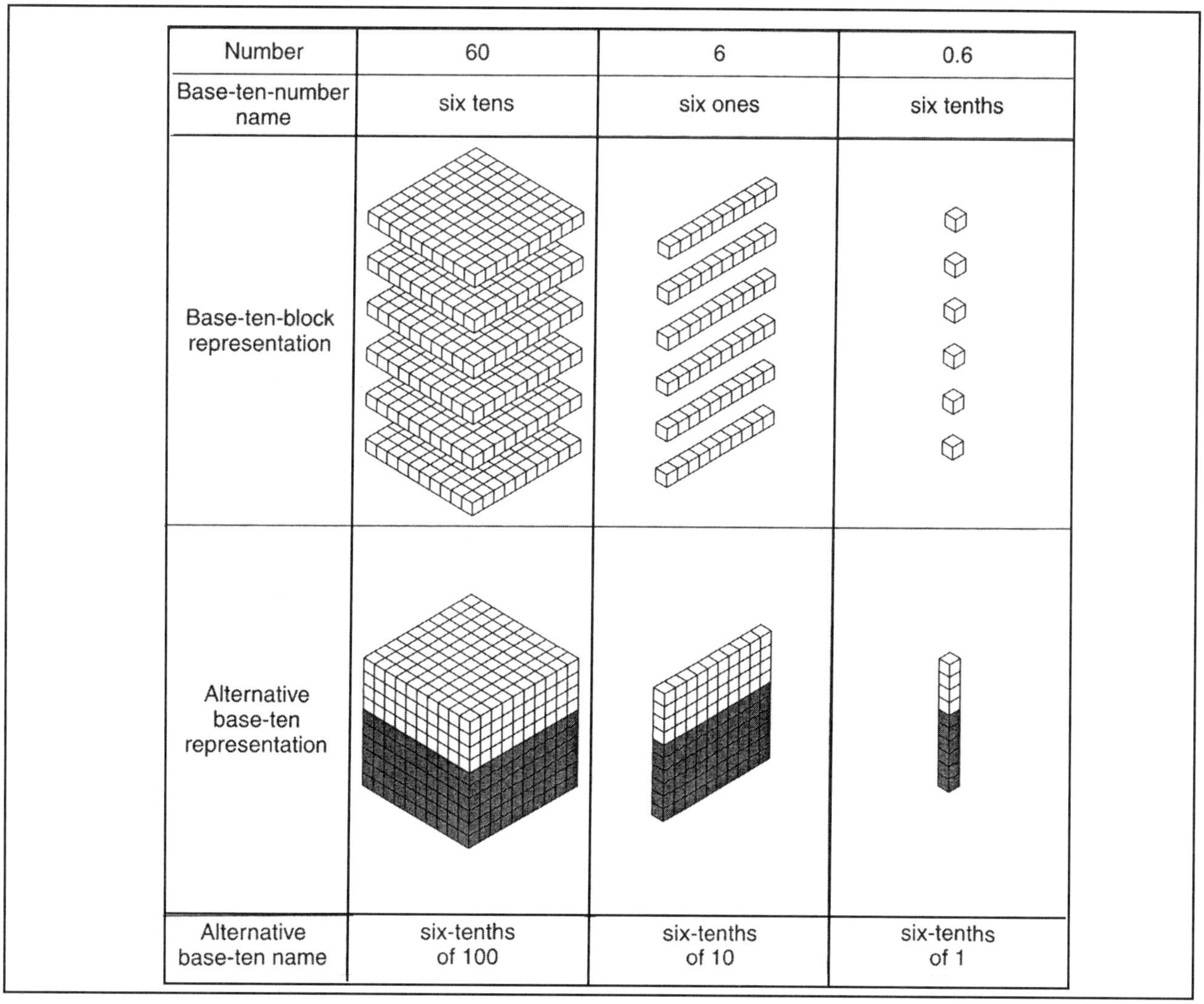

Number	60	6	0.6
Base-ten-number name	six tens	six ones	six tenths
Base-ten-block representation			
Alternative base-ten representation			
Alternative base-ten name	six-tenths of 100	six-tenths of 10	six-tenths of 1

Fig. 2. Alternative number names and representations when a long represents one unit

told me that they thought the students who completed this instructional unit had a much better grasp of decimal numbers than did their students in previous years.

The first lessons focus on gaining familiarity with base-ten materials, which can be ordered from most catalogs of mathematical aids. The materials consist of individual centimeter cubes, long blocks that are marked to look as though ten cubes have been glued in a row; flat blocks that are marked to look as though ten longs have been glued into a ten-by-ten block, and large blocks that are marked to look as though ten flat blocks have been glued to form a ten-by-ten-by-ten cube. Note that we do not call the smallest block a unit as is commonly done, because in this instruction we change the naming of the unit so that other blocks can represent one, that is, one unit.

Students must play with the blocks and learn relationships to answer such questions as the following:

- How many longs are in a flat?
- How many small blocks are in 3 longs?
- Where do you think there will be more longs, in 3 flats or in 1 big block?
- I have 6 longs and 3 small blocks. What do I have to add in order to have a flat?
- Which is bigger, that is, has more wood, 1 block or 10 flats? Four flats or 48 longs?

In the next lesson we begin to use the blocks to represent numbers. The small blocks are used to represent the number 1. Students then are asked what numbers are represented by various sets of blocks: two big blocks, three flats, and two little blocks; one flat and two longs; and so on. They must also represent numbers with blocks; for example, they show 404 with blocks. Two-dimensional drawings can later be used for the blocks, and these drawings can be used on assignments for problems like the one in **figure 3**.

Alternatively, students can be asked to show with blocks the larger of 99 and 111, which of the numbers 204 and 258 is closer to 235, and so on. Students should be asked questions that indicate the limitations of block representations of numbers when the small cube represents 1:

- Can you represent 46,321 with the blocks you have? Why or why not?
- Can you represent 8 1/2 with the blocks you have? Why or why not?

The next lessons should focus on changing the unit. First let one long represent one whole, or one unit. Students can then be asked to represent 76. (They would do so with seven flats and six longs.) After many such questions, they can again be asked, "Can you represent 8 1/2 with the blocks you have? Why or why not?" (Yes, with eight longs and five small cubes.) It is then worthwhile to ask a few questions—remaining in the whole-number system—where the flat represents one unit.

It is then natural to begin decimal instruction. If the flat represents one whole, then what does a long represent? It is obviously less than 1. What part of 1 is it? Since ten longs are in a flat, one long represents 0.1. Several questions should follow:

- How would you represent 0.3? 4.3? (**See fig. 4.**)
- How many tenths are in four wholes?
- What do you have to add to 0.9 to have one whole?
- 4.5 is ___ ones and ___ tenths, or ___ tenths.
- Which of the following are equivalent to one flat and four longs: 14? 1.4? 140? 14 longs? 41 longs? 41?

Likewise, children can come to understand that a small block in this context represents one-hundredth,

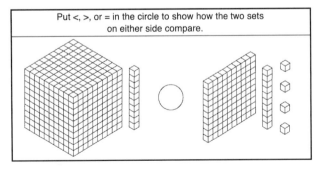

Fig. 3. Substituting a two-dimensional drawing for blocks

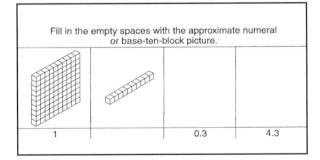

Fig. 4. Representing numbers with base-ten blocks

and many questions similar to the previous questions can be asked. Teachers can also present such problems as the following:

In 6.40 are ___ tens and ___ ones and ___ tenths and hundredths.

In 6.4 are ___ tens and ___ ones and ___ tenths and hundredths.

In 6.04 are ___ tens and ___ ones and ___ tenths and hundredths.

Are any of these numbers the same? Why?

A great deal of practice is needed in each of the lessons described here; the questions indicated are only a small sample. Try using the big block as the unit and going through all the exercises again, this time introducing one-thousandth. Ask students to describe how they could cut up the blocks to represent one ten-thousandth and one hundred-thousandth. When students feel very secure with the blocks, with changing units, and with problems involving decimals, it is time to switch to another representation. A day or two spent with money—dollars and cents—will work well. Finally, a lesson or two should focus directly on decimal numbers without using another representation (although many children will naturally answer in terms of "blocks" or "wood"). Questions like the following can be asked:

- Is 0.1 closer to 0 or to 1?
- Is 1.72 closer to 1 or to 2?
- I am a number. I am bigger than 0.5 and smaller than 0.6. Who am I?
- Are there decimals between 0.3 and 0.4? How many do you think there are?
- Are there decimals between 0.35 and 0.36? How many?
- Are there decimal numbers between 0.357 and 0.358? How many?

Draw baskets and label them "Numbers smaller than 0.5," "Numbers bigger than 0.5 but smaller than 1," "Numbers between 1 and 3," and "Numbers bigger than 3." Then give the students the following numbers and ask them to place each number in the appropriate basket: 0, 0.03, 1.01, 5.03, 2.63, 0.49, 0.93, 0.60, 1.19, and so on. This type of problem can be made more difficult with baskets labeled "Numbers between 0.4 and 0.5," "Numbers between 0.7 and 0.8," and "All other numbers."

If desired, these lessons could be interrupted before decimal numbers are introduced, and addition and subtraction of whole numbers could be introduced using the blocks. But when addition and subtraction of decimal numbers are introduced, lessons with the blocks should come first.

The two questions at the beginning of this article are trivial for students who have had this instruction. The students see the part of the number to the right of the decimal point as a natural extension of the place-value system, and they treat the entire number as one quantity. Such students also develop a good feeling for the sizes of decimal numbers and can compare them with one another. It did not occur to any of the students who received this instruction to round 0.52 to 0 or to 1 when estimating a product—0.52 was simply seen as "about a half."

When students understand what they are doing, they tend to enjoy doing mathematics. It is worth the time needed to build strong foundations. The time will be easily made up in future lessons, and students are much more likely to be successful.

Action Research Ideas

1. For each of the following pairs of decimal numbers, ask students to tell which is smaller. Then analyze their answers to see if any are making the rule-1 or rule-2 errors identified in the Sacker-Grisvard and Léonard (1985) study.

Number Pair	Use of Rule 1	Use of Rule 2	Correct
3.17 or 3.4	3.4	3.17	3.17
14.285 or 14.19	14.19	14.285	14.19
6.43 or 6.721	6.43	6.721	6.43
11.01 or 11.002	11.01	11.002	11.002
9.642 or 9.99	9.99	9.642	9.642
15.134 or 15.12	15.12	15.134	15.12
156.1 or 156.012	156.1	156.012	156.012

If you find evidence of systematic rule-1 or rule-2 errors, use some of the instructional ideas in this article and then reassess your students to determine whether their understanding of place value has improved. In addition to rule-1 or rule-2 errors, look for other systematic errors that students make. What are the misconceptions that underlie these errors?

2. Assess your students' understanding of place value by asking such questions as the following.

   (a) The Sweet Candy Company places 10 pecan clusters in each box they sell. The cook just made 262 pecan clusters. How many boxes can be filled with the fresh pecan clusters?

   (b) There is $2148 in the bank, ready to be used for prizes for the state science fair. If each prize is $100, how many prizes can be given?

Students with good place-value understanding will not need to do any division. Some students may solve *(a)* and *(b)* by using division. Some may not solve them at all. In either case try numbers like 260 or $2100 to see if easier numbers allow them to use their more limited place-value knowledge. If you find some students making large numbers of errors, use some of the instructional ideas in this article. Then reassess them using similar questions to determine whether their knowledge of place value has improved.

References

Hiebert, James. "Mathematical, Cognitive, and Instructional Analyses of Decimal Fractions." In *Analysis of Arithmetic for Mathematics Teaching,* edited by Gaea Leinhardt, Ralph Putnam, and Rosemary A. Hattrup, pp. 283–322. Hillsdale, N. J.: Lawrence Erlbaum Associates, 1992.

Markovits, Zvia, and Judith T. Sowder. "Developing Number Sense: An Intervention Study in Grade 7." *Journal for Research in Mathematics Education* 25 (January 1994): 4–29.

Sackur-Grisvard, Catherine, and François Léonard. "Intermediate Cognitive Organizations in the Process of Learning a Mathematical Concept: The Order of Positive Decimal Numbers." *Cognition and Instruction* 2 (1985): 157–74.

Sowder, Judith T. "Instructing for Rational Number Sense." In *Providing a Foundation for Teaching Mathematics in the Middle Grades,* edited by Judith T. Sowder and Bonnie P. Schappelle, pp. 15–29. Albany, N. Y.: SUNY Press, 1995.

Threadgill-Sowder, J. "Computational Estimation Procedures of School Children." *Journal of Educational Research* 77 (July–August 1984): 332–36.

Decimal Fractions

James Hiebert

This report summarizes what research says about the learning and teaching of decimal fractions and suggests some instructional strategies and activities to improve learning. Decimal fractions ordinarily are introduced in grade 4 or 5 and are treated intensively in grades 6 and 7. Research has focused on students at these grade levels, although some studies have included high school students. The following presents some of the significant findings obtained primarily from studies of students in conventional instructional programs in the United States and other countries.

1. Most students do not recognize that decimal fractions are just a different way of writing familiar (common) fractions (Carpenter et al. 1981; Ekenstam 1977; Hiebert and Wearne 1986), nor do they recognize that decimal fractions involve extensions of place value from whole numbers (Bell, Swan, and Taylor 1981; Brown 1981; Carr 1983). In other words, most students believe that decimal fractions represent an entirely new quantitative system. Few students see connections between the new (decimal fraction) symbols and other numeric symbols that they already know.

2. Most errors that students make in decimal problems are caused by their confusion about which rule, of all they have memorized, applies to which problem (Bell, Swan, and Taylor 1981; Brown 1981; Hiebert and Wearne 1985, 1986; Sackur-Grisvard and Léonard 1985). Some students make up their own rules by slightly changing the rules they were taught.

3. For many students, errors made in working with decimal fractions, unlike those made in working with whole numbers, do not correct themselves as students move through school. Low levels of performance on decimal-fraction problems are common through high school and beyond (Brown 1981; Carpenter et al. 1981; Carr 1983; Grossman 1983; National Assessment of Educational Progress 1983).

The major findings of research, together with what can be read between the lines (and found in the sources referenced) tell the following story. Most students do not develop sufficient meanings for decimal symbols when they are introduced. Soon students are asked to learn rules for manipulating decimals. Because they do not know what the symbols mean, they have no way of figuring out why the rules work. They must memorize each rule and hope that they remember on which problems to use them, a method that works for the simplest routine problems that are practiced heavily but not for problems that are even a little different. Without knowing what the symbols mean, students are unable to judge whether their answers are reasonable or whether they are on the right track. Because decimals are the last number system with which many students work, the initial errors they make persist and are difficult to remediate.

Some important changes must be made in the way decimals are commonly taught if we hope to improve students' understanding and performance:

1. More time must be devoted to developing the meaning of decimal fractions symbols *when they are introduced.* Concrete materials, such as Dienes base-ten blocks or square pieces of paper marked into tenths and hundredths, can be used as referents for the symbols. Such activities as writing the decimal fraction (e.g., 2.31) for a set of objects (e.g., 2 wholes, 3 tenths, and 1 hundredth), and vice versa, help students establish initial connections between symbols and referents. Additional activities, such as representing a written number (e.g., 1.83) with the materials in as many ways as possible (e.g., 18 tenths and 3 hundredths, 1 whole and 83 hundredths) enrich the connections and develop important concepts of the decimal system.

2. Many of the rules students are taught for working with decimal numbers should be developed from the meanings of the symbols. Often this can be done in a very natural way by using the referents for the symbols. For example, if Dienes blocks serve as referents, adding and subtracting can be introduced using the blocks and then symbol rules discussed as consequences of block activity. "Line-up-the-decimal-points" is simply a way to ensure

that same-sized pieces are combined. The extra time spent developing meanings of symbols is saved because students remember with little practice the rules they understand and remember on which problems to use them.

3. Students should be asked to estimate the answers to decimal computation problems before calculating, especially those for which the algorithm is more complicated—multiplication and division. Estimation helps students understand what the algorithm is doing for them and judge the reasonableness of their calculated answers.

All these suggestions are designed to improve performance *in the long run*. Introducing decimal addition by saying "These are just like whole-number problems after you line up the decimal points" is a quick way to get good performance *in the short run*, but it is counterproductive *in the long run*. The findings from research encourage us to adopt a long-term view and take the time to develop meaning at the outset.

References

Bell, Alan, Malcolm Swan, and Glenda Taylor. "Choice of Operation in Verbal Problems with Decimal Numbers." *Educational Studies in Mathematics* 12 (1981): 399–420.

Brown, Margaret. "Place Value and Decimals." In *Children's Understanding of Mathematics: 11–16*, edited by Kathleen M. Hart, pp. 48–65. London: John Murray, 1981.

Carpenter, Thomas P., Mary Kay Corbitt, Henry S. Kepner, Mary Montgomery Lindquist, and Robert E. Reys. "Decimals: Results and Implications from National Assessment." *Arithmetic Teacher* 28 (April 1981): 34–37.

Carr, Ken. "Student Beliefs about Place Value and Decimals: Any Relevance for Science Education?" *Research in Science Education* 13 (1983): 105–9.

Ekenstam, Adolph. "On Children's Quantitative Understanding of Numbers." *Educational Studies in Mathematics* 8 (1977): 317–32.

Grossman, Anne S. "Decimal Notation: An Important Research Finding." *Arithmetic Teacher* 30 (May 1983): 32–33.

Hiebert, James, and Diana Wearne. "A Model of Students' Decimal Computation Procedures." *Cognition and Instruction* 2 (1985): 175–205.

———. "Procedures over Concepts: The Acquisition of Decimal Number Knowledge." In *Conceptual and Procedural Knowledge: The Case of Mathematics*, edited by James Hiebert, pp. 199–223. Hillsdale, N.J.: Lawrence Erlbaum Associates, 1986.

National Assessment of Educational Progress. *The Third National Mathematics Assessment: Results, Trends, and Issues*. Denver, Colo.: Education Commission of the States, 1983.

Sackur-Grisvard, Catherine, and François Léonard. "Intermediate Cognitive Organizations in the Process of Learning a Mathematical Concept: The Order of Positive Decimal Numbers." *Cognition and Instruction* 2 (1985): 157–74.

Section 6

Fractions

Using Sharing Situations to Help Children Learn Fractions

Susan B. Empson

You might be surprised to learn that first graders know a lot about fractions. That is what two first-grade teachers and I discovered when we collaborated on a five-week fraction unit. This article describes the highlights of a case study of fractions in a first-grade class, then presents some preliminary findings suggesting that third-, fourth-, and fifth-grade children can learn fractions in similar ways.

Equal-Sharing Problems in the First Grade

Therese Kolan knew at the beginning of the year that her first graders already had a lot of intuitive knowledge of addition, subtraction, multiplication, and division with whole numbers. Her mathematics curriculum was based on creating story problems that incorporated her students' knowledge and encouraging the students to invent their own strategies for solving the problems. In a typical mathematics lesson, the class spent most of the time solving problems and discussing the children's various strategies. Kolan had found that she never needed to show a child how to solve a problem. She wanted to use this same approach when teaching fractions.

Most young children have experienced equal sharing and have quite a bit of intuitive knowledge of those situations. Research shows that kindergartners can invent their own strategies for solving story problems that are about equal sharing (Carpenter et al. 1993). We believed that first graders might be able to use their intuitive knowledge of equal sharing to solve story problems about equal-sharing situations to produce fractional quantities: "Four children want to share ten cupcakes so that each child gets the same amount. Show how much one child can have." Fourteen out of seventeen children in Kolan's class solved this problem before any instruction in fractions, suggesting that this instruction could begin by giving the children other similar problems and asking them to devise their own equal-sharing strategies.

Although we planned on making counters available and letting children use paper and pencil to solve problems, we decided not to introduce fraction manipulatives, such as preformed plastic pieces that fit together to form circles. When children use preformed manipulatives to represent fractions, they may not realize the significance of all the pieces' being of the same size. We thought the context of equal sharing would help children understand that pieces *need* to be of the same size. Rather than letting manipulatives dictate the situation, we wanted children's thinking to dictate the use of manipulatives and other tools for thinking.

Although we thought some children might have difficulty drawing circles or other shapes, and partitioning them exactly into a given number of segments, we were both surprised and pleased with the variety of ways in which children used drawings and counters to solve equal-sharing problems.

Kolan decided not to introduce fraction symbols to the children, although a few children already knew how to write the symbol for 1/2. She encouraged children to record their answers by showing how much one child might receive or by writing it out (**fig. 1b**). Older students learn the symbol more readily. The guiding principle in introducing symbols to children is to make sure that the symbols are related to concepts and situations that are already meaningful to children.

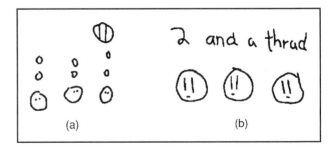

Fig. 1. Two examples of children sharing seven candy bars among three children by partitioning the last candy bar into thirds

Kolan started fraction instruction by giving children equal-sharing problems. Since children's earliest partitions are based on repeated halving, the first fraction problems Kolan gave her students involved partitioning leftover items in sharing situations with two and four children: "Four children want to share fourteen apples so that each child gets the same amount. Show how many apples one child can have." A typical strategy for this problem involved using fourteen counters to represent the apples, dealing them out one by one to each of four people until two remained. Children then had to decide what to do about the two leftover apples. Some children described cutting each apple in half, giving out a half to each child, or drew pictures of the two leftover apples, partitioning them into halves, for a total share of three and one-half. Other children showed how the two leftover apples could be cut into fourths for the four people, giving each person one-fourth from each of the two apples, or three and two-fourths. Other children decided just to give the remaining two apples to someone else. Kolan encouraged these children to share the apples among the four children by saying, "The children really want to have as much as they can. Can you think of anything they could do to share these two apples, too?" Other teachers may use a different prompt: "How would you share these apples with your three sisters if you were at home?"

Soon after giving her class problems that resulted in halves and fourths, Kolan decided to move beyond fractions that could be solved entirely by repeated halving by posing equal-sharing problems that involved three sharers: "Three children want to share seven candy bars so that everyone gets the same amount. How much would each child get?" Interestingly, only six children made thirds at first to share the extra candy bar in the problem (**fig. 1**). Most children partitioned the extra candy bar into fourths, using repeated halving, and indicated that they would give out three of the pieces. Several children stopped there, deciding that each sharer would get two and one-fourth candy bars. Sometimes children will continue to partition, dividing the last fourth into thirds; in that case, each sharer gets two and one-fourth, and a third of one-fourth of a candy bar.

When the children got two different answers for the same problem, Kolan pursued a discussion about the differences in how children had partitioned the last candy bar. Through reflecting on strategies that involved partitioning the last candy bar into thirds, several of the children who had made fourths were able to see how to construct thirds.

During the third week of fraction instruction, Kolan gave this problem: "Six children have ordered blueberry pancakes at a restaurant. The waiter brings eight pancakes to their table. If the children share the pancakes evenly, how much can each child have?" Most children got either one and one-third (**fig. 2a**) or one and two-sixths (**fig. 2b**) for their answer. Kolan made sure to call on one child who got each answer to share his or her strategy. After they had shared, one girl raised her hand and said that she had partitioned her pancakes into sixths, but "I would rather do the thirds... because thirds gets bigger pieces." Although she was correct about the relationship between thirds and sixths, she did not realize that each sharer got *two*-sixths, not just one. This comment evolved into a discussion that one-third was the same amount of pancake as two-sixths. Children argued that sixths could be made from thirds by cutting each third in half. This information helped other children conceptualize the relationship between thirds and sixths: it takes two-sixths to make one-third.

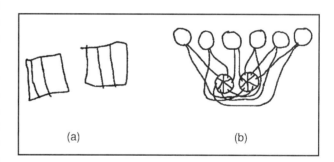

Fig. 2. Children shared eight pancakes among six children by partitioning the last two pancakes into (a) thirds or (b) sixths.

It is important to note that the children did not develop an understanding of equivalence in just one lesson. The class spent many lessons discussing how two different fractions actually represented the same quantity. As with most significant mathematical concepts, children's understanding of equivalence developed gradually over the course of many discussions, not all at once.

After five weeks of instruction, these first graders had learned a lot of about fractions. Sixteen out of seventeen children could exhaust the sharing material in situations that required partitions other than repeated halving (**fig. 3**). Even more significant, about half the children were able to use their new knowledge of fractions to solve problems that were different from those they had solved during instruction.

Fig. 3. Two examples of strategies for sharing eight candy bars among five children that did not rely exclusively on repeated halving: (a) each child gets "one (not shown) and a half"; (b) each child gets "one whole candy bar (not shown) and three-fifths."

For example, eight children solved the following problem that involved adding fractions with unlike denominators:

> Tina and Tony painted pictures this afternoon. Tina used half a jar of blue paint for her picture. Tony used three-fourths of a jar of blue paint for his picture. How much blue paint did Tina and Tony use altogether for their paintings?

All problems were read to the children as many times as they needed. All the children used pictures, rather than symbols, to solve the problem (**fig. 4**). Seven of the children quantified their answer as one and one-fourth, or five-fourths. The eighth child recognized that he had an extra quarter that would not fit into one jar but was not sure what to call the two amounts together. Further, ten children were able to solve a subtraction-of-fractions problem: "Robert had

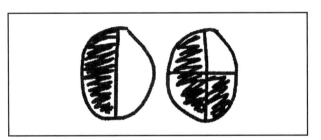

Fig. 4. To add one-half of a jar and three-forths of a jar, this child knew that two-forths would fill up the empty one-half of the first jar and that an extra one-fourth would not fit.

six giant peanut-butter cookies. He ate one-fourth of a cookie and decided he didn't want to eat any more. How many cookies did Robert have left?"

Readers may be wondering how first graders could solve problems that involved addition and subtraction of fractions without prior instruction. The key lies in students' solving and discussing equal-sharing problems, which can develop an understanding of equivalence. Equal-sharing problems fostered a sound conceptual foundation for understanding how fractional quantities are related by prompting children to reflect on different ways to partition the same amounts. Since the first graders were not trying to remember procedures that had been taught, they had to use conceptual knowledge to devise solutions. If teachers spend a great deal of time developing concepts of equivalence through equal-sharing problems, children will eventually be able to solve addition-and-subtraction-of-fractions problems using their own strategies based on concepts that make sense to them.

Equal Sharing in the Upper-Elementary Grades

I am now working with a group of researchers and teachers to understand how equal-sharing situations and their extensions can be used to help older elementary children in grades 3, 4, and 5 develop fraction concepts. These teachers approach instruction in ways similar to Kolan's. The majority of their curriculum revolves around problem solving and comparing the various strategies that their students invent. They all believe that children are capable of inventing their own strategies for solving problems and virtually never show children how to solve a problem. While working with fractions, they have encouraged children to create their own representations to solve problems without using fraction manipulatives.

Our preliminary work has confirmed that equal-sharing problems constitute a rich context in which to develop fraction concepts for older children (see also Streefland [1993]). The next step, to go beyond the first graders' work, is to investigate the important fraction topics in greater depth in the upper elementary grades.

Developing an understanding of equivalence takes place over several weeks through solving several kinds of equal-sharing problems. Several discussions of equivalence have been prompted when children get two or more different fractions for their answers. In one third-grade classroom, the teacher gave this problem on the day she wanted to introduce fractions: "Matthew has thirteen licorice sticks. He wants to share them with eight people. How much would each person get?" In **figure 5a**, the problem solver numbered the licorice sticks that each sharer received: one whole licorice stick, one-half, and one-eighth. Several children who did not know what to call an eighth called it "one-half of one-quarter," which the teacher accepted. One girl even called it "a tiny bite." After giving out one licorice stick to each sharer, other children decided to partition each of the remaining five licorice sticks into eighths (**fig. 5b**). One child who used this strategy expressed his answer as "1 whole + 5 eigths [sic]." The class had gotten two different answers: one and one-half and one-eighth (or a half of a quarter), and one and five-eighths.

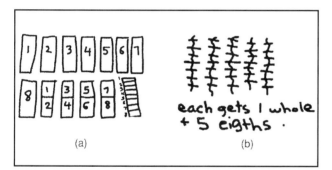

Fig. 5. Two different strategies for sharing thirteen licorice sticks among eight children: (a) each child gets one and one-half and one-eighth licorice sticks; (b) each child gets one licorice stick (not shown) and five-eighths.

The teachers usually handle this kind of instructional scenario by asking something like "Are these two answers the same amount?" or "Do you get more licorice with one answer than the other?" One of the first equivalence relationships to which children can relate on their own is how one-half is related to another fraction, such as one-eighth. In the foregoing example, children used a drawing, rather than symbols, to reason that one-half was the same amount as four-eighths. This equivalence helped them understand that one and one-half and one-eighth was the same amount of licorice as one and five-eighths.

As children develop strategies for solving equal-sharing problems with smaller numbers of sharers, such as two, three, four, six, and eight, some teachers have moved on to higher numbers of sharers. These problems potentially involve several equivalence relationships. In one fourth-grade classroom, after the children had spent about two weeks solving equal-sharing problems with smaller numbers, the teacher posed this question: "Twenty friends were sharing eight cakes. How much cake does each person get?" One child drew eight cakes and decided to partition each one into tenths. Note that every two cakes would then have twenty equal pieces, enough for one per child. She then gave each sharer one-tenth of a cake out of every two cakes, or four-tenths of a cake altogether (**fig. 6a**). Another child reasoned that since twenty people were involved, he would cut each of the eight cakes into twenty pieces and give each sharer one-twentieth out of each cake for a total of eight-twentieths. He used symbols to express his reasoning (**fig. 6b**). As these strategies were shared with the class, one child observed that eight-twentieths was the same as four-tenths. He explained that if each cake were cut into twenty pieces, each sharer would get one piece from each cake, or eight-twentieths, but if two one-twentieths were put together to make bigger pieces, then each cake would be cut into ten pieces and each sharer would get four pieces, or four-tenths. Although this child had some formal knowledge of fraction equivalence, the way he used the problem context to describe equivalence helped other children understand.

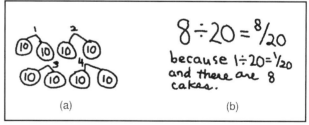

Fig. 6. To share eight cakes among twenty children, (a) one child partitions every two cakes into ten pieces and gives each child one piece from each pair, for a total of four-tenths; and (b) one child reasons that if each of the eight cakes was partitioned into twenty pieces, each child would get one piece of each, for a total of eight-twentieths.

Another way in which the teachers have elicited and developed students' thinking about equivalence is through comparisons of equal-sharing situations: "In art class, at one table four students were sharing three containers of clay so that everyone got the same amount. At another table eight students were sharing six containers of clay. At which table does a child get more clay?" This kind of problem involves two sharing situations that need to be compared. Other research on children's fraction thinking (Streefland 1993) and our preliminary work have documented two main ways that children think about similar comparison problems: (1) figure out how many pieces individual sharers at each table receive, and compare the resulting fractional quantities; (2) compare the ratios of sharers to containers at each table. For instance, a child might say that at the first table, four children get three containers. At the second table, the children and containers can be split up so that every four children get three containers here, too (**fig. 7**). Since the second example can be thought of as two sharing situations that are exactly like the first, the two situations are equivalent.

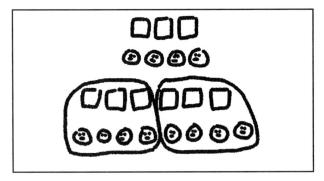

Fig. 7. One way to compare four children's sharing three things to eight children's sharing six things is to decompose the second sharing situation to mirror the first; the two situations are equivalent.

Conclusion

Whether you are a teacher of first or fifth graders, you can use equal-sharing problems to help children learn and understand fractions. After finishing the five-week unit on fractions with Kolan and her colleague, I asked them what they thought about first graders' working with fractions. The colleague, who had taught fifth grade before teaching first grade, said that she would have used the very same kinds of problems with her fifth graders. She continued, "There's no way they understood what they were doing, when I think about it.... We were just following the textbook. I mean, we had a lot of other activities that we did, but I don't think they understood as profoundly what fractions are all about as these first graders do now.... It just became a bunch of symbols to them at that point, and they knew how to manipulate symbols, but I don't think they really knew what they were doing." Her first graders had been encouraged to build their fraction understanding on their intuitive knowledge of equal sharing, and they understood fractions better than many children whose instruction is based on simply learning symbolic procedures.

Action research idea 1

Ask children to solve an equal-sharing problem where the amount each person gets is not a whole number, for instance, four children sharing seven cookies. A problem such as four children sharing twenty-two cookies is a little more difficult but is also a good beginning problem, since the cookies can be split in half. Note which children devise ways to divide up the leftovers. Ask several to share their strategies with the rest of the class. Pose a similar problem several days later. Note whether some additional children now propose a way to divide up the leftovers.

Action research idea 2

Ask children to solve a problem, such as four children sharing fifteen apples. Different students may get different answers, such as three and three-fourths; three and one-half and one-fourth; three and one-fourth and one-fourth and one-fourth; or fifteen-fourths. Question the students about whether these answers describe the same amount, and have them give the reasons for their answer. Note which children understand that two-fourths is the same amount as one-half. Repeat these problems several times a week over several weeks. Note whether more students understand this equivalence. Use a problem such as eight people sharing fourteen apples to create an opportunity to explore additional equivalences, or pose problems such as six people sharing eight apples to allow students to explore thirds and sixths. Note whether children's understanding of equivalence expands to include other relationships.

Action research idea 3

After evidence shows that children understand equivalence in the context of equal-sharing problems, explore their ability to use that understanding to solve addition and subtraction problems. One example: Anna wants one-half yard of fabric to make a pillow. Her brother Jason wants one-sixth yard of the same fabric to make a lunch sack. How much fabric should their father buy? Students who understand equivalence will be able to solve this problem without instruction. If few students can solve it, spend more time on developing equivalence relationships rather than on addition and subtraction.

More difficult problems would include those that add five-sixths yard and one-half yard; one-fourth yard and one-eighth yard; one-third yard and one-ninth yard; or one-third yard and one-fourth yard. Note which problems are more difficult than others. Consider why some addition and subtraction problems are more difficult for the students.

Bibliography

Carpenter, Thomas P., Ellen Ansell, Megan F. Franke, Elizabeth Fennema, and Linda Weisbeck. "Models of Problem Solving: A Study of Kindergarten Children's Problem-Solving Processes." *Journal for Research in Mathematics Education* 24 (November 1993): 428–41.

Empson, Susan. "Equal Sharing and Shared Meaning: The Development of Fraction Concepts in a First-Grade Classroom." Paper presented at the American Educational Research Association, San Francisco, 1995.

Streefland, Leen. "Fractions: A Realistic Approach." In *Rational Numbers: An Integration of Research,* edited by Thomas P. Carpenter, Elizabeth Fennema, and Thomas Romberg, pp. 289–326. Hillsdale, N.J.: Lawrence Erlbaum Associates, 1993.

The Problem of Fractions in the Elementary School

Leslie P. Steffe and John Olive

Researchers in Finland have warned for some years that children learn mathematics too mechanically in our comprehensive school. They learn rules and tricks, but not mathematical thinking. It is rote learning without meaning. I think this is what happens often with the fraction concept.

Tuula Strang (1990)

Strang's findings are based on the results of testing nearly three thousand students, aged nine to twelve years, in the Finnish comprehensive school. But are the implications of the findings restricted to students living in Finland? Kerslake (1986) found that English students of thirteen to fourteen years relied on rote memory of previously learned techniques when working with fractions. She believes the underlying problem is that "with the exception of certain simple examples such as 1/2 and 1/4, fractions do not form a normal part of a child's environment, and the operations on them are abstractly defined" (p. 87). Students of similar ages in the United States also learn rules and tricks for fractions and rely on rote memory of these rules and tricks (Hunting 1980; Kieren 1988; Nik Pa 1987; Payne 1976).

How can we teach fractions so that students develop concepts of fractions, gain a sense of fractional number, use models to find equivalent fractions, and apply fractions to problem situations, all of which are recommended by the *Curriculum and Evaluation Standards for School Mathematics* (NCTM 1989, pp. 57–59)? Such goals for teaching fractions are not revolutionary because they form the basis of the current curriculum. But these goals are not being achieved by the way in which fractions are currently being taught. Hunting (1980) suggests that students are forced to learn rules and tricks and rely on rote memory of those rules and tricks in part because they lack connections between what is taught and their informal ways and means of operating. To make such a connection, a major transformation is needed in teaching and learning (Steffe and Wood 1990). We would add that a major shift is needed in how most of us view the nature—the very essence—of fractions.

Students' Knowledge of Fractions

In a study of the evolution of mathematical concepts, Wilder (1968) once said that "mathematics is not something that was handed down by divine revelation to some mathematical 'Moses' in bygone times. Mathematics is something that man himself creates . . ." (p. 4). If fractions are viewed as something that students themselves create, then we as teachers need to become more aware of how they create them. Consider, for example, the following dialogue involving Teresa (Tr), thirteen years of age, and her teacher (T) (Ning 1990).

T: Can you make one-third using one candy bar?

Tr: (Draws a picture like the following.)

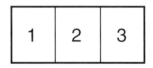

T: What is the first one?

Tr: Third.

T: How about the second?

Tr: Third.

T: And the third one?

Tr: Third.

T: Are they equal?

Tr: Yes.

T: Okay, how many thirds make a whole candy bar?

Tr: One and one-third!

T: Tell me what that means.

Tr: Because, if I put them two together (the first and second pieces), it will be a whole and then I'll have one-third of this right here (the third piece), and that will be one-third. So one and one-third!

Rather than find how many thirds make a candy bar, Teresa appeared to be finding a way to express how much candy she had. Because she usually interpreted fractional situations in terms of halving and doubling, she combined two pieces to make a whole and then combined this whole with the remaining piece, which she had already named as *one-third*. *One-third* for Teresa had no meaning with respect to the fractional whole. It was simply one of three pieces that she had made.

On the basis of the operations that we infer Teresa performed, we would accept her answer of one and one-third as being a legitimate result of those operations. For us as mathematics teachers, this attitude is very liberating because it allows us to suspend our own ways and means of operating and to focus on making sense of how Teresa operated. Focusing on the positive aspects of her mathematical activity enables us to see that she operated in a rational manner that we should accept, at least for the moment.

Students like Teresa have constructed only what we call *prefractional concepts* (Steffe and Olive 1990). She could not find, say, *one-fifth* of ten items because in her mind one-fifth referred to five single elements in a collection or to five parts of a continuous whole. *One-fifth* did not refer to the mental operations of making five composite units of indefinite numerosity and then distributing an equal number of the ten units among them.

Part-whole concepts of fractions

Viewing the mathematical knowledge of students as being of primary interest leads to a general change in an assumption permeating mathematics teaching. No longer should mathematics teachers assume that students' concepts of fractions are all the same. For example, given six pieces of a candy bar and told that this portion was three-fourths of the bar, Karla, a fifth-grade student, said, "If there are six pieces in three-fourths, then there must be two more in the whole bar," which is a solid indication that she indeed had constructed part-whole knowledge of fractions (Olive and Steffe 1990). Karla's concept was very different from Teresa's. The crucial difference between prefraction and part-whole fraction concepts resides in numerical part-whole operations. A hallmark of mastery of these operations is that given a part of a whole, a student can produce the whole (Hunting 1980; Piaget and Szeminska 1960).

Equivalent fractions. Students with part-whole concepts of fractions can develop strategies for comparing fractions. For example, Michael, a nine-year-old, could operate in a very sophisticated way when asked to compare eight-tenths and four-fifths (Saenz-Ludlow 1990, pp. 55–56). Michael (M) had already put 100 pennies into ten cups, with 10 per cup, and 100 pennies into five cups, with 20 per cup, without the help of his teacher (T).

T: Which is bigger, eight-tenths or four-fifths?

M: (After a pause) they are the same.

T: How did you find that out?

M: Because there are twenty cents left here (showing two of the ten cups), and here (showing four of five cups) there are 20, 40, 60, 80. So there are twenty left in here too. So they are the same.

Michael's strategy was to compare the numerosities of the complements of the two fractional parts of 100 pennies. His strategy of comparing complements was based on his meaning for the two fractions, which was the "result of carrying out the operations symbolized by *eight-tenths* and *four-fifths*." These results consisted of a numerosity (80 pennies) and an awareness that the 80 pennies were a part of the whole 100 pennies. Michael's strategy of comparing complements was constructed for comparing fractions close to 1, and it has been observed in other students, as well (Behr et al. 1984).

Michael had not constructed the conventional concept of equivalent fractions that serves as a basis for the current objectives concerning equivalent fractions in school mathematics. The conventional concept is based on a strategy of transforming a given fraction into equivalent forms. For example, if after partitioning a whole into five equal parts, students can put these five separated parts together, reconstituting the fractional whole, they might be able to use the five parts in the fractional whole as a basis for further partitioning. Students might split each of the five parts into two equal parts and on that basis establish that four of the five equal parts constitute the same part of the fractional whole as eight of ten equal parts. Partitioning the results of a partitioning is essential to establishing equivalent fractions in a mature and meaningful way. In this example, eight-tenths is *identical to* four-fifths, whereas for Michael the two fractions were *different objects* that were equivalent only because their results produced the same numerosity.

It would be very unfortunate if Michael's teacher decided to present equivalent fractions to Michael using the conventional concept, because it involves reasoning one stage beyond comparing complements. Michael could perform partitioning operations, but he had yet to use those operations to partition the results of partitioning to achieve a new goal. Directly teaching the conventional concept would require that Michael memorize the conventional process for establishing equivalent fractions, because it is unlikely at his conceptual level that he would modify his method of comparing complements into the more abstract operation of partitioning a result of partitioning. We need to learn how students modify their methods in mathematical learning; nothing less is acceptable in our reform efforts.

Active Mathematics Learning

In the ebb and flow of students' mathematical activity, teachers should expect rarely to encounter mathematical reasoning identical to their own. For Michael, equivalent fractions *as his teacher viewed them* were not accessible. Nor could he be guided to look around the proverbial corner and discover them as if they were already lying there like a shiny coin. Rather, through his mathematical activity, he would have to modify his strategy for comparing fractions in such a way that the modified results would lead to the conventional concept. The following episode concerning Karla's construction of improper fractions illustrates how one teacher induced such a modification in a student's means of operating.

Constructing improper fractions

Karla interpreted all fractional language using part-whole concepts. In the following exchange between Karla (K) and her teacher (T), she interpreted *two over one* as two pieces of one candy bar.

T: How much of a candy bar is two over one?

K: That's one whole candy bar.

T: Is it?

K: Two over one, two pieces of one candy bar.

The teacher then attempted to introduce a cognitive conflict.

T: What is two over two?

K: It's one whole candy bar.

T: So what do you think two over one is?

K: Uh, it's . . . one-half of a candy bar.

T: What's one over two?

K: That's half of a candy bar.

T: So are they the same? Is two over one the same as one over two?

K: (Nods her head in the affirmative.)

After her teacher posed the questions designed to promote cognitive conflict, Karla assimilated the phrase *two over one* using her part-whole concept of one-half—one of two pieces of a candy bar—without modifying the concept. Nevertheless, she was aware of an ambiguity because she independently told a computer with which she was working to make "two over one" using a command that she knew. The computer produced the display in **figure 1**, which her teacher took for an occasion for further intervention.

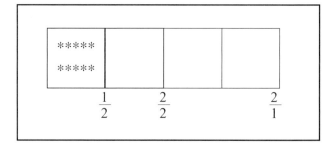

Fig. 1

T: Do you see where the two over one is? What is the two over one marking?

K: It's marking two pieces, and it's different from two over one (meaning one over two).

Karla's attempt to use the computer to confirm her current belief that *two over one* and *one over two* have the same meaning lead to a disconfirmation. In the face of the computer output, she became aware that *two over one* and *one-half* were different, but she still had no idea why. So her teacher asked more questions.

T: What is two over one?

K: Two pieces of one candy bar— one-half (a classic part-whole interpretation).

T: OK. Here's two over one. The whole thing from here (the left edge of the display) to here (pointing at "2/1") is two over one. How much of a candy bar is that?

K: That's going to be two candy bars.

T: Right—two of one. Two of one candy bar. If I give you two of one, how much have I given you?

K: You've given me two whole candy bars.

T: Right. If I give you one of two, how much have I given you?

K: Half of a candy bar.

T: Good! If I give you two of two, how much have I given you?

K: The whole thing!

Karla finally understood what *two of one* (not *two over one*) means. The modification that she made was to take the two pieces of a candy bar together as a unit and use this unit as content for making a replicate unit. A candy bar had previously served as a framework inside of which Karla made two separate but equal units. She now "stepped back," took these two units as one thing, and replicated the whole composite unit in the same way that she could replicate one-half of a candy bar to produce a whole candy bar.

Learning environments

Although students may not construct fractions in out-of-school contexts, they do construct a wealth of informal knowledge on which we can base the teaching of fractions. Learning this informal knowledge is crucial but not sufficient. We must also learn how students actively construct fractions in school-based learning environments.

In the example of Karla's constructing her concept of "two of one," we see that interactive mathematical communication served as a basis for her experiencing doubt that *two over one* and *one over two* have the same meaning. When she tested her belief by telling the computer to generate a display, the computer display was unexpected and disconfirming. Karla's problem was then to find out how to produce the desired computer output, and she was highly motivated to do so. She was deeply and experientially involved in doing mathematics.

To simply tell Karla "that is wrong" or "that is not how to do it" or to show her the "right way" would not have helped. A primary function of learning is to help students develop constraints on how they operate so they will be motivated to modify their concepts. The teacher has a responsibility to create and monitor environments in which students live mathematics (Kieren 1988). Establishing these environments in the classroom is a basic challenge for reform in teaching and learning fractions.

References

Behr, Merlyn, Ipke Wachsmuth, Thomas Post, and Richard Lesh. "Order and Equivalence of Rational Numbers: A Clinical Teaching Experiment." *Journal for Research in Mathematics Education* 15 (November 1984): 323–41.

Hunting, Robert. "The Role of Discrete Quantity Partition Knowledge in the Child's Construction of Fractional Number." Ph.D. diss., University of Georgia, 1980. *Dissertation Abstracts International* 41 (1981): 4380–81A. University Microfilms no. 8107919.

Kerslake, Daphne. *Fractions: Children's Strategies and Errors.* Windsor, England: NFER-Nelson, 1986.

Kieren, Thomas E. "Personal Knowledge of Rational Number: Its Intuitive and Formal Development." In *Number Concepts and Operations in the Middle Grades,* edited by J. Hiebert and Merlyn Behr, pp. 163–81. Hillsdale, N.J.: Lawrence Erlbaum Associates, 1988.

National Council of Teachers of Mathematics (NCTM). *Curriculum and Evaluation Standards for School Mathematics.* Reston, Va.: NCTM, 1989.

Nik Pa, Nik Azis. "Children's Fractional Schemes." Ph.D. diss., University of Georgia, 1987.

Ning, Tzyh Chiang. "A Case Study of Teresa." Athens, Ga.: University of Georgia, 1990.

Olive, John, and Leslie P. Steffe. "Constructing Fractions in Computer Microworlds." In *Proceedings of the Fourteenth PME Conference,* edited by George Booker, Paul Cobb, and Teresa N. de Mendicuti, pp. 59–66. Mexico City: Program Committee for the International Group for the Psychology of Mathematics Education, 1990.

Payne, Joseph. "Review of Research on Fractions." In *Number and Measurement,* edited by Richard Lesh, pp. 145–87. Columbus, Ohio: ERIC/SMEAC, 1976.

Piaget, Jean, and Szeminska, Alina. *The Child's Conception of Geometry.* New York: Basic Books, 1960

Saenz-Ludlow, Adalira. "Michael: A Case Study of the Role of Unitizing Operations with Natural Numbers in the Conceptualization of Fractions." In *Proceedings of the Fourteenth PME Conference,* edited by George Booker, Paul Cobb, and Teresa N. de Mendicuti, pp. 51–58. Mexico City: Program Committee for the International Group for the Psychology of Mathematics Education, 1990.

Steffe, Leslie P., and John Olive. "Children's Construction of Rational Numbers of Arithmetic." 1990. NSF Grant no. MDR-8954678.

Steffe, Leslie P., and Terry Wood. *Transforming Children's Mathematics Education: International Perspectives.* Hillsdale, N.J.: Lawrence Erlbaum Associates, 1990.

Strang, Tuula. "The Fraction Concept in Comprehensive School at Grade Levels 3–6 in Finland." In *Proceedings of the Fourteenth PME Conference,* edited by George Booker, Paul Cobb, and Teresa N. de Mendicuti, pp. 75–80. Mexico City: Program Committee for the International Group for the Psychology of Mathematics Education, 1990.

Wilder, Raymond L. *Evolution of Mathematical Concepts: An Elementary Study.* New York: John Wiley & Sons, 1968.

Fractions: In Search of Meaning

Mary Lou Witherspoon

Results of the National Assessment of Educational Progress indicate that "although most students could perform simple whole-number calculations, many evidenced little knowledge of the most fundamental concepts of fractions, decimals, or percents" (Carpenter et al. 1988, p. 40). This statement proved quite accurate during a project to compile a videotape to help preservice teachers learn how elementary school students think about fractions. Students who had completed the fifth grade struggled—and often failed—to give appropriate pictorial representations of such "simple" situations as one-third of a pie and one-fourth of a set of eight marbles. (For convenience, the word *fraction* is used to mean a fractional number.)

Interpretations and Representations of Fractions

It is not surprising that students have such problems when one considers the complicated nature of rational numbers. Kennedy and Tipps (1988, pp. 396–98) identify five interpretations of a fraction, such as one-half:

1. A unit subdivided into equal-sized parts. This part-whole interpretation gives rise to the familiar "half of a pie" model. Generally the unit is a region.

2. A set subdivided into equal-sized parts. In this part-whole interpretation the unit is a set: "half of the children are girls."

3. A ratio. This use describes the relationship between two quantities: I have half as much lemonade as you do."

4. An indicated division. This more abstract notion is the quotient of two integers: one-half is the result when one is divided by two.

5. An expression of rational numbers. One-half is the point on a number line that is midway between zero and one.

Furthermore, a fraction such as one-half can be represented in a variety of ways: with symbols (e.g., 1/2 or 3/6), with concrete models (e.g., fraction bars), with real-life situations (e.g., sharing a piece of gum), with pictures, or with spoken language (Lesh, Post, and Behr 1987). Hence, we see that giving meanings to fractions is no small task.

The fraction interpretation that students probably encounter most frequently in elementary school is that of part of a region. Pictures of subdivided regions to be shaded to indicate some fractional part accompany discussion of the real-life example of sharing a pizza. Commercially produced fraction manipulatives and games also usually involve premarked-region models. Although these devices are worth exploring, if these are the only contexts in which students encounter fractions, they learn only a small part of the underlying concepts. Their repertoire for problem solving involving fractions becomes quite limited. Two specific areas are problematic for upper elementary school students in their ability to deal with fractions: the geometry of unmarked-region models and the application of knowledge of regions to other interpretations.

The Interviews

During the videotaping project mentioned earlier, the author conducted interviews of seven students who had completed the fifth grade. The students, selected by a sixth-grade teacher and the principal of the school they attended, were chosen to represent a cross section of abilities. Some of the interviews were conducted in the summer, and others occurred at the beginning of the sixth-grade year. Each interview was based on three interpretations of fractions: part of a region, part of a set, and ratio. The students were asked to illustrate on a chalkboard three statements:

1. One-third of the pie is gone. (region)

2. There are eight marbles. One-fourth of the marbles are white. (set)

3. There are nine students in the class. There are half as many boys as girls. (ratio)

Of the seven students, only four gave an appropriate representation of "one-third of the pie is gone." In other words, each successful student drew a circular region and had a strategy involving using its center for making reasonable-looking subdivisions. Two of the others tried to make thirds by repeatedly halving the circular region. Yet another student divided the pie into three pieces that were obviously not the same size and then shaded one piece (**fig. 1**). Although he did portray three pieces, he did not use an appropriate strategy for partitioning a circular region into thirds. Thus, the geometry of the situation could have interfered with his making a suitable model. In a situation involving unmarked regions, even what was probably the most familiar interpretation of a fractional number seemed difficult for several of these students.

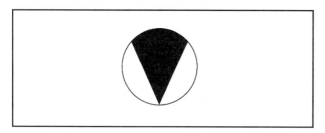

Fig. 1. A student's representation of "One-third of the pie is gone."

In representing the marble and ratio problems, several of the students tried to apply what they knew of regions. One student, in response to the problem "There are eight marbles. One-fourth of the marbles are white," colored in only a portion of four of the marbles (**fig. 2**). Apparently she had no strategy for finding a fractional part of a set; hence, she tried a familiar approach—subdividing. Knowledge of regions may actually have interfered with her ability to represent the set problem correctly. Another student first wrote "8/4" on the chalkboard. When asked to draw a picture of the marbles, he drew eight circles, colored one white, and then stepped back to think. Next, he draw a larger circle, partitioned it into eighths, and gave the resulting figure some consideration. He was trying to use a region model to arrive at an answer for a set problem. Although the student exhibited insight into transformation from one interpretation to another and although he had a "correct answer" when he wrote "8/4" (one-fourth of eight marbles would be 8/4, or 2, marbles), he did not draw an appropriate representation of the situation. Hence, he was unable to use what he knew of regions to solve a problem involving a set.

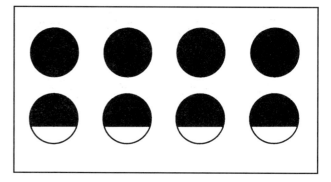

Fig. 2. A student's response to "One-fourth of the marbles are white."

Only one of the seven students correctly pictured the problem "There are nine children. There are half as many boys as girls." (Although it is true that specifying nine children was an added constraint to the problem, it did give the students a starting point. A strategy of guess and check would easily have revealed the proper situation had the students understood "half as many.") Two of the students went so far as to picture "half-persons." Once again, the region-model strategy of subdividing was applied in an incongruous situation. One of these pictured nine stick people and then carefully drew a line down the center of the middle person (**fig. 3**). The other student drew seven and one-half stick figures. When asked which were the boys and which were the girls, she said, "Oh, yes," drew nine more figures, and indicated that the seven and one-half people were boys,

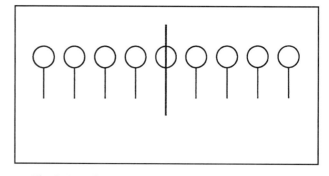

Fig. 3. A student's representation of "There are half as many boys as girls."

whereas the nine were girls (**fig. 4**). The phrase "half as many boys" seems in these students' minds to include half of a boy in its representation. The unreasonableness of their pictures did not appear to bother either of them. Although one-half is supposedly a simpler idea than either one-third or one-fourth (Lesh, Behr, and Post 1987), these students had less success with the problem involving one-half than with those involving one-third or one-fourth. Apparently one-half was not "understood" when its interpretation was a ratio.

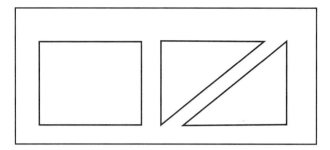

Fig. 5. Students were doubtful that two triangular regions could be halves of a rectangular region.

Fig. 4. Another student's response to "There are half as many boys as girls."

Teaching Suggestions

Thus, we see evidence of two areas of difficulty in giving reasonable representations of situations involving fractions. The first is that of the geometry of an unmarked region. Students must experience subdividing regions in various ways so that they become familiar with the geometry of various shapes. In a second-grade lesson, the teacher showed students examples and nonexamples of halves of a rectangular region. When she held up two halves that were formed by cutting along a diagonal, the students said that those pieces could not be halves. Their rationale was that the two would not fit onto the rectangle (**fig. 5**). Many "oohs" were heard when the teacher assembled the triangular region into a rectangular region. Had they seen a picture with the diagonal drawn in,

probably few disagreements would have arisen about its showing halves, but the restriction of having to picture in their minds how the pieces could be positioned to form the rectangular region was more challenging.

Curriculum and Evaluation Standards (NCTM 1989) states, "It is important that children connect ideas both among and within areas of mathematics" (p. 32). Certainly this assertion is true with geometry and the region interpretation of fractions. Experiences such as those suggested by Payne and Zawojewski in the "IDEAS" section of the December 1986 issue of the *Arithmetic Teacher* give students the opportunity to relate the geometry of regions to fractional concepts.

The other aspect that proved troublesome was the students persistent attempts to use a region interpretation by directly applying the "cutting" strategy to situations that represented other fractional-number concepts. The set interpretation of fractions is quite useful in a number of situations, such as exploring probability and learning to multiply fractions by whole numbers.

To understand how to use the set interpretation, students must be able to understand that the "one" is a set, not a single object. A dozen eggs is a useful set model for students to use initially for two reasons. One is that the students are accustomed to thinking of twelve eggs as a single unit—a dozen. Another advantage is that because twelve has a relatively large number of factors, one can show halves, thirds, fourths, sixths, and twelfths. Equivalent fractions are easily seen in this situation, as well. Discussing how half a pizza is like half a dozen eggs and how the two are different would be fruitful so that the students could make connections between the two interpretations.

Conclusions

We should be careful not to assume that students "understand" fractions merely because they are able to carry out an algorithm or recite definitions. Even such a basic and "simple" idea as one-half carries much deeper meaning that must be applied in myriad situations. Certainly the successful application of rules to symbols in fraction problems is an exercise in futility if the student cannot interpret the result of his or her labors.

The region model is frequently encountered in textbooks, but it is by no means the only context in which a fraction is used. Some students seem to feel more comfortable with that model, almost certainly because it is the one they see most frequently in instruction. Experience with only region models could limit students' thinking when solving problems involving other interpretations.

Furthermore, seeing premarked examples probably does not allow students to develop strategies for handling the geometry of various figures. Specifically, students must understand a variety of fractional-number interpretations and representations. They can do so only after having experiences that allow them to explore many different situations.

References

Carpenter, Thomas P., Mary M. Lindquist, Catherine A. Brown, Vicky L. Kouba, Edward A. Silver, and Jane O. Swafford. "Results of the Fourth NAEP Assessment of Mathematics: Trends and Conclusions." *Arithmetic Teacher* 36 (December 1988): 38–41.

Kennedy, Leonard M., and Steve Tipps. *Guiding Children's Learning of Mathematics.* 5th ed. Belmont, Calif.: Wadsworth Publishing Co., 1988.

Lesh, Richard, Merlyn Behr, and Tom Post. "Rational Number Relations and Proportions." In *Problems of Representation in the Teaching and Learning of Mathematics,* edited by Claude Janvier, pp. 41–58. Hillsdale, N. J.: Lawrence Earlbaum Associates, 1987.

———. "Representations and Translations among Representations in Mathematics Learning and Problem Solving." In *Problems of Representation in the Teaching and Learning of Mathematics,* edited by Claude Janvier, pp. 33–40. Hillsdale, N.J.: Lawrence Earlbaum Associates, 1987.

National Council of Teachers of Mathematics (NCTM). *Curriculum and Evaluation Standards for School Mathematics.* Reston, Va.: NCTM, 1989.

Payne, Joseph N., and Judith S. Zawojewski. "IDEAS." *Arithmetic Teacher* 34 (December 1986): 18–25.

Making Connections to Understand Fractions

Nancy K. Mack

Suppose you asked your students to solve this problem: If you have 3/8 of a pepperoni pizza and I give you 2/8 more of the same pepperoni pizza, how much of a pepperoni pizza do you have? Then suppose you asked them to explain how they got their answers. What do you think they would say? Several of them might respond. "I'd have '5/8' because there's a pizza cut into eight pieces, and I got three of them and then you gave me two more of the eight pieces, so I have five of the eight pieces, or five-eighths of the pizza." Several of them might respond. "The answer is '5/8' because when the bottom numbers are the same, you add the top numbers and just put the bottom number down." Which of these two explanations infers a deeper understanding of fractions?

NCTM's *Curriculum and Evaluation Standards for School Mathematics* (1989) suggests that all students should learn fractions with understanding. When students explain their answers to a problem such as 3/8 + 2/8 in a manner similar to the first of the foregoing explanations, focusing on the number of pieces in a whole, they connect the mathematical meaning of fractions, the definition of addition, and meaningful real-world applications. However, researchers and national assessment data suggest that many students in grades 3–8 have little understanding of fraction symbols and procedures (Kouba et al. 1988). This research shows students solving computation problems like 3/8 + 2/8 by adding numerators together and denominators together to get 5/16 or by rotely "adding the top numbers together and putting the bottom number down" to get 5/8.

How can students construct an understanding of fraction symbols and procedures? Recent research shows that many students come to instruction with a rich store of practical knowledge related to fractions that they are able to draw on to solve a variety of real-world word problems (Kieren 1988: Leinhardt 1988: Mack 1990). For example, many students are able to solve in meaningful ways the foregoing problem involving 3/8 and 2/8 of a pepperoni pizza before receiving formal instruction on addition and subtraction of fractions.

NCTM's *Professional Standards for Teaching Mathematics* (1991) suggests that one way teachers may be able to help students learn mathematics with understanding is by relating mathematical symbols and procedures to real-world problems that draw on students' knowledge of mathematical situations outside of school. This knowledge that students bring to instruction is sometimes called *informal knowledge*. It refers to that real-life circumstantial knowledge, constructed by individual students, that may be either correct or incorrect and can be drawn on by students in response to problems posed in the context of familiar situations (Leinhardt 1988).

As we strive to help our students understand fractions, it may be helpful to examine characteristics of students' informal knowledge of fractions and ways in which students can use that informal knowledge to give meaning to fraction symbols and procedures. The following observations and dialogues are based on the author's teaching experiences with third-, fourth-, and sixth-grade students of diverse mathematical abilities in both individualized and whole-class settings.

Characteristics of Students' Informal Knowledge of Fractions

One of the primary characteristics of students' informal knowledge of fractions is that students' informal solutions involve separating units into parts and dealing with each part as though it represents a whole number, as opposed to dealing with each part as a fraction (Mack 1990). For example, consider the following problem: If you have 5/6 of a cake and I eat 2/6 of the cake, how much of the cake do you have left? Students often refer to the fractions in this problem in terms of "the number of pieces" (e.g., five pieces or five pieces of six). However, the use of fraction names (e.g., five-sixths) refers to the fractions as specific parts of a whole.

Although this "number of pieces" interpretation of fractions appears to draw more on students' knowledge of whole numbers than on knowledge related to a part-whole definition, it enables students to solve problems in meaningful ways that are relatively error free. The following dialogue, which is taken from a typical instructional session with my students, illustrates how students can use informal knowledge to solve fraction problems in meaningful ways:

Ms. M: Suppose you have two lemon pies and you eat 1/5 of one pie, how much lemon pie do you have left?

Ned: You'd have 1 4/5. First of all you had 5/5 to start with. Then if you ate one, you'd have four pieces left out of five, and you still have one whole pie left.

Ned's response suggests that he was thinking of "two lemon pies" as one whole pie and 5/5 of a pie and that he thought of 5/5 as "one" piece and "four pieces . . . out of five" pieces. Although Ned thought of 5/5 in terms of numbers of pieces, his response further suggests that solving problems in this way was meaningful to him, just as it is to many other students as they draw on their informal knowledge of fractions.

Although students are able to draw on their informal knowledge to solve numerous real-world fraction problems, recent research suggests that students' informal knowledge of fractions is initially disconnected from their knowledge of fraction symbols and procedures (Mack 1990). Thus, at the beginning of instruction on fractions, students often are able successfully to solve numerous problems presented in the context of real-world situations (e.g., eating pizzas, cakes) and consistently to explain their solutions in terms of their informal knowledge of fractions (e.g., "three pieces of four"). However, students often are unable successfully to solve problems represented symbolically (e.g., 1/4 + 2/4) that are similar to the real-world problems. Further, students often explain their solutions to these symbolic problems by using faulty knowledge related to fraction symbols and algorithmic procedures.

The following dialogue illustrates how students' knowledge of fraction symbols and procedures is not initially connected to their informal knowledge.

Ms. M: Suppose you have two pizzas of the same size and you cut one of them into six equal-sized pieces and you cut the other one into eight equal-sized pieces. If you get one piece from each pizza, which one do you get more from?

Julie: The one with six pieces doesn't have as many pieces, so each piece is bigger.

Ms. M: Tell me which fraction is bigger, 1/6 or 1/8 (showed Julie a card with 1/6 and 1/8 printed on it).

Julie: One-eighth is bigger. Eight is a bigger number, I think. Eight is bigger than six.

Julie's response "The one with six pieces . . . is bigger" suggests that she possessed informal knowledge related to the size of fractions. However, her response "One-eighth is bigger" suggests that she had not made a connection between the fraction symbols and her informal knowledge.

Julie's response is typical of the way that many students respond when they have not made connections among fraction symbols, procedures, and their informal knowledge. Furthermore, students often suggest that it is acceptable to obtain different answers to corresponding problems "because on this (real-world problem) you're talking about pieces and on this (symbolic representation) you're talking about numbers, and they're not the same."

Building on Informal Knowledge

Although students' knowledge of fractions is initially disconnected, recent research shows that students are able to use their informal knowledge to give meaning to fraction symbols and procedures (Mack 1990). The following dialogues illustrate explanations offered by students during typical instructional sessions. In these dialogues, the students use their existing informal knowledge to help them solve symbolic problems.

First, Greg is asked to compare 3/4 and 3/5. Greg draws on his informal knowledge of fractions as pieces of a cake to give meaning to symbolic representations for 3/5 and 3/4.

Greg: See this (3/5) is like this (3/4), except that on this (3/5) you have five pieces in the whole cake or whatever and you want three of them, but on this (3/4) you only have four pieces in the whole cake.

Teresa is asked to convert 3 5/8 to an improper fraction and to express 14/3 as a mixed numeral. Teresa draws on her informal knowledge of 1 as 8/8 or 3/3.

Teresa: [Three and five-eighths is] 29/8, 8 goes into 3, I mean 8/8 goes into 1, so it's 8, then 16, then another 1 is 24, plus 5 is 29.... [Fourteen-thirds is] (writes 3/3 3/3 3/3 3/3 2/3) 4 2/3. I had to write it down or else I'd get it mixed up in my head.

Research indicates that when students draw on their informal knowledge, they often invent their own algorithms to solve problems. These algorithms differ from those that we traditionally teach students (Kieren 1988; Mack 1990). For example, Teresa's method for converting mixed numerals and improper fractions is to count by 8s or 3s, the number in the denominator, keeping track of the number of counts. This approach differs from the traditional method of dividing the numerator of an improper fraction by the denominator to find a corresponding mixed numeral. Teresa's algorithm, however, enabled her to convert mixed numerals and improper fractions in a meaningful way. Other students may invent other alternative algorithms to solve symbolic fraction problems in meaningful ways.

Implications for Instruction

As I work with students in instructional situations, I attempt to encourage them to build on their informal knowledge by presenting them with real-world problems in whatever contexts the students choose. I also present them with corresponding symbolic problems or ask them to record symbolically the number sentence for the problem and their solution on their paper. As we proceed in this manner, the students experience times when they quickly make connections between informal knowledge and symbols on their own. However, they also experience times when connections do not immediately occur. Then I try to present students with symbolic problems that are more closely matched to their informal knowledge.

This tactic often involves clarifying various concepts (e.g., the 3 in 3 − 1/5 refers to 3 wholes rather than 3 of 5 pieces), moving back and forth between problems represented symbolically and those in the context of real-world situations, and removing symbolic representations when students are not yet ready to work with them. Generally this approach enables students to make the appropriate connections.

The following dialogue, which is taken from an instructional session, illustrates ways in which students and teachers can work together to make connections among fraction symbols, procedures, and informal knowledge.

Ms. M: Suppose you have one whole pepperoni pizza and you eat four-fifths of the pizza for lunch, how much of the pizza do you have left?

Ted: One-fifth.... I ate four pieces of five pieces and there's one piece left.

Ms. M: (Wrote 1 − 4/5 on Ted's paper) What's the answer to that problem?

Ted: I don't know.

Ms. M: Is this (symbolic representation) the same as the pizza problem?

Ted: No, 'cause [the 1] is 1/5 (in 1 − 4/5). You only said one on my problem.... That "1" means one piece, and when it was a pizza you said one whole pizza.

Ms. M: What if I said that the "1" (in 1 − 4/5) means one whole, what would you get then?

Ted: One-fifth.... I'm going to put "5/5" for the one whole pizza 'cause it's the same thing just cut up, so it's 1/5....

Ms. M: (Wrote 2 − 3/4 on Ted's paper) Can you read that problem?

Ted: (Got a surprised look on his face and gasped) We haven't done this before 'cause we're doing the twos and we were doing the ones.... Two whole pizzas take away three-fourths of a pizza . . . Eight-thirds, no wait (pause) . . . I don't know.

Ms. M: Let's turn your paper over for now. If you had two whole pizzas and you ate three-fourths of one pizza, how much pizza would you have left?

Ted: One and one-fourth.

Ms. M: (Turned over Ted's paper and pointed to 2 − 3/4) Can you think of how to solve that problem now?

Ted: No.

Ms. M: Okay, let's try something different. Can you write 7/8 + 5/8? What's that equal to?

Ted: (Wrote problem correctly) Twelve-eighths.

Ms. M: If you have twelve-eighths of a pizza, do you have more than one whole pizza or less than one whole pizza?

Ted: You have more . . . four more pieces.

Ms. M: (Showed Ted how to write 1 4/8) Can you write 4 − 3/5?

Ted: (Wrote 20/5 − 3/5) Seventeen-fifths.

Ms. M: Can you solve the problem another way?

Ted: Three and two-fifths (on his own wrote 3 5/5 − 3/5 = 3 2/5) . . . five-fifths is the same as one whole, and I had three that weren't cut up, so I have four; it's just that one's cut up.

Ted's response "I'm going to put '5/5' for the one whole pizza" suggests that he drew on his informal knowledge to give meaning to 1 − 4/5. Furthermore, Ted's solution for 4 − 3/5 suggests that after we worked together to address concepts related to fractions greater than 1, he was able to extend his knowledge on his own to give meaning to the symbolic representation for 4 − 3/5. Although other ways can be used to encourage students to make connections among fraction symbols, procedures, and their informal knowledge, this dialogue demonstrates that matching symbolic representations to problems that drew on Ted's informal knowledge aided him in making connections.

This dialogue, as well as the observations concerning students' informal knowledge of fractions, suggests that students can learn fractions with understanding. As we strive toward this goal, research suggests that we encourage our students to build on their informal knowledge. We can do so by presenting them with symbolic problems that are closely matched to real-world problems that draw on their informal knowledge. As we do so, we may not only hear evidence of our students' making connections among fraction symbols, procedures, and their informal knowledge, we may also hear our students comment, "I like fractions. They make sense!"

References

Kieren, Thomas E. "Personal Knowledge of Rational Numbers: Its Intuitive and Formal Development." In *Number Concepts and Operations in the Middle Grades,* edited by James Hiebert and Merlyn J. Behr, pp. 162–81. Hillsdale, N.J.: Lawrence Erlbaum Associates, 1988.

Kouba, Vicky L., Catherine A. Brown, Thomas P. Carpenter, Mary M. Lindquist, Edward A. Silver, and Jane O. Swafford. "Results of the Fourth NAEP Assessment of Mathematics: Number, Operations, and Word Problems." *Arithmetic Teacher* 35 (April 1988): 14–19.

Leinhardt, Gaea, "Getting to Know: Tracing Student's Mathematical Knowledge from Intuition to Competence," *Educational Psychologist* 23 (Spring 1988): 119–44.

Mack, Nancy K. "Learning Fractions with Understanding: Building on Informal Knowledge," *Journal for Research in Mathematics Education* 21 (January 1990): 16–32.

National Council of Teachers of Mathematics (NCTM). *Curriculum and Evaluation Standards for School Mathematics.* Reston, Va.: NCTM, 1989.

———. *Professional Standards for Teaching Mathematics.* Reston, Va.: NCTM, 1991.

Children's Strategies in Ordering Rational Numbers

Thomas Post and Kathleen Cramer

Our comments here are based on interviews from two teaching experiments with fourth and fifth-grade children. The students were taught many aspects of fractions using a variety of manipulative materials including circular and rectangular pieces, Cuisenaire rods, number lines, and chips. Each student had his or her own materials. Each worked independently, in small groups, and as a whole class, spending much time talking about and performing a wide variety of fraction-related tasks.

One of the more fundamental rational-number notions that we explored with students was that of fraction order and equivalence. The following comments reflect the nature of children's thinking as they grapple with these concepts.

For all children, their previous whole-number schemas have influenced their ability to reason about the order relation for fractions. Children reasoned that one-third is greater than one-half because three is greater than two. Children often regressed to additive or subtractive strategies when comparing fractions. Two-fifths equaled five-eighths because $2 + 3 = 5$ and $5 + 3 = 8$. They considered three-fourths and two-thirds to be equal because the difference between the numerator and denominator in each fraction was one. For some children the influence of whole-number ideas on their thinking about rational numbers was persistent. Perhaps the fourth and fifth graders we worked with were hesitant to use multiplicative strategies because the concept of multiplication had not yet been fully developed. In one sense it seems logical that they would use the arithmetic ideas most familiar to them.

We believe many experiences with physical models of fractions are essential to overcome the influence that whole-number schemas have on children's rational-number reasoning.

Although one strategy is sufficient to order whole numbers, multiple strategies (depending on whether fractions have the same numerators, the same denominators, or nothing the same) are needed to order fractions. This idea was new to children and caused considerable difficulty. In the identical numerator situation (2/3, 2/5), the ordering decision is made by comparing the size of pieces. (Thirds are larger than fifths so 2/3 > 2/5.) In the identical denominator situation (4/11, 7/11) the ordering decision is made by comparing the number of pieces. (The pieces are the same size, and four pieces are fewer than seven pieces. Therefore, 4/11 < 7/11.) Children initially confused these issues. When asked to compare 3/9 and 4/9, some argued, "3/9 is greater than 4/9 because in fourths the pieces are smaller and it would take more of them to equal the whole unit," or similarly, "because in 3/9 the pieces are larger because there are fewer of them."

The words *more* and *greater* can lead to misunderstandings. *More* can mean more pieces in the partitioned whole, or *more* can mean more area covered by each part. *Greater* can mean a greater number of parts in the partitioned whole or a greater fraction size. When asking a student to select the greater of two fractions or which of two fractions is more, the correctness of the answer will depend on how the student interprets the words *greater* or *more*. For example, many children responded that 1/3 is greater than 1/2 because you have more pieces when you divide the whole into thirds than when you divide the whole into halves. Children often followed up our ordering questions with their own question: "Do you mean size of piece or number of pieces?" They needed clarification of the word *more* before they could give an answer. If the focus is on the size of piece, 1/2 is greater than 1/3; if the number of pieces is the variable considered, then 1/3 is greater than 1/2.

Students with strong mental images developed through extensive experiences with concrete aids employed ordering strategies that had not been taught. An example of a student-created strategy that we later named the "residual strategy" is the following: When comparing 5/6 and 7/8, the student observed that both fractions have one piece left over (to make a whole). Since 1/6 is greater than 1/8, more is left over (to make a whole) for 5/6; thus, it is the smaller fraction. Students also created what we called the transitive strategy: "3/7 is less than 5/9 because 3/7 doesn't cover half the unit and 5/9 covers over half."

Students who created their own strategies for comparing fractions and those who mastered the multiple strategies for comparing fractions demonstrated a quantitative understanding of fractions that would enable them to think reasonably about them. These students would not make the common error of adding numerators and denominators in, for example, the problem 1/2 + 1/3. They could reason that a sum of 2/5 would be unreasonable because it is less than 1/2, and the answer must be greater than 1/2 because one of the addends is 1/2.

Thinking quantitatively about fractions depends on the internal images children have of fractions and their ideas of order and equivalence. Our experience has shown that extended instruction with a variety of manipulative materials can help children overcome the initial influence that whole-number ideas have on their thinking and enable them to think quantitatively about fractions.

Activities for Ordering Rational Numbers

The following samples of activities from our teaching experiments are designed to develop order and equivalence concepts using fraction circles similar to the ones shown in **figure 1**. Fraction circles can easily be constructed by duplicating the pattern on construction paper, laminating the paper for durability, and then cutting them apart.

To compare the size of fractions, such as 1/2 and 1/3 or 3/6 and 3/9, students need to understand the inverse relationship between the number of pieces into which the whole unit is divided and the size of the resulting pieces. By using the fraction circles as a frame of reference, one-third can be shown to be less than one-half because when a circle is divided into three equal parts, the size of each piece is smaller than either of the two pieces that will result when the same size circle is divided into two equal parts. With fractions, "more" pieces does mean "less" size.

Children can be led to discover this relationship. **Figure 2** presents an activity to help children see this relationship and to come to understand how to order fractions with identical numerators. In the first example, the brown and red pieces are compared. Three brown pieces are needed to cover the whole circular area, whereas twelve red pieces are needed to cover the same circular area. More red pieces are needed than brown ones, and the red pieces are smaller in size. This information is recorded on the chart. Three more comparisons are shown in **figure 2**.

Notice the last question in this activity. After several identical-numerator comparisons, students are asked to make an identical-denominator comparison. Children need to realize that different ordering strategies are used for different types of fractions. They often want to apply the same strategy in all situations. For this reason, it is helpful to include identical-denominator comparisons among identical-numerator comparisons in all ordering activities that you do with children.

The next activity introduces children to the idea of equivalent fractions. By using fraction circles, equivalent fractions are initially defined as fractions that take up the same amount of area. For example, by superimposing two blue pieces over one yellow piece, children can see that 2/4 is the same amount as 1/2. By "creating" the equivalence chart (**fig. 3**) cooperatively, children will construct for themselves many of the common equivalences.

The activity should start by finding fractions equal to 6/12. Start by defining the whole circular area as the unit. Ask children to show 6/12 of the whole circular area with their pieces. Now ask if they can find any other ways, using pieces of another color, to cover those six reds exactly. Students may try to cover

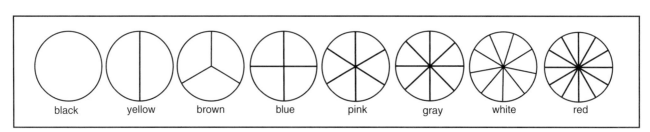

Fig. 1. Fraction circles

Ordering Fractions

Use your circular pieces to complete the table.

Color	Number of pieces to cover the whole unit	Which color takes <u>more</u> pieces to cover the whole?	Which color has the <u>smaller</u> pieces?
Brown	3		
Red	12	✔	✔
Pink			
White			
Brown			
Yellow			
Gray			
Pink			

??????

1) It takes 29 equal silver pieces to cover the whole circle. It takes 15 equal green pieces to cover the whole circle. Which is smaller, 1 silver piece or 1 green piece? _____

2) It takes 100 equal gold pieces to cover the whole circle. It takes 50 equal violet pieces to cover the whole circle. Which is bigger, 3 gold pieces or 3 violet pieces? _____

3) It takes 10 maroon pieces to cover the whole circle. Which covers the smaller area, 2 maroon pieces or 3 maroon pieces? _____

Fig. 2

red	$\frac{1}{12}$	$\frac{2}{12}$	$\frac{3}{12}$	$\frac{4}{12}$	$\frac{5}{12}$	$\frac{6}{12}$	$\frac{7}{12}$	$\frac{8}{12}$	$\frac{9}{12}$	$\frac{10}{12}$	$\frac{11}{12}$	$\frac{12}{12}$
white												
gray						$\frac{4}{8}$						
pink						$\frac{3}{6}$						
blue						$\frac{2}{4}$						
brown												
yellow						$\frac{1}{2}$						

Fig. 3. Equivalence chart

6/12 with the whites and see that it cannot be done. They will easily see that one yellow covers 6/12, and by exploring they will find that four grays and two blues also do the job. Ask students to name the fractions for these pieces and record that information on the chart. The rest of the chart can be completed in a similar manner, perhaps starting with 12/12 next. Once the chart is completed, a large classroom model of it can be posted and children can refer to it when the need arises.

Using just these two fraction-circle ideas, teachers can create a variety of activities that use these circles to teach fraction ideas. Many other physical models, such as fraction bars and Cuisenaire rods, are very useful. What is important is that initial fraction concepts be introduced concretely so children will be able to operate meaningfully on fractions represented abstractly.

References

Behr, Merlyn J., Ipke Wachsmuth, Thomas R. Post, and Richard Lesh. "Order and Equivalence of Rational Numbers: A Clinical Teaching Experiment." *Journal for Research in Mathematics Education* 15 (November 1984): 323–41.

Post, Thomas R., Merlyn J. Behr, and Richard Lesh. "Research-based Observations about Children's Learning of Rational Number Concepts." *FOCUS: On Learning Problems in Mathematics* 8 (winter 1986): 39–48.

Post, Thomas R., Ipke Wachsmuth, Richard Lesh, and Merlyn J. Behr. "Order and Equivalence of Rational Numbers: A Cognitive Analysis." *Journal for Research in Mathematics Education* 16 (January 1985): 18–36.

Building a Foundation for Understanding the Multiplication of Fractions

Nancy K. Mack

I was preparing to teach the multiplication of fractions to a class of fifth-grade students. My initial goal was to help the students understand situations involving taking a part of a part of a whole, such as eating two-thirds of three-fourths of a cookie or giving a friend one-half of one-half of a pizza. I also wanted my students to solve these problems in meaningful ways.

My students had already studied the addition and subtraction of fractions. They had given meaning to concepts, symbols, and procedures related to these two operations by building on their informal knowledge of fractions (Mack 1993). I thought that my students might also be able to build on their informal knowledge to understand and solve problems involving taking a part of a part of a whole. I decided to begin our study of the multiplication of fractions by helping my students build a foundation on the basis of their informal knowledge.

To begin, I needed to address three questions related to understanding the multiplication of fractions:

- What mathematics do I want the students to learn?
- What informal knowledge do the students already have?
- How might I help the students build on their informal knowledge to lay a foundation for their understanding?

Later I would help them draw on this foundation to give meaning to number sentences involving the multiplication of fractions.

The following observations and dialogues are based on the results of a study that examined the development of students' understanding of the multiplication of fractions and on my experiences teaching multiplication of fractions to average-ability fifth-grade students.

What Mathematics Do I Want Students to Learn?

Understanding the multiplication of fractions involves understanding ideas about fractions and understanding ideas about multiplication. When working with fractions, students need to understand that equal-sized parts are needed and that the size of a part is based on the size of the unit. They also need to be able to solve problems that involve equal sharing. Furthermore, students need to understand that each fraction has many equivalent representations; 9/12 and 6/8 are forms of 3/4, for example.

Although multiplication can be viewed in several ways, such as repeated addition or arrays, many situations with fractions involve taking a part of a part of a whole. For example, the following problem involves finding one-fourth of one-half of a whole cookie and can be represented mathematically by "1/4 of 1/2 of 1 whole" or by "1/4 × 1/2."

> You have one-half of a giant chocolate-chip cookie. You give your friend one-fourth of the piece you have. How much of the whole cookie did you give your friend?

This "taking a part of a part of a whole" interpretation can be applied to the multiplication of fractions by whole numbers greater than 1, by 1, and by fractions smaller than 1. For example, 3/4 × 6 can be thought of as starting with six items or units; partitioning the collection into four equal parts, or one-fourths; and selecting three of the one-fourth parts. The answer, 4 1/2, is expressed in terms of the original unit. Similarly, 3/4 × 1 can be thought of as starting with one item or unit; partitioning it into four equal parts, or one-fourths; and selecting three of the one-fourth parts. Here again, the answer, 3/4, is expressed in terms of the original unit. Finally, 3/4 ×

This article is dedicated to the memory of Merlyn J. Behr, who helped the author understand the multiplication of fractions. The author wishes to thank Denine Pruszynski Burkett for her assistance with data collection. This study was supported by a grant from the University of Pittsburgh School of Education.

1/2 can be thought of as starting with one-half of a unit, partitioning the one-half into four equal parts, or one-fourths; and selecting three of the one-fourth parts. The answer, 3/8, is expressed in terms of the original unit.

I wanted my students to learn to solve "taking a part of a part of a whole" problems involving a wide range of fractions. First I needed to determine whether these ideas were already a part of their informal knowledge.

What Informal Knowledge Do Students Have?

Before we began to study the multiplication of fractions, I asked my students to solve problems involving equal-sharing situations (Empson 1995). Many students solved a problem such as the following by first partitioning each cookie into thirds, or "three equal pieces."

> There are eight cookies on a plate. Three people want to share all the cookies. How many cookies will each person get?

They then distributed the pieces of each cookie among the people sharing them, one cookie at a time. Other students distributed as many whole cookies as they could among the three people. They then partitioned any remaining cookies so that each cookie could be shared among the three people. I could see that the students understood that the pieces of the partitioned cookies had to be of the same size.

I also asked my students to explain the meaning of a fraction. For example, they explained that 3/4 meant three out of four pieces. When I asked if 6/8 or 9/12 could be thought of as 3/4, they responded, "No, there's more than four pieces." From these responses I learned that my students did not understand that two different fractions could represent the same amount, that is, could be equivalent.

Next I asked my students to solve problems involving taking a part of a fractional quantity. The students were able to solve problems that involved finding one-fourth of one-half and finding one-half of one-half. To find one-fourth of one-half of a cookie, Sara first drew one-half of a cookie and partitioned it into four equal-sized parts. Sara then related one of the four parts to the whole cookie and determined that one-eighth of the whole cookie is the same amount as one-fourth of one-half of the cookie (see **fig. 1**). Many students used Sara's strategy.

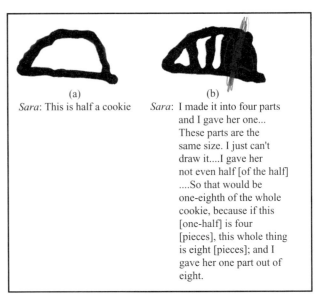

(a)
Sara: This is half a cookie

(b)
Sara: I made it into four parts and I gave her one... These parts are the same size. I just can't draw it....I gave her not even half [of the half] ....So that would be one-eighth of the whole cookie, because if this [one-half] is four [pieces], this whole thing is eight [pieces]; and I gave her one part out of eight.

Fig. 1. Sara's solution to a problem involving one-fourth of one-half of a cookie. Please note: her solution is presented in a way that attempts to show its development.

From my students' solutions to finding one-fourth of one-half of a cookie, I learned that they could do more than find a fractional part of one whole. They could find a part of a part of a whole. I also learned that they understood to what quantity each fraction referred in a problem and that they could state the answer in terms of the original unit. More specifically, the students understood that in 1/4 of 1/2 of a cookie, the 1/2 referred to the whole cookie and the 1/4 referred to the one-half cookie and that one-eighth of the whole cookie was equivalent to one-fourth of one-half of a whole cookie.

Although they were able to find one-fourth of one-half and one-half of one-half of a cookie, they were unable to solve any other problems that involved taking a part of a fractional quantity, and they were unable to connect their solutions to the operation of multiplication. When I asked them what mathematical operations they could use to solve problems like sharing eight cookies among three people and giving a friend one-fourth of one-half of a cookie, they responded, "Division." As Sam explained, "In both of them you gotta divide it up." When I asked the students if multiplication could be used for problems like taking one-fourth of one-half, they responded, "No." The students initially thought of multiplication in terms of repeated addition only. Abby explained, "Multiplication is like five times two, like five plus five.... You can't add one-fourth one-half times. It doesn't make sense."

It was clear that my students did not think of taking a part of a part of a whole in terms of multiplication. They viewed these situations in terms of division, and this conception was very strong. I wondered whether it would be possible for me to help these students build on their informal knowledge to understand a wider variety of situations involving taking a part of a part of a whole. I also wondered whether I could help my students learn to view these situations as involving the multiplication of fractions.

Helping Students Build on Their Informal Knowledge

As we began our study of the multiplication of fractions, I encouraged my students to draw on their informal knowledge by asking them to solve problems based on real-world situations. Our first problems involved equal-sharing situations, such as sharing ten cookies among four people. Although equal-sharing situations are usually considered division problems, these problems helped my students deepen their understanding of the need for equal-sized parts when working with fractions. The problems also helped them understand the meaning of partitioning a whole cake or one-half of a pizza into thirds, fourths, and other fractional parts.

After working with equal-sharing situations, we moved on to finding a fraction of a whole-number amount, such as two-thirds of twelve and one-fourth of nine. As the students tried to understand and solve these problems, both the students and I frequently referred to the equal-sharing problems, the need for equal-sized parts when working with fractions, and the meaning of partitioning something into thirds or fourths. To help the students focus on mathematical ideas and solve the problems in meaningful ways, I asked such questions as "How many groups do you need if you are going to eat one-fourth of these?" "Remember when we were sharing cookies? What did we have to do with some of the cookies?" and "Could each person get more than one cookie?"

My students continued to draw on their knowledge of the need for equal-sized parts and the meaning of partitioning a quantity into thirds or fourths as we moved on to problems that involved taking a part of a part of a whole. I first asked my students to solve problems similar to "finding one-fourth of four-fifths of a cookie," problems in which the denominator of the first fraction was the same as the numerator of the second fraction. Such problems encouraged my students to begin thinking of the meanings of fourths and thirds from a broader perspective. They also helped my students realize that they could find a portion of a fractional quantity other than one-half. As the students focused on these ideas, they solved problems like one-fourth of four-fifths by seeing one fractional quantity as being embedded within another. The following vignette illustrate how Lee was able to understand and solve one-fourth of four-fifths. Lee's problem was inspired by photographs from a recent visit to the zoo.

Ms. M: You have four-fifths of a chocolate cake left to feed the camel after feeding the meerkats one-fifth of the cake. The camel is not very hungry. It is only going to eat one-fourth of what you have to feed it. How much does the camel eat of the whole cake?

Lee: [Drew a circle, divided it into five supposedly equal-sized parts, and put a dot on the one part fed to the meerkats. See **fig. 2**.] That one [points to one of the four unmarked pieces]. That's one-fifth [of the whole cake] . . . 'cause there's five of these [pieces in the whole cake] and I gave one to the camel of these four there [indicated the four unmarked pieces].

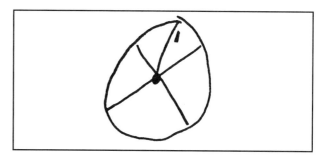

Fig. 2. Lee's solution to a problem involving one-fourth of four-fifths of a chocolate cake

Lee's comment "That's one-fifth 'cause there's five of these" suggested that he drew on his informal knowledge related to equal-sized parts and to partitioning a quantity into fractional amounts as he solved the problem. Lee's explanation and diagram suggested that as he drew on this knowledge, he was able to understand the problem, solve the problem in a meaningful way, and determine how to name the result.

After my students solved such problems as finding one-fourth of four-fifths, I asked them to solve increasingly complex problems involving taking a part of a part of a whole. Next I asked the students to solve problems similar to "finding 3/4 of 2/3 of a cookie," in which the denominator of the first fraction is a multiple of the numerator of the second fraction. Problems of this nature encouraged the students to further broaden their idea of the meaning of partitioning a quantity into a fractional amount. The students solved the problem of finding three-fourths of two-thirds by drawing a picture showing two one-thirds and partitioning each of the one-third pieces in half to obtain four parts. They then selected three of the parts and determined that the answer was three-sixths, or one-half.

Following this activity, I asked the students to solve problems similar to "finding 2/3 of 9/10 of a cookie," in which the denominator of the first fraction is a factor of the numerator of the second fraction. Such problems encouraged the students to think of the meaning of fractions in terms of ideas of equivalence. Some students found two-thirds of nine-tenths of a cookie by forming three groups of three pieces from the nine pieces in nine-tenths. They then selected two of the three groups and determined that the answer was six-tenths. Other students solved this problem by partitioning each of the nine pieces in nine-tenths into three equal-sized parts. They then selected two pieces from each group of three.

Last, I asked the students to solve problems similar to "finding 3/4 of 7/8" of a cookie, in which the greatest common factor of the denominator of the first fraction and the numerator of the second fraction is 1. Problems of this nature encouraged the students to further broaden their conception of the meaning of partitioning a quantity into a fractional amount. They also helped the students realize that they could both repartition a quantity and group the pieces when solving problems that involved taking a part of a part of a whole. The following vignette, also based on the zoo pictures, illustrates how Abby was able to understand and solve in a meaningful way a problem involving finding three-fourths of seven-eighths.

Ms. M: You have seven-eighths of a gigantic chocolate-chip cookie [after feeding the flamingoes one-eighth of the cookie]. You go to feed the bear, and the bear is not real hungry. The bear already ate some fish. The bear eats three-fourths of the remaining piece of cookie. How much of the whole cookie does it eat?

Abby: [Drew a circle.] I gotta split it into eight. [Divided the circle into eight supposedly equal parts and shaded one part that fed the flamingoes. See **fig. 3a**.]

Ms. M: You have to feed the bear three-fourths of what you have. How many parts do you need if you need to feed the bear three-fourths?

Abby: Four.

Ms. M: Is there some way you could make this [seven-eighths] so you would have four parts . . . of the same size? There could be more than one piece in each part.

Abby: [Divided each of the seven unshaded pieces in half, found fourteen pieces, and divided each piece in half again. See **fig. 3b**.] I got twenty-eight pieces.

Ms. M: Can you divide that into four parts?

Abby: [Counted out four groups of seven and indicated groups of seven with heavy lines. See **fig. 3c**.] There's my four parts. That's four-fourths. Seven [in each part]. . . . so the bear gets that much [indicated three of the four parts].

Ms. M: Do you know what the fraction name is for that? What did you do to each of these [pointed to the seven one-eighth pieces]?

Abby: Split 'em into four.

Ms. M: The question is, How much of the whole cookie did the bear eat? How are you going to find out how much of the whole cookie the bear ate?

Abby: Split this [the shaded one-eighth piece] into four [see **fig. 3d**]. The bear ate twenty-one

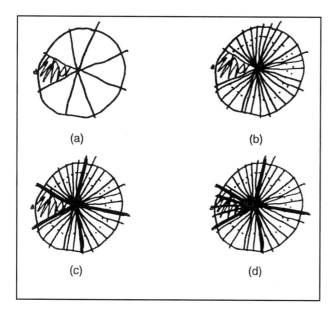

Fig. 3. Abby's solution to a problem involving three-fourths of seven-eighths of a cookie. Please note: Her solution is presented in a way that attempts to show its development.

pieces and there's thirty-two pieces in the whole cookie, ... so twenty-one thirty-twos.

Such questions as "How many parts do you need?" and "Is there some way you could make this so that you would have parts of the same size?" appeared to encourage Abby to draw on her informal knowledge related to equal-sized parts and the meaning of partitioning a quantity into fractional amounts. The way that Abby drew her diagram suggested that she realized that she could both repartition and group pieces as she solved the problem in a meaningful way.

After about six days of instruction, my students were able to understand and solve a variety of problems involving taking a part of a part of a whole. They had developed this understanding by building on their informal knowledge of fractions. Although we had not yet worked with number sentences, I was confident that the students had developed a foundation that could support their understanding of the multiplication of fractions.

Action Research Ideas

1. As you prepare to work on the multiplication of fractions, assess your students' ability to solve problems that involve equal-sharing situations. If they can solve such problems as how to share eight cookies among three people, have them try a sharing problem like how to share one-half a cookie among four people. If they are successful, try having them take a part of a part of a whole, such as one-fourth of one-half of a cookie. As your students work on these problems, note and record which students understand (1) that equal-sized parts are needed; (2) that whole-number problems can result in fractional amounts; (3) what it means to partition a quantity into a fractional amount, such as thirds; and (4) to what quantity each fraction in a problem refers. Have students who do not understand these ideas solve more equal-sharing problems. Help them by asking questions, such as "Are the pieces the same size?" and "When you say that each person gets two and one-half, what do they get two of and what do they get one-half of?" and "If there are three people, what fractional part of all the cookies does each person get?" Such questions will help the students focus on the important mathematical ideas in the problem.

2. Students who understand the need for equal-sized parts and the meaning of partitioning a quantity into fractional amounts may be successful at taking a part of a part of a whole. As they work, note which students draw diagrams and note the nature of their diagrams. Which ones draw diagrams similar to those of Lee and Abby? Do any of your students draw rectangular arrays that are frequently used to represent multiplication? Whatever diagrams they draw, note whether they represent taking a part of a part of a whole. Also, note ways in which students might later connect their solutions with diagrams to number sentences involving the multiplication of fractions.

References

Empson, Susan B. "Using Sharing Situations to Help Children Learn Fractions." *Teaching Children Mathematics* 2 (October 1995): 110–14.

Mack, Nancy K. "Making Connections to Understand Fractions." *Arithmetic Teacher* 40 (February 1993): 362–64.

How Children Think about Division with Fractions

Mary Ann Warrington

When children are allowed to create and invent, their fertile minds enable them to solve problems in a variety of original and logical ways. When their minds have not been shackled by rules and conventions, children are free to invent procedures that reflect their natural thought processes.

Much of the research and documentation concerning children's inventions focuses on their approaches to the addition, subtraction, multiplication, and division of whole numbers. Research and practice in these areas have shown that children can develop sophisticated and meaningful procedures in computation and problem solving without explicit instruction in the use of conventional algorithms. These invented procedures have been reported not only in the United States (Kamii 1989, 1994; Madell 1985) but also in Brazil (Carraher, Carraher, and Schliemann 1985), Holland (Heege 1978; Treffers 1987), and South Africa (Olivier, Murray, and Human 1991). In the United States, some leading educators, such as Burns (1994) and Leinwand (1994), have renounced the teaching of algorithms to young children, and some researchers, including Kamii (1994), have even shown the practice to be harmful. Despite compelling evidence about children's procedures with fractions (Mack 1990; Streefland 1993), most educators still believe that to handle the more complex mathematics of the middle grades and beyond, children need to learn specific procedures, or algorithms.

As a teacher of fifth and sixth grade, I am passionately committed to reform in mathematics education and firmly believe in the merits of children's constructing their own knowledge. Yet when I began working in the middle grades four years ago, like many I was ambivalent about whether students could continue to progress in mathematics without instruction in procedures, operations, and algorithms. Despite my convictions about constructivist teaching in the primary grades, the thought of tackling a middle school curriculum involving fractions, decimals, and percents with such an approach seemed overwhelming. However, the thought of playing it safe and teaching in a traditional manner contradicted everything I knew about how children learn mathematics. Furthermore, I was well aware of the alarming statistics regarding mathematics achievement in our schools and the apparent ineffectiveness of many currently used methods. Thus, I set forth teaching mathematics to a class of fifth and sixth graders using a constructivist approach in which the students were encouraged to think deeply about mathematics concepts and invent their own methods of solving problems.

The purpose of this article is to let others know that contrary to popular belief, children *can* indeed construct knowledge about sophisticated and abstract concepts in mathematics without the use of algorithms. I chose to focus on the students' division with fractions because the rule of "invert and multiply" had always puzzled me, and I was concerned about students' ability to find meaning in an area that few adults understand. Most adults "invert and multiply" without any notion of why they are doing so, and students usually cannot explain the reasoning behind this frequently used and widely accepted procedure.

The following account provides insight into how children naturally think when they are encouraged to do their own thinking. The students whose ideas are expressed in this article were fifth and sixth graders in a self-contained, mixed-age, mixed-ability classroom. The children had been exposed to a variety of teaching practices in mathematics before entering my class. Some had been taught algorithms, some had worked extensively with manipulatives, and some had had two years of a constructivist approach to mathematics. The culture of the classroom and school is one that values the process of learning, and children are accustomed to sharing their ideas openly.

Children's Thoughts on Dividing Fractions

By the time this topic was introduced, the students had constructed considerable knowledge about fractions and were quite confident and proficient with respect to equivalent fractions. The work documented here began in February, so I had been working with these students for at least five months. The children had already proved capable of inventing ways to add, subtract, and multiply fractions without direct instruction on procedures or algorithms.

Because I believe that children construct knowledge on the basis of what they aledy know, I have always taught from this perspective; when introducing a new concept, I begin with a familiar topic and move forward. Thus, the initial discussion about dividing fractions began with a general question about division. I asked the class to think about the expression.

$$4 \div 2$$

and what it meant to them. Their responses ranged from "It means if you have four things and you divide them into two groups, how many are in each group?" to the most common response, which was "It means how many times does two fit into four or how many groups of two fit into four?" This brief discussion of division informed me about how the children were thinking about this mathematical principle. It is essential to learn what your students know and how they are thinking before proceeding to new territory. Their prior knowledge about division must be used as a base or a starting point.

Next, I presented them with the problem

$$2 \div \frac{1}{2}.$$

Within seconds many children were eager to respond. When called upon, one student responded, "Four [pause] because one-half goes into two four times." Another child followed up with "I think it's four also [pause] because if you had two candy bars and you divided them into halves, you'd have four pieces." I was pleased with their thinking thus far and inspired by their willingness to attempt to solve the problems using their own devices. The students were used to relying on themselves for solutions, so the proverbial "I don't know how to do this" did not surface. The students seemed confident about their reasoning, and I was eager to move into other problems, knowing that halves tend to be easier for children than other fractions, such as thirds and fifths.

The next problem I wrote on the chalkboard was

$$1 \div \frac{1}{3}.$$

Without hesitation the children responded, almost in unison, "Three [pause] because one-third goes into one three times."

I then wrote

$$1 \div \frac{2}{3}.$$

Two responses were forthcoming. About one-third of the class said that the answer was 6, and the rest believed the answer to be 1 1/2. The reasoning of those claiming 1 1/2 was that 1 ÷ 1/3 is 3, so 1 ÷ 2/3 must be 1 1/2 because 2/3 is twice as big as 1/3 and so fits into 1 half as many times. Since they had already determined that 1 ÷ 1/3 is 3, 1 ÷ 2/3 must be half of that, or 1 1/2. This explanation seemed to convince others that 6 was not feasible. Many of those who had initially responded with 6 quickly retracted their answer, whereas others took some time to debate before noticing the flaw in their reasoning. The exchange of ideas is an important aspect of a constructivist classroom. Although constructivists believe that children learn from one another, they do not believe that children acquire mathematical knowledge *from* other people. Such knowledge has to be *constructed* by each individual from the inside. Social interaction stimulates critical thinking, but it is not the source of mathematical knowledge (Kamii 1994). In this case, the students who believed that the answer was 6 had to think about their own reasoning as well as that of their peers and determine who was correct. In deciding that 6 was incorrect, they had to *modify* their thinking. The social interaction undoubtedly stimulated these children to question their thinking; however, the actual construction of knowledge—determining the answer to be 1 1/2—was internal.

From that point the students solved 3 ÷ 1/3 with relative ease, and when asked to try

$$\frac{1}{3} \div 3,$$

they again constructed their reasoning from what they knew. One child said, "It's one-ninth because one divided by three is a third, so if you want to divide it [one-third] by three, you have to take a third of a third, which is one-ninth." Another child's explanation went like this: "I think it's one-ninth because if you had one-third of a pie left and you were sharing it with three people [two friends], each person would get one-ninth." What intrigued me about that argument was that the child took a straight computation problem and assigned meaning to it by creating a word problem.

After working through several more problems, I gave the class the following:

> I purchased 5 3/4 pounds of chocolate-covered peanuts. I want to store the candy in 1/2-pound bags so that I can freeze it and use it in smaller portions. How many 1/2-pound bags can I make?

The students were used to estimating first, so they quickly gave estimates ranging from 10 to 12. They then set out to find the exact answer. One child responded, "Eleven bags, and you would have a quarter of a pound left over, or half a bag." When I asked how she obtained that answer, she replied, "You get ten bags from the five pounds because five divided by one-half is ten, and then you get another bag from the three-fourths, which makes eleven bags, and there is one-fourth of a pound left over, which makes half of a half-pound bag." Another student solved the problem by changing the 5 3/4 to 6 pounds and then dividing that by 1/2 to get 12 bags. He then took the 1/4 he had added and divided that by 1/2 to get the 1/2, which he subtracted from 12 to arrive finally at 11 1/2 bags.

Perhaps the invention that startled me the most on this particular problem came from a child who nonchalantly raised her hand and said, "I just douobled it [five and three-fourths] and divided by one." Her peers responded, "Can you do that?" She went on to explain that it did not change the problem. She cleverly cited how the answer remains the same when an equation is doubled, as in $10 \div 5 = 2$ and $20 \div 10 = 2$. She astutely used a mathematical relationship, without direct instruction about proportions. This sort of inventive thinking and intellectual risk taking are simply not present in classrooms where teachers impose methods on children. I also found it fascinating, although not surprising, that not one child considered converting the mixed numeral 5 3/4 to an improper fraction of 23/4, which would be the first step in a traditional algorithm. Although many advocates for teaching algorithms assert that children do not invent efficient methods to solve complex computations, the evidence cited here does not support such claims. What could be more efficient than doubling 5 3/4 to make 11 1/2 and dividing by 1?

During the next week the children continued to work with division of fractions. I was amazed by the thinking that was taking place and thrilled with how much I was learning from the children. Their work had exceeded my expectations thus far, yet I was still curious about whether they could continue this upward spiral as the problems and computation became more difficult. One day I asked them to solve the following problem:

$$4\frac{2}{5} \div \frac{1}{3}$$

By now the entire class could estimate that the answer was "a little more than twelve." After estimating, they set out to calculate the exact answer. As expected, this problem took longer than previous ones, and more head scratching, frowning, and exchanges of ideas ensued than usual. After a while we gathered as a group to discuss the outcome.

Several children remarked that they had an answer that was close but were not sure if it was exact. I assured them that I was more interested in hearing their strategies. Various children volunteered answers, which I recorded on the chalkboard. (All the answers were slightly more than 13.) One child's explanation was the following: "I got thirteen and one-fifteenth [pause]. I started with four divided by one-third, and that's twelve because one-third goes into four [pause] twelve times . . . Then I changed two-fifths to six-fifteenths and one-third to five-fifteenths." I interrupted her at this point and asked her to explain why she did that. She continued, "Because it is easier for me to divide them now, and they are still the same number [pause]. Then I figured five-fifteenths goes into six-fifteenths one more time, which makes thirteen, and there is one-fifteenth left over, so it's thirteen and one-fifteenth." Several children nodded with approval; some children exclaimed, "That's what I did!" Other asked to have the thinking repeated.

After a lengthy discussion about this problem, everyone seemed convinced that the answer was indeed 13 1/15 except for the child who had "doubled and divided by one" in the previous example. She raised her hand and claimed to agree with everything "except the last part." She said that "six-fifteenths divided by five-fifteenths is one, and there is one-fifteenth left over, which still has to be divided by five-fifteenths. One-fifteenth divided by five-fifteenths is one-fifth because five-fifteenths could fit in [to one-fifteenth] one-fifth of a time, so the answer is thirteen and one-fifth." This response was not only logical and mathematically correct but also a shining example of the autonomy (Kamii 1985, 1994) that children develop when encouraged to think for themselves. Here was a child standing alone, disagreeing with an entire class of peers and perhaps even her teacher. She was not willing to accept any thinking or procedure that did not make sense. This is one of the glorious products of constructivism: intellectual autonomy.

Needless to say, the debate over the answer to this problem went on for some time, and for many it carried over into recess. During the heated debate, one child looked to me and said, "Well, which answer is right?" When the class realized that I was typically not giving answers, the debate resumed. This sort of intellectual bantering among children is a desirable and typical occurrence in a constructivist classroom. This type of social interaction or debate engages children in critical thinking. It does not contribute to the sort of confusion that many people experienced in mathematics class, which resulted merely in frustration. It is a processing of ideas that results in deeper understanding. Piaget attributed great importance to social interaction. In his theory, social interaction is absolutely essential to the construction of knowledge, and it is indispensable in childhood for the elaboration of logical thought (Kamii 1989).

We continued to work on dividing with fractions, and the problems became more involved and had more complexity with respect to computation. The children continued to thrive, and eventually almost all of them had resolved their confusion about remainders (such as the 1/15 in the previous problem). Some students continued to struggle with the more difficult problems, yet they were able to give extremely close estimates, which indicates a developing understanding and excellent number sense.

Conclusion

When I first began teaching children, I was fascinated with their ability to think and reason. And their strategies for solving problems have never ceased to amaze me. In many instances and with many different topics, such as the one described in this article, students have taught me what it means to be a teacher and what it means to "think and communicate mathematically."

I want to stress to readers that children *can* and *do* invent ways to do sophisticated mathematics; however, the culture of the classroom must be one that truly values and encourages thinking. Children must feel safe if they are to take the intellectual risks necessary to construct knowledge. They must be given ample time to think and reflect about numbers and to exchange ideas with peers, and they must be developmentally ready for the material being presented. Furthermore, a child who has been fed a strict diet of algorithms and has viewed mathematics as simply calculating using a system of memorized rules cannot suddenly begin to think deeply about numbers and to invent procedures. Such children have learned to be dependent on teachers for methods and solutions, and it is extremely difficult to change such behavior.

As teachers it is our duty to provide learning environments that allow children to be successful. This obligation means that we must look carefully at the traits we value. Are we merely interested in the correct answer, and is that all we assess? Do we as educators truly value mathematics, and if so, how do we communicate that regard to students? Have we taken the time and energy to learn the mathematics we claim to teach? Do we really value and encourage intellectual autonomy?

Finally, one should not assume that the teacher's role in a constructivist classroom is one of a passive observer who sits idly waiting for children to construct knowledge. Setting up a classroom environment in which the children invent methods to solve problems is not an easy task. The teacher must strive to understand each child's thinking and must carefully determine just when and how to guide a child to a deeper and higher level of understanding. Creating such a classroom and formulating appropriate questions to probe children's thinking and lead them to new intellectual heights is perhaps the subject of a subsequent article.

References

Burns, Marilyn. *About Teaching Mathematics: A K–8 Resource.* Sausalito, Calif: Marilyn Burns Education Associates, 1992.

———. "Arithmetic: The Last Holdout." *Phi Delta Kappan* 75 (February 1994): 471–76.

Carraher, Terezinha Nunes, David William Carraher, and Analucia Dias Schliemann. "Mathematics in the Streets and in Schools." *British Journal of Developmental Psychology* 3 (March 1985): 21–29.

Heege, Hans ter. "Testing the Maturity for Learning the Algorithm of Multiplication." *Educational Studies in Mathematics* 9 (February 1978): 75–83.

Kamii, Constance. *Young Children Reinvent Arithmetic.* New York: Teachers College Press, 1985.

———. *Young Children Continue to Reinvent Arithmetic, Second Grade.* New York: Teachers College Press, 1989.

———. *Young Children Continue to Reinvent Arithmetic, Third Grade.* New York: Teachers College Press, 1994.

Leinwand, Steven. "It's Time to Abandon Computational Algorithms." *Education Week* 9 February 1994, p. 36.

Mack, Nancy K. "Learning Fractions with Understanding: Building on Informal Knowledge." *Journal for Research in Mathematics Education* 21 (January 1990): 16–32.

Madell, Rob. "Children's Natural Processes." *Arithmetic Teacher* 32 (March 1985): 20–22.

Olivier, Alwyn, Hanlie Murray, and Piet Human. "Children's Solution Strategies for Division Problems." In *Proceedings of the Thirteenth Annual Meeting of the North American Chapter of the International Group for the Psychology of Mathematics Education,* vol. 2, edited by Robert G. Underhill, pp. 15–21. Blacksburg, Va.: Virginia Polytechnic Inst., 1991.

Streefland, Leen. "Fractions: A Realistic Approach." In *Rational Numbers,* edited by Thomas P. Carpenter, Elizabeth Fennema, and Thomas A. Romberg, pp. 289–325. Hillsdale, NJ.: Lawrence Erlbaum Associates, 1993.

Treffers, A. "Integrated Column Arithmetic According to Progressive Schematisation." *Educational Studies in Mathematics* 18 (May 1987): 125–45.

Section 7

Geometry, Geometric Reasoning, and Spatial Sense

Concept Learning in Geometry

David J. Fuys and Amy K. Liebov

How well do primary students understand such geometric concepts as angle, rectangle, and cube? How might your students show that they understand them? These are important questions, since the geometry strand for K–4 programs contains a host of concepts involving two- and three-dimensional shapes. More generally, the *Standards* states that "the K–4 curriculum should be conceptually oriented" (NCTM 1989, 17). However, NAEP testing (Kouba et al. 1988) indicates that many students demonstrate a lack of understanding of underlying concepts in mathematics. Surveys of classroom practice show "a heavy emphasis on skill development and slight attention to concepts" (Porter 1989, 11). Also, children exhibit various types of misconceptions, such as undergeneralization, which can occur because they include irrelevant characteristics; overgeneralization, which can occur because some key properties are omitted; and language-related misconceptions (e.g., that "diagonal" means slanty). Research on concept learning sheds some light on how children form concepts (and misconceptions), with implications for ways to teach geometric concepts and also concepts in other areas of mathematics (e.g., even/odd, factor) and in other subjects.

Best Examples

One way that children form a concept is to begin with a few cases and by averaging their features develop an "average representation" or prototype, which they use for categorizing new examples (Reif 1987). Children may form prototypes that include extraneous or even erroneous features that can lead to misconceptions (Clements and Battista 1992; Fuys, Geddes, and Tischler 1988; Tennyson, Youngers, and Suebsonthi 1983), such as thinking of a shape only in terms of cases in standard orientation, as usually found in the everyday world. However, beginning with "best examples," that is, "clear cases demonstrating the variation of the concept's attributes" (Tennyson, Youngers, and Suebsonthi 1983, p. 281), can help develop correct concepts (**fig. 1**). We need to provide "best examples" that are rich in imagery of familiar objects and manipulative models with which children have worked. Research (Lehrer et al. 1989) indicates that children should develop multiple representations of concepts (e.g., everyday objects, manipulatives, diagrams, verbal definitions). Some children encode information better in a verbal format than visual; others, vice versa. Also, they should link representations for a concept, such as by drawing examples and giving a list of properties.

Primary pupils are better able to learn geometry concepts when they handle models and use diagrams (Clements and Battista 1992; Mitchelmore 1982; Prigge 1978). They should experience both region/ surface models and side/edge models (**fig. 2**). Children may find some aspects of shapes easier to grasp by using one type of model than another. Here, too, they can connect representations for a concept—

There are many examples of *regular polygons.* Here are some "best examples."

Fourth graders used these best examples to learn about regular polygons (Tennyson, Youngers, and Suebsonthi 1983, L82). What other examples could we have used? We might investigate which examples work best for our students.

Fig. 1. Best examples of regular polygons

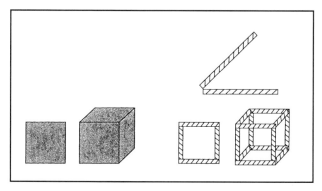

Fig. 2. Region/surface models and side/edge models

for example, interrelating angle as a "wedge," made with two fastened straws that open, a turn of their body, or a Logo turtle turn.

Concept Cards

It is important to present both examples and nonexamples of a concept (Clements and Battista 1992; NCTM 1989; Tennyson, Youngers, and Suebsonthi 1983) as in the concept card (**fig. 3**). Nonexamples have proved more effective than examples for "difficult" concepts and when familiar prototypes for a concept frequently have irrelevant features. Nonexamples should vary all irrelevant features. Carefully chosen nonexamples help children eliminate irrelevant features and identify crucial ones.

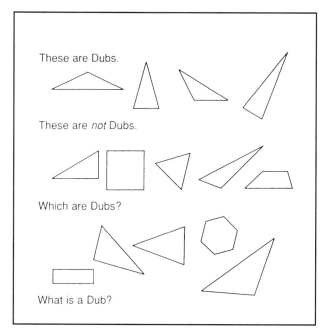

Fig. 3. Concept card with examples and nonexamples

The concept-card approach helps children learn how to formulate a correct definition. Initially children can use their own language for definitions, although it will be imprecise at times, and then through class discussion formulate a working verbalization using geometric terms. Youngsters may need to touch parts of shapes or concrete models when explaining verbally. Children sometimes memorize verbal definitions and can spout them with ease, so we are cautioned not to rely solely on verbal definitions to see if children understand a concept. Ask them to draw examples and nonexamples or to explain which cases are or are not examples and why.

Defining Shapes

Children can identify examples of a concept by thinking in terms of prototypes, which has advantages (e.g., speed in recognizing familiar examples) and disadvantages (e.g., misidentification of unusual cases). Another way is to use a rule or definition for the concept, that is, to think in terms of key properties of the concept. Rule-based concepts are powerful in that they enable you to identify and generate examples and nonexamples and also to deduce conclusions from them (Reif 1987).

One way to express a rule is by a verbal definition or list of properties called a "declarative specification" of the concept—that is, a statement of "what it is" in terms of its characterizing features (Lehrer et al. 1989; Reif 1987). Most textbook definitions are of this type. Another way is through a "procedural specification," which enables one to construct or identify examples of the concept. A child might give a procedural description for "square" by telling how to make it with manipulatives (e.g., "Take four sticks the same size, put two like this for a square corner, and then . . .") or by a Logo procedure. Children can use one type of specification to understand another—for example, a third grader who came to understand her textbook's declarative definition for square by thinking about her procedural Logo notion.

We can incorporate examples and nonexamples, best examples, and different representations of concepts into discovery and expository lessons for introducing geometric concepts.

Formulating a declarative definition from a procedural one

Confused by her textbook's definition for square (a quadrilateral with right angles and equal sides), Jenny reflects on her Logo experiences. "I need four FORWARD 100 and four RIGHT 90 . . . so, oh, . . . four sides . . . equal sides . . . right angles . . . yeah, I got it now." How might Jenny think about rectangles? Might she think of squares as being special rectangles?

Discovery Lessons

Present a concept card on the chalkboard, and guide children to formulate and test a rule for the shape. One variation is to present the examples and nonexamples one at a time (on the chalkboard or with computer guess-my-rule software) and challenge children to predict where they go and why (**fig. 4**). Another variation is to have children work individually or in small groups on concept cards in a learning center and later share their definitions.

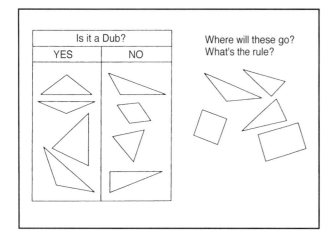

Fig. 4. Guess-my-rule sort of example and nonexamples

Expository Lessons

First present two "best examples" (leaving them out for viewing), next give the crucial features of the concept as they relate to the examples, and then ask questions structured to have children check each crucial property for several cases. This approach proved more effective for teaching third and fourth graders geometric concepts than did those not having "best examples" or not using examples and nonexamples (Tennyson, Youngers, and Suebsonthi 1983), suggesting why the expository material in many textbooks may not suffice for developing concepts—namely, examples are seldom "best examples"; nonexamples are rare; and exercises usually ask only for identification, not "why."

Conclusion

The research gives us perspectives for reflecting on the geometry-learning environment currently in our classroom and also for planning for the future. It challenges us to offer a richer, more continuous activity-oriented program in geometry. It also suggests principles and effective approaches for classroom practice. Finally, as we help youngsters become explorers of geometry, the research invites us to raise and address questions about ways to improve the teaching and learning of geometry in our classrooms.

Reflecting on the geometry-learning environment in her classroom, coauthor Amy Liebov wonders about ways to create a more global and integrated geometry program. She is concerned about using time wisely and "fitting in" all the important topics. She knows she needs to evaluate the activities her students currently use and identify additional resources for a more stimulating and enriched program. She wonders where she can find such resources and how she can implement them using research-based approaches. Finally, she feels the need to interact with colleagues about concerns, resources, and innovations involving the teaching and learning of geometry in their classes. These are concerns of all reflective teachers.

Action Research Ideas

1. Develop a "concept card" (see **fig. 3**) for each two- and three-dimensional shape in your curriculum. Be sure to include examples, nonexamples, and undesignated figures that are not in standard position. To assess your students' understanding, duplicate the cards and ask each student to respond in writing on the card. After collecting the students' work, have them explain their responses to the class. Use their written work to identify the shapes that each student recognizes and the shapes that each student can accurately describe. Develop new

cards for each shape, and engage the students in written activities and discussions of the shapes and their properties at various times throughout the year. You may want to use guess-my-rule activities as a variation. Keep a record of each student's progress.

2. Develop some best examples (see **fig. 1**) for each two- and three-dimensional shape included in your curriculum. Sequence the shapes in a way that makes sense (polygons before regular polygons, for example). For each shape, have the students identify the common features of the best examples. At a later time use concept cards or guess-my-rule activities to assess understanding of the shapes introduced through best examples.

3. Have your students create some best examples for shapes from your curriculum. Have them present their examples to the class, along with their reasons for selecting these particular examples. Members of the class should be given an opportunity to comment on each student's examples. Note what examples each student uses, and see if you can determine what they reveal about the student's understanding.

4. If you have used both concept cards and best examples, reflect on whether students learn more from one or the other or whether a combination of both approaches works best.

References

Clements, Douglas H., and Michael T. Battista. "Geometry and Spatial Reasoning." In *Handbook of Research on Mathematics Teaching and Learning,* edited by Douglas A. Grouws, pp. 420–64. New York: Macmillan Publishing Co., and Reston, Va.: National Council of Teachers of Mathematics, 1992.

Fuys, David, Dorothy Geddes, and Rosamond Tischler. "The van Hiele Model of Thinking in Geometry among Adolescents." *Journal for Research in Mathematics Education* Monograph Series No. 3. Reston, Va.: National Council of Teachers of Mathematics, 1988.

Jenson, Robert J., ed. "Geometry and Spatial Sense." In *Research Ideas for the Classroom: Early Childhood Mathematics.* Reston, Va.: National Council of Teachers of Mathematics, 1993.

Kouba, Vicky L., Catherine A. Brown, Thomas P. Carpenter, Mary Montgomery Lindquist, Edward A. Silver, and Jane O. Swafford. "Results of the Fourth NAEP Assessment of Mathematics: Measurement, Geometry, Data Interpretation, Attitudes, and Other Topics," *Arithmetic Teacher* 35 (May 1988): 10–16.

Lehrer, R., W. Knight, R. Sancilio, and M. Love. "Software to Link Action and Description in Pre-Proof Geometry." Paper presented at the meeting of the American Educational Research Association, San Francisco, California, April 1989.

Mitchelmore, M. *Final Report of the Cooperative Geometry Research Project.* Kingston, Jamaica: UWI School of Education, 1982.

National Council of Teachers of Mathematics (NCTM). *Curriculum and Evaluation Standards for School Mathematics.* Reston, Va.: NCTM, 1989.

Porter, A. "A Curriculum Out of Balance: A Case Study of Elementary School Mathematics." *Educational Researcher* 18 (1989): 9–15.

Prigge, Glenn R. "The Differential Effects of the Use of Manipulative Aids on the Learning of Geometric Concepts by Elementary School Children." *Journal for Research in Mathematics Education* 9 (November 1978): 361–67.

Reif, F. "Interpretation of Scientific and Mathematical Concepts: Cognitive Issues and Instructional Implications." *Cognitive Science* 11 (1987): 395–416.

Tennyson, R. D., J. Youngers, and P. Suebsonthi. "Concept Learning of Children Using Instructional Presenting Forms for Prototype Formation and Classification-Skill Development." *Journal of Education Psychology* 75 (1983): 280–91.

Spatial Abilities

Douglas T. Owens

Look at **figure 1** (Del Grande 1987). Which quadrilateral is different from the others? In **figure 2** (Ben-Chaim, Lappan, and Houang 1988), you are first given the back view of a building. Which of the pictures to the right of it represents the front view?

Fig. 1. Which quadrangle is different from the others?

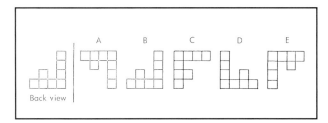

Fig. 2. You are given the back view of a building. Find the front view.

These are two examples of tasks that involve spatial perception.

Spatial perception is important "for its relationship to most technical-scientific occupations and especially to the study of mathematics, science, art, and engineering" (Ben-Chaim, Lappan, and Houang 1988) or even for such everyday activities as accurately understanding road signs and maps. The ability to perceive spatial relationships is the focus of this article.

What has been described as spatial *ability* might be better described as spatial *abilities*. These abilities appear to comprise two main factors (Bishop 1983; Halpern 1986):

1. A visualization factor, which includes the ability to imagine how pictorially presented objects will appear when they are rotated, twisted, or inverted "or how a flat object will appear if it is folded or how a solid object will appear if it is unfolded" (Halpern 1986, p. 49)

2. An orientation factor, which includes the ability to detect arrangements of elements within a pattern and the ability to maintain accurate perceptions in the face of changing orientations.

Figure 3 shows an example of a visualization task (Halpern 1986). Imagine folding a piece of paper and then punching a hole through it. The task is to identify the correct representation of how the paper would look when unfolded again. According to Halpern (1986), **figure 4** is an example of an orientation task. The task is to choose, from the pictures of three cubes on the right, the one that could represent a different view of the cube on the left. Individuals "are not necessarily good or poor at both types of tasks" (Halpern 1986, 49). That is, some people are good at visualization tasks but not orientation tasks; others are good at orientation but not visualization.

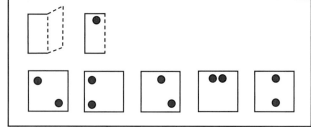

Fig. 3. Fold a piece of paper, and punch a hole through it. How would the paper look when it is unfolded?

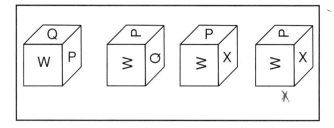

Fig. 4. Which of the three cubes on the right represents a different view of the cube on the left?

Findings from Research

Although different tests of spatial abilities will often lead to different results, one of the most consistent findings from research is that males tend to score higher than females on tests of spatial abilities, especially from adolescence onward (Halpern 1986). Ben-Chaim, Lappan, and Houang (1988) found sex differences in spatial abilities among sixth, seventh, and eighth graders, but not fifth graders, prior to a three-week unit of instruction on spatial tasks. The unit was then implemented in three schools, involving twenty-one teachers and 1000 students. The researchers arrived at the following conclusion (p. 66):

> The most important result of this investigation was that after the instruction intervention, middle school students regardless of sex gained significantly from the training program in spatial visualization tasks. The students made across-the-board gains in posttest item types.

Sex differences were present before and after instruction, but boys and girls responded similarly and positively to the instructional program. Moreover, the sex differences observed were small when compared with the magnitude of the gains in test scores. Researchers continue to struggle with the reasons for sex differences in spatial abilities. However, regardless of the differences that seem to exist among people, everyone should be given the opportunity to develop spatial abilities.

Spatial settings are obviously all around us, not just in geometry class! For example, research involving spatial relations in a geography setting has been conducted in the Soviet Union. In one study (Kabanova-Meller 1968), ten students in grade 4 were involved in a small-group "teaching experiment" in which they were taught the following rule: Stand facing down river. The right bank will be on your right, and the left bank will be on your left. The students were then given four tasks. In the first task, they were shown a map of two rivers they knew ran north and asked to locate the right and left banks. All ten children responded correctly. This task was easy because simply by looking at the map, the students were facing north (as labeled on the map), the direction both rivers flowed. The second task involved using a map of rivers that the students knew ran south. They were again asked to locate the right and left banks. This time, eight of the ten students answered correctly on their own; the other two needed the teacher's help. To solve the problem, some of the students actually rotated their bodies to help determine the relationship among the flow of the rivers, their bodies, and the location of the banks. In the third task, the students were presented a picture of a river rapidly running down a mountain slope. Only five of the ten students were able to use the rule to determine the right and left banks. One student turned his head, mentally looking down stream, then correctly said, "This will be the right bank, and this, the left." The other five students apparently had difficulty reconstructing the situation and were unable to solve the problem on their own. For example, one student quickly said, "I'm looking downstream" and incorrectly identified the right and left banks. In the fourth task, the students were shown a picture of a plains river. They were shown the right bank and asked, "Can you show which way the river is flowing?" Only two of the ten children independently answered correctly. One of these two said, "I'll stand so my right arm will be by the right bank. Then I'll stand facing this way, downstream." One other student solved the problem with the teacher's help, and the other seven students were unable to solve it. Clearly, this task was the most difficult of the four because the students had to both mentally reconstruct the situation and apply the rule in reverse.

In a different teaching experiment, other grade-4 students were taught another rule: If you face north, east will be on the right, west will be on the left, and south will be behind. The students were then assigned the following task: Facing south, show the directions. The students had significant difficulties with this task. As was true in the previous tasks, when fourth graders had to adapt the rule and mentally reconstruct the situation, the task was difficult or impossible to complete. When Kabanova-Meller (1968) gave the same two rules and two sets of tasks to ten seventh graders, they had no difficulties moving from direct relationships to reverse ones. In fact, using the second rule, the seventh graders also were able to determine the other directions when facing northwest, northeast, southwest, and southeast.

Whereas the two previous examples involved the use of spatial abilities in other school subjects, the next example is a school geometry activity. **Figure 5** was given to the fourth graders after they successfully distinguished between squares and other rectangles and identified trapezoids and general quadrilaterals (Kabanova-Meller 1968). The students were asked to pick out and draw figures from a complicated drawing of overlapping figures. The following response of one student is typical of those reported. He first drew triangle *BDT,* which stands out and has no other lines

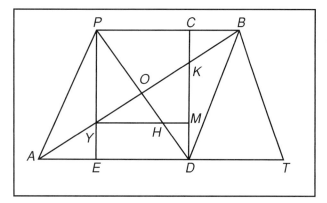

Fig. 5. Find squares, rectangles, trapezoids in the figure.

crossing through it. Then he proceeded left to right, identifying figures with a common side: triangles *APY* and *YPO*, quadrilateral *OPCK*, triangle *YOH*, and so on. Out of twelve students, only two were able to identify overlapping figures. For example, one of the two identified the quadrilateral *APCD* and then the triangle *ABD* after singling out several figures with one common side.

Suggestions for Practice

Students need many more experiences like those that have been described here if they are to develop their spatial abilities. Spatial activities can be used in teaching many subjects, ranging from geography to physical education. In fact, according to Musick (1978), motor activity and full body movement are important in developing spatial abilities.

Within the mathematics curriculum itself, teachers need to give more time to geometry, where spatial concepts and relationships can be studied explicitly. Activities suggested for levels of the van Hiele model of thinking in geometry often involve not only geometrical concepts but spatial perception as well. For example, the diagrams in **figure 6** can be used in the following activity described by Fuys, Geddes, and Tischler (1988, p. 34):

> Students look at a right triangle grid (see diagram A . . .), and are asked to identify shapes and lines in it. They are then given an acetate sheet with a ladder drawn on it (diagram B), and asked if they can find it in the grid. They are asked what it looks like, and are shown pictures of "ladder-like" objects (diagram C). Two more ladders drawn on acetate are presented (diagrams D and E), to be identified in the grid. The instructor then

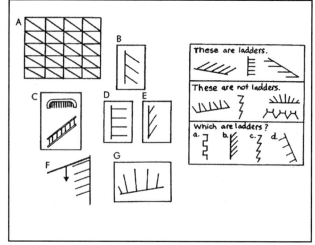

Fig. 6. What shapes can be found in grid A?

demonstrates how on a ladder, one stick can be placed on the "side," and another stick slid down it to make the "rungs" (diagram F). Students are asked to demonstrate this on another ladder. Students are shown a picture of a nonladder (diagram G): "Do you think this looks like the ladders you've seen? Why not? Actually this one is not a ladder. Do you see why?" If students have difficulty forming this concept, a "creature card" is available which shows more examples and nonexamples, and asks the student to decide on some others. Finally students are asked to describe a ladder.

Other activities that involve spatial abilities are based on suggestions by Young (1982). For example, **figure 7** involves different designs that can be made with the same tangram pieces. One student could create a design, trace its perimeter, and challenge a classmate to create the same design. **Figure 8** shows how to fold and cut paper to create a design; before unfolding, the students should draw what they predict the design will look like. Or the student could be shown an unfolded figure and asked to predict what it will look like when folded in certain ways. Using an arrangement of six squares like that shown in **figure 9**, the teacher could ask, Can this arrangement of six squares be folded into a cube? Can you create any other arrangements of six squares (hexominoes) that can be folded into a cube?

Finally, **figure 10** illustrates an activity involving building with cubes and drawing pictures of solids on dot paper (Lappan, Phillips, and Winter 1984). The student is asked to build the pictured solid, take away certain cubes, and then draw a picture of the remaining solid.

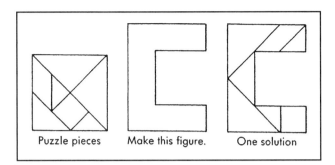

Fig. 7

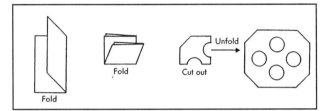

Fig. 8. Fold and cut paper to make a design.

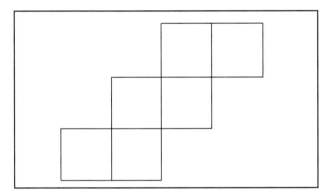

Fig. 9. Can these squares be folded to make a cube?

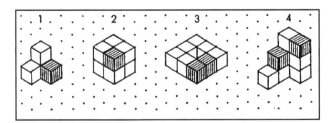

Fig. 10. Build the solid, then take away the shaded cube(s).

Conclusion

All students can and should develop spatial abilities. Spatial abilities are not just an important part of learning geometry. They are involved in other parts of the mathematics curriculum (e.g., when we use certain representations to help students learn fraction concepts), in other parts of the school curriculum beyond mathematics, and in many parts of people's lives and careers. The kinds of activities presented in this article have been shown to improve the spatial abilities of boys and girls alike.

References

Ben-Chaim, David, Glenda Lappan, and Richard T. Houang. "The Effect of Instruction on Spatial Visualization Skills of Middle School Boys and Girls." *American Educational Research Journal* 25 (spring 1988): 51–71.

Bishop, Alan J. "Space and Geometry." In *Acquisition of Mathematics Concepts and Processes,* edited by Richard Lesh and Marsha Landau, 175–203. New York: Academic Press, 1983.

Del Grande, John J. "Spatial Perception and Primary Geometry." In *Learning and Teaching Geometry, K–12,* 1987 Yearbook of the National Council of Teachers of Mathematics (NCTM), edited by Mary Montgomery Lindquist and Albert P. Shulte, pp. 126–35. Reston, Va.: NCTM, 1987.

Fuys, David, Dorothy Geddes, and Rosamond Tischler. "The van Hiele Model of Thinking in Geometry among Adolescents." *Journal for Research in Mathematics Education* Monograph Number 3. Reston, Va.: National Council of Teachers of Mathematics, 1988.

Halpern, Diane F. *Sex Differences in Cognitive Abilities.* Hillsdale, N.J.: Lawrence Erlbaum Associates, 1986.

Kabanova-Meller, E. N. *Students' Mental Development and the Formation of Techniques of Mental Activity.* Moscow: Prosveshchenie Publishers, 1968.

Lappan, Glenda, Elizabeth A. Phillips, and Mary Jean Winter. "Spatial Visualization." *Mathematics Teacher* 77 (November 1984): 618–23.

Musick, Judith Smith. "The Role of Motor Activity in Young Children's Understanding of Spatial Concepts." In *Recent Research Concerning the Development of Spatial and Geometric Concepts,* edited by Richard Lesh, pp. 85–104. Columbus, Ohio: ERIC, 1978.

Young, Jerry L. "Improving Spatial Abilities with Geometric Activities." *Arithmetic Teacher* 30 (September 1982): 38–43.

Enhancing Mathematics Learning through Imagery

Grayson H. Wheatley

Mathematics is often seen as a subject in which rules are followed and symbols manipulated to achieve "correct" answers. Fortunately, this characterization of mathematics is changing, in large part because of NCTM's initiatives. As we move into the decade of reform in school mathematics, we should explore all options for enhancing mathematics learning. This article considers the role of visual imagery in doing mathematics.

Although a few teachers may supplement mathematics instruction with spatial activities, most students rarely have opportunities to use imagery in school mathematics classes. In fact, school curricula militate against such use. As Sommer (1978, p. 54) states, "School more than any other institution is responsible for downgrading visual thinking. Most educators are not only disinterested in visualization, they are positively hostile to it." The absence of spatial activities in textbooks is clear evidence that imagery is thought to be nonessential in school mathematics. But is it?

As described in Wheatley (1990), imagery involves (1) constructing an image from pictures, words, or thoughts: (2) re-presenting the image as needed: and (3) transforming that image. A student might decide that the two shapes shown in **figure 1** are "the same" by mentally rotating her or his image of the one on the left until it is in the same orientation as the one on the right. The need to make such comparisons occurs frequently in mathematics and is facilitated by the use of visual imagery.

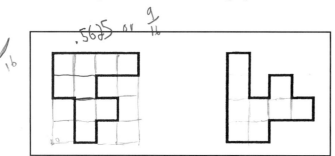

Fig. 1. Comparison of two figures using imagery

Imagery is not just a process of taking a mental picture and retrieving the picture from memory (Wheatley and Cobb 1990). As Lakoff (1987, p. 129) states. "Different people, looking upon a situation, will notice different things. Our experience of seeing may depend very much on what we know about what we are looking at. And what we see is not necessarily what is there." A nine-year-old girl was asked to draw the right triangle shown in **figure 2a**. Each time she tried to draw it, the segment on the right slanted out as shown in **figure 2b**. She knew her drawing did not look right, but she was so influenced by her concept and the images she had constructed of a triangle that she could not bring herself to draw one side vertical: for her, the sides of triangles "slant." Understanding geometric concepts and relationships is, in large part, a matter of constructing images.

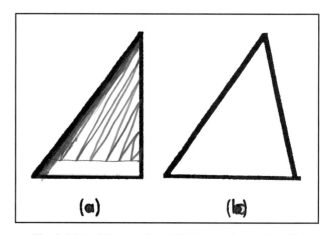

Fig. 2. (a) A right triangle and (b) Tammy's drawing of it

In a recent study (Brown and Wheatley 1989), fifth-grade students of high and low spatial ability were interviewed about various mathematics tasks. The goal was to assess students' meaningful mathematical knowledge rather than their facility for manipulating symbols. Six students were interviewed, three with high spatial ability and three with low spatial ability. The results could not have been more dramatic. The performance of the students with low spatial ability was average or above on standardized mathematics tests and in fifth-grade mathematics class. The results of the clinical interviews, however, showed that these students attached little meaning to mathematical ideas and could not solve nonroutine problems. For example, when asked to arrange thirty-six Unifix cubes in the shape of a rectangular region, one girl struggled for some time and finally made a five-by-seven region with one cube left over. She was unable to make a rectangular region using all thirty-six cubes even though she was attempting to do so. Contrastingly, the students with high spatial ability whose performance was average or below on standardized mathematics tests and in school mathematics class had an excellent grasp of mathematical ideas and were able to solve nonroutine problems, often creatively. These students with high spatial ability were not being recognized for the gifts they possessed. We interpret these findings to indicate that spatial ability lies at the heart of meaningfulness. Students can memorize facts and become proficient in demonstrated procedures, but they may not be giving meaning to their mathematical activity.

The use of imagery on a mathematics task is a function of the instructional setting. If mathematics tasks are presented in a familiar setting, students have a greater opportunity to use their prior experience in giving meaning to the tasks. For example, rather than just pose the problem 36 – 29 in abstract form, the teacher could present the task in a potentially meaningful setting, for example,

> 36 chairs are needed for a school party, 29 chairs are already in the room. How many more are needed?

In this way students can construct images for the numbers and use this imagery in developing a procedure for subtracting. Some might imagine moving one chair in with the twenty-nine to make five rows of six (30 chairs), then six more chairs to make seven chairs moved in. Posing tasks in a setting meaningful to students facilitates the use of imagery and the construction of mathematical relationships.

The use of imagery is also influenced by the student's intentions. Wheatley and Lo (1989) reported on a third grader who could determine the number of dots in a four-by-five rectangular pattern shown briefly by using subpatterns but who had great difficulty solving 20 – 5 because she tried to count back from twenty rather than use the imagery of 2 tens and break up 1 ten into five and five. She was intent on using a particular procedure—counting back from twenty to fifteen—instead of imaging a set of dots to be transformed. School experiences can contribute in significant ways to the powerful use of imagery in mathematical settings.

Learning new ideas and solving nonroutine problems are situations in which imagery is particularly useful. Since dealing successfully with novelty is one of the most important competences in a rapidly changing and complex technological society, we need to cultivate the development of essential tools that allow us to deal effectively with new situations. Imagery is one of those essentials.

Just as images of landscapes, houses, and people are formed when we listen to a conversation or read a novel, images are formed when we give meaning to our mathematical activity. Consider the following problem:

> Paul has a swimming pool in the shape of a rectangle. It is 31 feet long and 23 feet wide. A 3-foot-wide walkway surrounds the pool. What is the length of a fence around the walkway?

An experienced problem solver might form an image of the swimming pool with the walkway and fence around it and make a sketch based on the image. But note that a sketch can be created only if a mental image has been constructed first. In fact, a sketch can prove useful in thinking through the problem. Perhaps the imaging occurs in steps. On first reading the problem, a student might image and then sketch the swimming pool; he or she might elaborate the constructions after rereading the problem, with first the walkway and then the fence added to the image and the sketch. Finally, the numerical information might be added to the sketch to help solve the problem.

Wheatley and Wheatley (1982) found that few sixth-grade pupils drew diagrams in solving the foregoing problem. It should also be noted that few students successfully solved the problem. Yet most of these students were doubtless capable of forming the necessary images. When students are encouraged and given opportunities to form mental images, most readily do so, and when they are encouraged to use imagery, their mathematical power is greatly increased.

Some students are particularly good at forming images and are able to use them effectively in solving problems. The following transcript of an interview with Mike, a beginning first grader, illustrates the use of imagery in mathematical reasoning. In flashcard fashion, he was shown a series of cards with various arrangements of dots; after each card was briefly displayed, Mike was asked how many dots he saw. Here is how he responded to three of the dot patterns. Each card was shown for about one second.

I: (After showing **figure 3**) How many dots did you see?

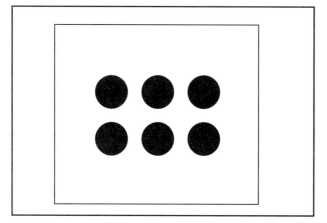

Fig. 3. Mike was shown this six-dot pattern and then asked, "How many dots did you see?"

Mike: Six

I: How did you know that one?

Mike: (Putting his hands up vertically, in front of his chest, as if to box in the two rows of three) They were all put together.

I: (After showing **figure 4**) How many dots did you see?

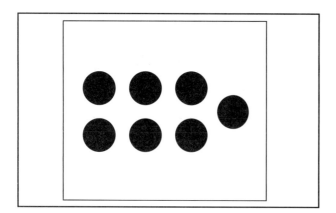

Fig. 4. Mike was shown this seven-dot pattern and then asked, "How many dots did you see?"

Mike: S-seven

I: (After showing **figure 5**) How many dots did you see?

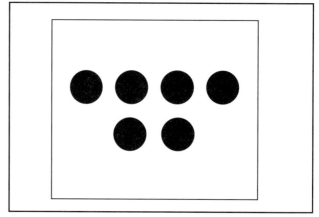

Fig. 5. A six-dot pattern that Mike mentally transformed into the pattern in **figure 3**

Mike: Six

I: How did you do that one?

Mike: (Holding his hands up in the same gesture as before) They weren't put together right, but I figured it out.

I: You tell me what you mean by "They weren't put together."

Mike: I mean by . . . those were right (pointing to the three dots at the upper left). Those . . . and it came down like that (indicating that he had made a linear pattern of the remaining three dots).

It was clear that Mike had transformed his image of the four-over-two pattern to a three-over-three pattern so as to create an image for which he knew the number. Under normal school conditions, Mike would likely be considered slow (he had to figure out the problem 3 + 1 by counting from one) when in fact he had exceptional visual reasoning and great potential as a mathematics student.

As the foregoing examples illustrate, imagery plays a prominent role in many if not all aspects of mathematics. Imagery is useful not only in coming to know a parallelogram and its properties but also in doing numerical tasks and especially in problem solving. Listed here are suggestions for incorporating imagery in mathematics instruction.

Instructional implications

Since dynamic imagery can be particularly useful as students give meaning to geometric and numerical relationships, it is important to give them opportunities to develop and use their imagery.

1. Encourage students to build mental pictures—to "imagine what it would look like." Design tasks that go beyond manipulating objects to transforming constructed images as Mike's task did.

2. Encourage students to communicate their images by building models, drawing pictures and graphs, and using verbal descriptions.

3. Design activities that promote the use of visual imagery. The *Arithmetic Teacher* offers many suggestions for building imagery into the mathematics curriculum (see especially Yackel and Wheatley [1990]).

4. Show students that you value imagery by allocating time for spatial activities, accepting imagery-based explanations, and describing your own imagery.

References

Brown, Dawn, and Grayson Wheatley. "Spatial Visualization and Arithmetic Knowledge." In *Proceedings of the Eleventh Annual Meeting of the North American Chapter, International Group for the Psychology of Mathematics Education,* edited by Carolyn A. Maher, Gerald A. Goldin, and Robert B. Davis. New Brunswick, N.J., September 1989.

Lakoff, George. *Women, Fire and Dangerous Things: What Categories Reveal about the Mind.* Chicago, Ill.: University of Chicago Press, 1987.

Sommer, Robert. *The Mind's Eye: Imagery in Everyday Life.* Palo Alto, Calif.: Dale Seymour Publications, 1978.

Wheatley, Grayson. "Constructivist Perspectives on Mathematics and Science Learning." *Science Education* 75 (January 1991): 9–21.

———. "One Point of View: Spatial Sense and Mathematics Learning." *Arithmetic Teacher* 37 (February 1990): 10–11.

Wheatley, Grayson, and Paul Cobb. "Analysis of Young Children's Spatial Constructions." In *International Perspectives on Transforming Early Childhood Mathematics Education,* edited by Leslie Steffe. Hillsdale, N.J.: Lawrence Erlbaum Associates, 1990.

Wheatley, Grayson, and Jane-Jane Lo. "The Role of Spatial Patterns in the Construction of Number Units." In *Proceedings of the Eleventh Annual Meeting of the North American Chapter, International Group for the Psychology of Mathematics Education,* edited by Carolyn A. Maher, Gerald A. Goldin, and Robert B. Davis. New Brunswick, N.J., September 1989.

Wheatley, Grayson, and Charlotte Wheatley. *Calculator Use and Problem Solving Strategies of Grade Six Pupils: Final Report.* Washington, D.C.: National Science Foundation, 1982. ERIC document no. ED 175720.

Yackel, Erna, and Grayson Wheatley. "Promoting Visual Imagery in Young Pupils." *Arithmetic Teacher* 37 (February 1990): 52–58.

Conquer Mathematics Concepts by Developing Visual Thinking

Rina Hershkowitz and Zvia Markovits

It seems that visual thinking will be the primary way of thinking in the future. In modern life the presentation of phenomena has changed from tables and formulas heavy with numbers and symbols to a dynamic visual presentation using computers. To understand, analyze, and predict, we will have to engage in some form of visual thinking. To communicate visually and to develop advanced visual thinking, we need certain linguistic elements—a language.

The Agam program is an example of an effort to interweave the development of a visual language with a process of developing visual thinking. This project is the vision of the artist Yaacov Agam that has become an educational reality through the work of a team of researchers and educators of the Science Teaching Department in the Weizmann Institute. The program was developed, tested, and implemented with several groups of students beginning in nursery school with three-to-four-year-olds and continuing with the same groups to the third grade. The development and implementation was accompanied by research and evaluation that showed that the "Agam children" can apply visual abilities and visual thinking in learning tasks more successfully than children in control groups (Razel and Eylon 1990).

Some of the program's thirty-six curriculum units introduce students to such basic visual concepts as the main geometric figures, directions, colors, and size relationships. These units make up a "visual alphabet" that forms the basis for more advanced concepts, such as symmetry, ratio and proportion, numerical intuition, and other concepts that serve as building blocks in scientific and mathematical thinking.

The following example demonstrates how thinking is developed in the Agam program. The first unit is on the circle, and the second, on the square. The three-to-four-year-old student becomes acquainted with these concepts by methods relying almost exclusively on nonverbal experiences, for example, an activity in which each student takes one end of a rope (the ropes being first of equal lengths [see **fig.1a**] and subsequently of different lengths [see **fig. 1b**]) whose other

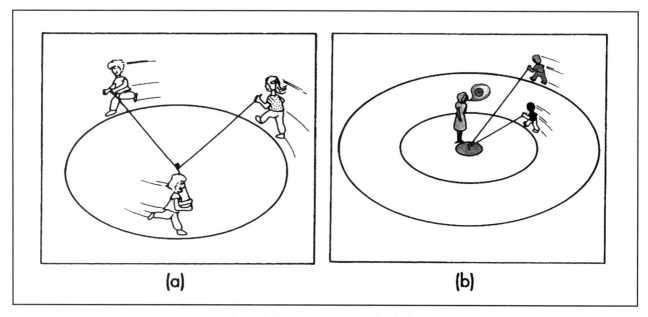

Fig. 1. Students use ropes to make circles.

end is fixed to some point (the center) and by walking to create circles, experiences visually and physically the properties of a circle.

From circles and squares as "visual letters," patterns, which are "visual words" or "visual sentences," can be created (patterns are treated in the third unit in the program). Pattern, in the sense used here, is a visual periodic series whose elements at this stage in the curriculum are circles and squares of different sizes, colors, and position with changing intervals between them (see **fig. 2**). Periodic series are the bases of many mathematical and scientific concepts, such as certain functions, waves, and movements. This unit tackles the concept visually in a way that is meaningful to the young student.

When students identify patterns furnished by the teacher, books, or the classroom environment or when they memorize—store various patterns and recall them—they internalize the concept of pattern and realize that it is the same irrespective of the changes in the periodic themes that create different patterns.

When students create patterns, they are problem solving with high-level thinking. They have to analyze the main characteristics of patterns—the building blocks that are to be used in their creation—and choose those that they would like to have in their own special pattern. Finally they have to synthesize all the aforementioned in the reproduction of their pattern. In the patterns unit all the activities deal with linear (one-dimensional) patterns, but children's creativity

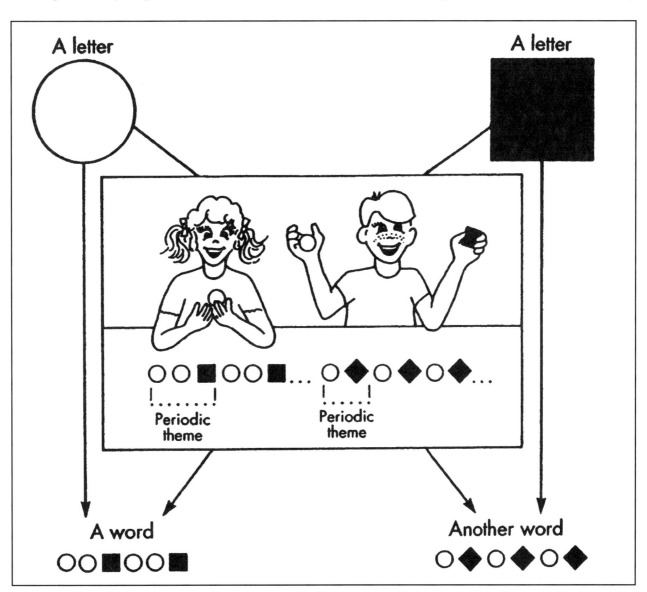

Fig. 2. Patterns of squares and circles make words.

is unbounded. See, for instance, the "sun" in **figure 3a** in which the periodic series has a "ray" as the periodic theme and each "ray" itself is a linear pattern. The matrix (see **fig. 3b**) is also a multidimensional pattern: each row, each column, and the two main diagonals exhibit a linear pattern, and the whole matrix is a pattern with the column as the periodic theme.

In the following example we describe two units—"Ratio and Proportion" and "Numerical Intuition"—in which third graders, who have completed many units of the program, come closer to these higher-level mathematical concepts. These concepts, which are usually taught in higher grades, are the basis of many other concepts in mathematics, science, and other areas of life. Considerable evidence exists that these concepts are difficult for even adolescents to grasp (Tourniaire and Pulos 1985). Our experience indicates that children can overcome many difficulties if these concepts are experienced visually.

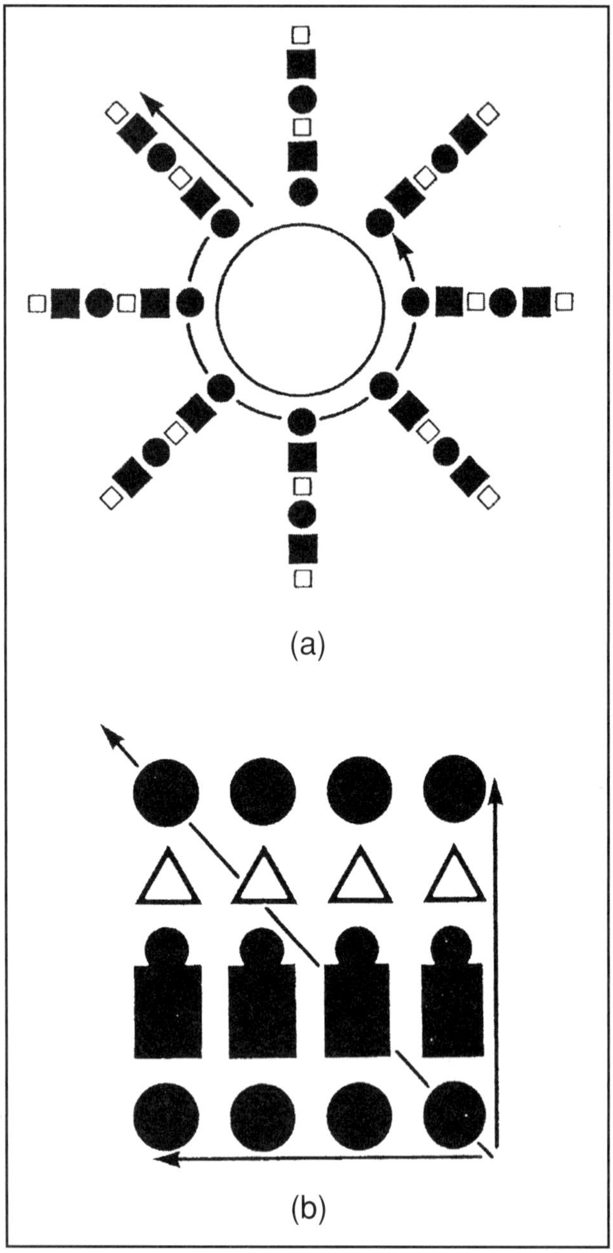

Fig. 3. Multidimensional periodic patterns created by four-year-old students

Ratio and Proportion

The following briefly describes the sequence of activities that are the basis of the visual presentation.

1. Sticks of various lengths are used, with a short one serving as a unit ("unit stick"). Students count how many times the unit can be placed beside two different sticks so as to equal their lengths and draw conclusions about the ratio between the lengths of these sticks.

2. Repetition of the same activity with sticks of different lengths leads the students to discover a ratio between certain sticks regardless of their particular lengths.

3. Students are asked to find different pairs of sticks related in the same way. In this way they discover the concept of proportional relationship.

4. Students do similar activities with towers of blocks and colored liquid in glasses. Towers of various heights demonstrate that two ratios can be the same even if they are expressed by different numbers, for instance, 1/3 = 2/6 (see **fig. 4a**). This comparison establishes a visual basis for equality of fractions.

5. Through similar activities with angles of different sizes (the unit on angle comes earlier in the program), students reinforce their intuition that the same ratios and proportions exist for different measures.

6. By measuring with sticks or paper strips, students find that the ratio of the lengths of the sides of similar shapes is constant.

7. Students discover that the ratio between the lengths of two vertical sticks is equal to the ratio

Fig. 4. Students build pairs of block towers that are related in the same way.

between their shadows, thus extending their concept of proportion (see **fig. 5**).

Fig. 5. Students discover the proportion between the lengths of sticks and the lengths of their shadows.

8. In the more advanced activities three values in a proportion are given and the student is asked to find the fourth. For example, a pair of towers and a single tower are given; the student estimates the height of the fourth tower that completes the proportion and then checks the estimate by building the tower (see **fig. 4b**). In a similar activity the student estimates the height of a vertical stick given the height of a second stick and the length of the two shadows.

Classroom observations. As is to be expected, the level of learning varied among students and according to the type of task. The following are some examples of our observations of the students' learning.

- In activities 1, 2, and 3, the students leaned heavily on the unit sticks; for example, after they measured the lengths of two sticks to find the ratio, they would not remove the "unit sticks" even though they had already found the ratio. After time and with accumulated experiences, most of them no longer needed the manipulative.
- In type-3 activities many students were satisfied with finding a single pair of sticks with a given ratio. With time, most of them succeeded in finding as many different pairs as could be obtained from the sticks at their disposal.
- Some students used trial-and-error strategies in finding a pair of sticks in a given ratio, whereas others planned their moves very carefully. For example, when finding a pair in the ratio 2 to 3, they first took five unit sticks, put two in a line and three in another line, and only then found suitable sticks.
- When presented with three sticks and asked to produce a drawing of sticks in the same ratios, some students explained their drawing qualitatively as big, medium, and small, whereas others succeeded in seeing and even expressing the ratios quantitatively.
- By the end of the unit, all the students were able to create different pairs of sticks, towers of blocks, and so on, for a given ratio (proportion), but few were able to deal successfully with activities in which

three values in a proportion were given and the fourth one had to be found.

Numerical Intuition

Numerical intuition is evidenced when a group of objects is shown briefly to students and they must determine the number of objects without counting. This activity occurs frequently in everyday life, but few of us bother to ask such questions as "How close was I to the exact number of objects?" "How did I get my number?" and "Do I use the same strategy each time?"

The following is a brief description of the activities in the numerical-intuition unit.

1. Students are presented with dot patterns (see **fig. 6**) for a very short time and then asked to write down how many dots they saw. As feedback the teacher allows them to count the exact number of dots and the students write the result next to their estimate.

2. Students play a game that requires the winner not to reach an exact number but to come as close as possible to it. They experience being more or less close to the exact number.

3. Students perform number-estimation activities similar to the first one with such concrete objects as circles, squares, or pencils spread randomly on the desk; with pictures in children's books; and with objects from the classroom.

4. Students experience numerical-intuition activities outside the classroom.

Classroom observations. Students were interviewed before and after the unit. In the interview they were asked first to estimate the number of objects in a certain collection and then to explain how they obtained their estimate. The following strategies were identified:

1. Counting. In the short time allowed, some students counted as many objects as they could. This strategy is, of course, not very efficient.

2. Estimation. We found a few different estimation strategies:

   - The student mentally divides the objects into groups with an equal number of objects in each. For example, in response to being shown **figure 6c**, Iair said, "Thirty—the points are spread, about four in each place. I circled groups of four in my head."

   - Any other kind of estimation with a reasonable explanation, for example, Avner responded to **figure 6c**, "It's complicated, . . . twenty-five . . . , it's a bit more than the picture on the left.

   - Some students used a kind of global estimation that does not use any algorithm or at least does not express the existence of such an algorithm, as demonstrated, for example, in responses like, "This is what it seems to me."

As a result of students' working through the unit, counting strategies dropped dramatically, whereas the use of all the aforementioned three estimation strategies increased, with the most dramatic increase in global estimation. We can conclude that the visual activities described developed the students' numerical intuition.

The activities in this unit lead directly to the question "How close am I to the exact number?" which is the basis of the concept of absolute and relative error.

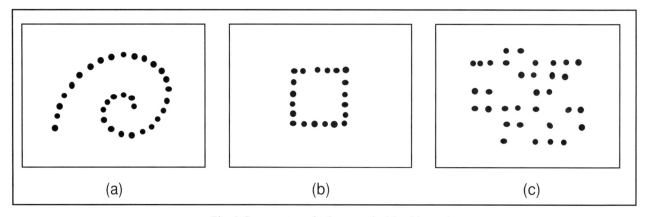

Fig. 6. Some patterns in the numerical-intuition unit

As in some ratio-and-proportion activities, students gave "additive" answers to the question, for example, that 11 dots is a better answer for a picture of 10 dots than 102 dots is for a picture of 100 dots. With a very short additional lesson, the students moved toward "relative" responses.

References

Razel, Micha, and Bat-sheva Eylon. "Development of Visual Cognition: Transfer Effects of the Agam Program." *Journal of Applied Developmental Psychology* 11 (1990): 459–84.

Tourniaire, Francoise, and Steven Pulos. "Proportional Reasoning: A Review of the Literature." *Educational Studies in Mathematics* 16 (May 1985): 181–204.

Using Spatial Imagery in Geometric Reasoning

Michael T. Battista and Douglas H. Clements

Two second-grade students are investigating geometry with Logo. They are working in a TEACH environment in which the turtle-movement commands they enter are not only executed but simultaneously recorded in a procedure (Battista and Clements 1991; Clements 1983–84). During earlier lessons, they had successfully maneuvered the turtle to draw squares of various sizes. For each of these squares, they had used 90-degree turns. When asked how they knew whether a figure was a square, however, neither student had mentioned 90-degree turns.

Classroom Episodes

These two students are then individually asked to use the turtle to draw a tilted square (they are shown a picture of a square whose sides are not vertical or horizontal). Each begins by turning the turtle and making it go forward. Megan uses a trial-and-error approach to find the turns, turning the turtle by increments until the heading looks right. She then makes the turtle go forward and examines the results, erasing and reissuing the turn command if necessary. As she completes her figure, she announces that it is a square, even though her turns were not of 90 degrees and though she has used forward inputs of 30, 5, and 5 for the last side instead of 30 as for the first three sides. (She had said earlier that squares have all their sides equal.)

Teacher: How do you know it's a square for sure?

Megan: It's in a tilt. But it's a square because if you turned it this way, it would be a square.

Megan's reasoning is clearly at the visual level in the van Hiele hierarchy (Battista and Clements 1988; Clements and Battista 1992). She thinks of shapes as visual wholes rather than as geometric objects possessing properties. According to van Hiele, for students at this level, "There is no why, one just sees it" (van Hiele 1986, p.83). As her remarks indicate, however, for Megan there *is* a "why." This student has reasoned that the figure in front of her is a square by using spatial imagery. That is, she has observed that squares that are tilted often do not look quite like squares. So perhaps because her intention had been to draw a square, she visualizes that this figure can be turned to look like a square in standard orientation. She does so even though the figure drawn does not have all the properties that she previously named for squares. The students' next task is to use the turtle to draw a tilted rectangle. Coletta, who has discovered that she needs 90-degree turns to draw a square, uses 90s on her first attempt at making the tilted rectangle, reasoning as follows:

Coletta: Because a rectangle is just like a square but just longer, and all the sides are straight. Well, not straight, but not tilted like that (makes an acute angle with her hands). They're all like that (shows a right angle with her hands), and so are the squares.

Teacher: And that's 90 (showing hands put together at a 90° angle)?

Coletta: Yes.

In other discussions, she has also stated that a square is a rectangle.

Teacher: Does that make sense to you?

Coletta: It wouldn't to my [4-year-old] sister, but it sort of does to me.

Teacher: How would you explain it to her?

Coletta: We have these stretchy square bathroom things. And I'd tell her to stretch it out and it would be a rectangle.

The research on which this paper was based was supported by the National Science Foundation under grant no. MDR-8651668. Any opinions, findings, conclusions, or recommendations expressed are those of the authors and do not necessarily reflect the views of the National Science Foundation.

Here again, we see how a student uses visual imagery to reason in a geometric context. First, she reasons that rectangles have 90-degree turns because of their visual similarity to squares. Second, it "sort of made sense" that a square is a rectangle because a square could be stretched into a rectangle. This response may be more sophisticated than one might initially think, for Coletta has just demonstrated her knowledge that squares and rectangles are similar in having angles made by 90-degree turns. Thus, she may have understood at an intuitive level that all rectangles could be generated from one another by certain *legal* transformations, ones that preserve 90-degree angles. That is, her conceptual knowledge may have influenced her imagery. Contrast Coletta's use of imagery with that of Megan, whose visual transformation contrasted with her verbal statement. We have additional evidence that Coletta's thinking about shapes was based more on properties than Megan's. So Coletta's imagery was more likely to be influenced by property-based considerations.

Another example of visual reasoning occurs as a fifth grader works with *Logo Geometry* (Battista and Clements 1991). Ryan is deciding whether he can draw various figures with a rectangle procedure that takes the lengths of the sides as input. He has successfully used the procedure to draw a tilted rectangle (labeled 4 in **fig. 1**) and is now reflecting on his unsuccessful attempt to make a nonrectangular parallelogram (labeled 7).

Teacher: Could you use different inputs, or is it just impossible?

Ryan: Maybe if you used different inputs. (Ryan types in a new initial turn. He stares at the picture of the parallelogram on the activity sheet.) No, you can't. Because the lines are slanted, instead of a rectangle going like that. (He traces a rectangle over the parallelogram.)

Teacher: Yes, but this one's slanted (indicating the tilted rectangle, labeled 4, that Ryan had successfully drawn with the Logo procedure).

Ryan: Yeah, but the lines are slanted (meaning nonperpendicular). This one's still in the size (shape) of a rectangle. This one (parallelogram)—the thing's slanted. This thing (rectangle) isn't slanted. It looks slanted, but if you put it back (shows a turn by gesturing, meaning to turn it so that the sides are vertical and horizontal), it wouldn't be slanted. Any way you move this (the parallelogram), it wouldn't be a rectangle (shaking his head). So there's no way.

Ryan has used three of the four classes of image processes defined by Kosslyn—generating an image, inspecting an image, and transforming an image (1983). First, after making the initial turn and trying to choose the inputs, he recognizes that the relationship between adjacent sides is not consistent with the implicit definition of a rectangle in the Logo procedure. He then generates an image of a rectangle that is consistent with this definition, tracing it on the activity sheet. Second, he inspects this image and compares it with the parallelogram, noting the differences.

Third, to elaborate the difference in the "slantiness" of the figures, Ryan mentally transforms his images of the rectangle and the parallelogram. His assertion that only the rectangle could be rotated to look like a rectangle is the capstone of his argument. His emerging knowledge of the properties of figures is supported by his visual reasoning.

In sum, we see a reciprocal effect. Work with Logo involving thinking about properties of figures has refined Ryan's visual reasoning. In turn, his visual reasoning has supported his analysis of the properties of figures.

Finally, an episode with a kindergarten student illustrates another natural role that imagery plays in geometric problem solving. John is asked to use Logo to draw an open path that has three bends in it. He draws a "box" in which the beginning and ending points do not touch. He explains that he "just thought of all the paths we made. One of the paths had the right number." John visually reviews solutions to similar problems, analyzing each to see if it meets the constraints for the new problem.

These episodes are consistent with the hypothesis (Battista 1990) that spatial visualization may be a more important factor in geometric problem solving for students who are at the visual, rather than higher, levels of geometric thinking. That is not to say, however, that visual imagery is unimportant at higher levels of thought. Many students that we have classified at the descriptive-analytic level (the level immediately following the visual level) are inhibited in their reasoning and problem solving because their spatial imagery is not coordinated with their knowledge of properties. For instance, many of these students know the properties of a rectangle but have difficulty recognizing whether a figure in a nonstandard spatial orientation (such as shapes 4 and 8 in **fig. 1**) possesses these properties. Moreover, because students at all levels often reason by considering specific images, their reasoning can go astray if they make a mistake in generating, inspecting, or transforming those images. In sum, visual imagery—when properly developed—can make a substantial contribution at all levels of geometric thinking. Thus, in teaching geometry, we should not focus solely on properties of figures and relationships among them. We should also help students develop vivid images and coordinate these images with their conceptual knowledge.

Instructional Implications

The foregoing discussion suggests two recommendations for teaching:

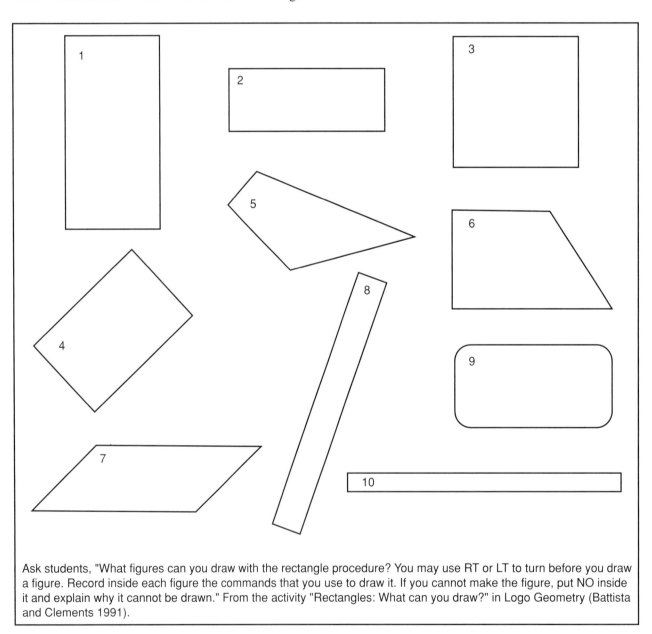

Ask students, "What figures can you draw with the rectangle procedure? You may use RT or LT to turn before you draw a figure. Record inside each figure the commands that you use to draw it. If you cannot make the figure, put NO inside it and explain why it cannot be drawn." From the activity "Rectangles: What can you draw?" in Logo Geometry (Battista and Clements 1991).

Fig. 1. Which figures can you draw using a rectangle procedure with two inputs?

1. Recognize that a great many students use visual imagery to reason. Respect this mode of thinking. Help students use it to analyze and convince themselves of the truth of geometric ideas.

2. Discuss visual reasoning with students. Ask questions that might help students incorporate conceptual knowledge into their visual-reasoning processes. For instance, after Coletta gave her bathroom-tile justification that squares are rectangles, the teacher might have drawn a trapezoid and asked, "I think I can stretch this figure into a rectangle. Is this geometric figure a rectangle?" This question would have prompted Coletta to clarify the nature of her "stretch" transformation. Also, when discussing a concept such as that of a triangle, make sure to use a variety of different examples, such as many different sizes and shapes of triangles in nonstandard orientations. This approach will help students evoke dynamic images when reasoning about concepts. When students have become familiar with mathematical transformations, they can be encouraged to describe their mental transformations using mathematical terms.

Many of the tasks that we have presented to students in our research and curriculum development projects can promote students' coordination of visual imagery and conceptual knowledge. Some examples follow. (In our project, students in grades K–1 used a single-key version of Logo, students in grades 2–4 used TEACH, and students in grades 5–6 used an enhanced version of regular Logo [Battista and Clements 1991].)

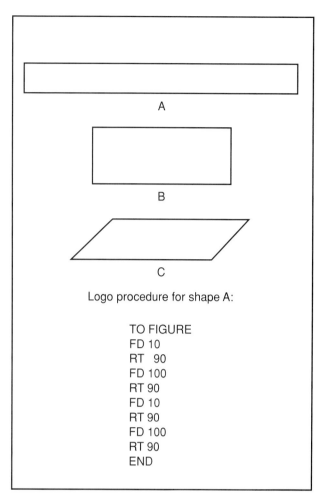

Fig. 2. Change the Logo procedure to make figures B and C.

Activity 1. Ask students to draw a square in Logo. Then ask them to draw a tilted square. Do the same for rectangles. (Grades 1–4)

Activity 2. Give students a procedure RECT that draws a rectangle when the two dimensions for the rectangle are given as input. Ask them which figures (as in **fig. 1**) they can draw with RECT. (Grades 2–6)

Activity 3. Show students the three figures in **figure 2**, two at a time. Ask how figure A is different from figure B. (The majority of students will give a global response, such as, "It's bigger or longer," not referring specifically to side lengths.) Give the students a Logo procedure that draws figure A, and let them test it on a computer. Then ask them how they can alter their procedure to draw figure B. Repeat with figures B and C. (Grades 2–6)

Note: If you do not have access to a computer, you can still use Logo-like commands. Have students work in pairs, one reading the commands and one acting as the turtle, following the commands with a metric ruler (FORWARD 50 means forward 50 mm) and a protractor. Students take turns acting as the turtle on paper.

References

Battista, Michael T. "Spatial Visualization and Gender Differences in High School Geometry." *Journal for Research in Mathematics Education* 21 (January 1990): 47–60.

Battista, Michael T., and Douglas H. Clements. "A Case for a Logo-based Elementary School Geometry Curriculum." *Arithmetic Teacher* 36 (November 1988): 11–17.

——— . *Logo Geometry.* Morristown, N.J.: Silver Burdett & Ginn, 1991.

Clements, Douglas H. "Supporting Young Children's Logo Programming." *Computing Teacher* 11 (December 1983–January 1984): 24–30.

Clements, Douglas H., and Michael T. Battista. "Geometry and Spatial Reasoning." In *Handbook of Research on Mathematics Teaching,* edited by D. A. Grouws. Reston, Va.: National Council of Teachers of Mathematics and Macmillan, 1992.

Kosslyn, Stephen M. *Ghosts in the Mind's Machine.* New York: W. W. Norton & Co., 1983.

van Hiele, Pierre M. *Structure and Insight.* Orlando, Fla.: Academic Press, 1986.

Estimation and Mental-Imagery Models in Geometry

John Happs and Helen Mansfield

Analogies are frequently used as mental models to introduce new concepts in a number of subject areas (Gentner and Gentner 1983). For instance, an electrical circuit can be likened to a water pipe so that students can visualize electricity as "flowing" through the wire in the electrical circuit.

Students' Construction of Mental Imagery

An examples such as the foregoing, the analogies are deliberately supplied to students so that they can think and reason in a more concrete way about concepts that may be new or inherently abstract. However, in situations where analogies are not given, the students' construction of mental images of their own may facilitate learning. In constructing these mental images. Students may draw on their previous experiences involving concrete tasks. These images may also need to be constructed spontaneously as the need arises in nonroutine tasks.

Estimation is an important skill in measurement. Standard 13 of *Curriculum and Evaluation Standards for School Mathematics* (NCTM 1989, p. 116) states that "in grades 5–8, the mathematics curriculum should include extensive concrete experiences using measurement so that students can estimate, make, and use measurements to describe and compare phenomena." Reasons for emphasizing estimation in measurement include the following: to help students develop a *mental frame of reference* [emphasis added] for the sizes of units of measure relative to each other (Bright 1976), to furnish students with activities that illustrate the basic properties of measurement (Bright 1976), to give students a means for determining whether a given measurement is reasonable (Coburn and Shulte 1986), and to give students the experience of estimating in different settings so they might appreciate when it is appropriate to overestimate or underestimate (NCTM 1989).

Estimation in measurement involves a form of judgment based on previous experience that is transferred to similar situations. The *National Statement on Mathematics for Australian Schools* (Australian Education Council 1990) suggests that estimation activities encourage students to recognize that estimation is not simply guessing but should draw on informed judgment.

We argue that learning to estimate can be difficult unless students have prior or contemporaneous experiences in measurement. Siegel, Goldsmith, and Madson (1982) have emphasized the importance of primary students' ability to draw on *mental images* as cues to facilitate their estimation. It appears that students move along a continuum from guesswork, with no reasons for the guess, to the use of benchmarks with nonstandard units to accurate estimation and the use of standard units.

Making a Good Estimate

We have found that a useful starting activity in lessons on estimation for sixth- and eighth-grade students is to ask the students to estimate the length of a room in meters. Students almost always construct their own mental image of a unit that they then imagine being used to measure the length of the room. These mental images are sometimes based on tangible objects located within the same room, such as light fixtures or carpet tiles or the student's own body length. In other instances, students have told us that they imagine the length of a meter independent of any object within the room.

This estimation technique for determining the length of a room is appropriate in that it involves iteration of a nonstandard or standard unit. However, answers have varied from six to eighteen meters for a room nine meters long. This discrepancy suggests that students' skills in constructing a mental image of a unit of an appropriate size might be inadequate for the task and could be improved by further experiences in carrying out additional measuring tasks in conjunction with estimating.

In our research into sixth- and eighth-grade students' understanding of angles, we have investigated the range of strategies these students use to estimate the sizes of angles shown in various contexts and with various materials, such as line drawings, cardboard sectors, and geostrips. The following examples in which eighth-grade students were asked to estimate the sizes of angles illustrate the strategies they employed in answering the questions.

The protractor as a mental image

Tonia was asked to estimate the measure of the angles shown as line drawings in **figure 1**. Tonia looked at a nearby protractor and then at the angles on several occasions before offering her estimates of 15 degrees and 30 degrees.

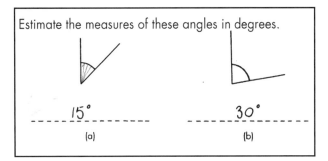

Fig. 1. Tonia's estimates of the measures of angles presented in line drawings

When Tonia was told that she appeared to be making a comparison with the protractor without actually picking it up to superimpose on the angle, her response indicated that she needed only to look at the protractor to make an image of it in her mind to assist with the estimation. Tonia repeated the same strategy with other angles during the interview.

Elizabeth took the mental image of a protractor a step further when she was asked to estimate specific angle measures. In each instance, she drew a picture of her mental image of a protractor on the angles. Additionally, she estimated and marked off her picture in 10-degree sectors and was able to do so whether the drawings were presented with "short arms" or "long arms." In each of these situations, she drew a protractor with an appropriate scale and with the 10-degree sectors drawn with relative accuracy.

The right angle as a mental image

Benn used a mental image of a right angle as a benchmark to estimate accurately the measures of the angles shown in **figures 1** and **2**. In the instance of the 45-degree angle shown in **figure 1(a),** he imagined a right angle and thought of the given angle as half of that right angle. For the 270-degree-angle in **figure 2 (b)**, rotation, he used his right-angle image actually to sketch three right angles, as shown, so as to arrive at his final estimate. Benn's initial estimate of the measure of the angle in **figure 2a** was 43 degrees, but he changed it to 87 degrees because "it's tilted a bit more, but it's not as big as ninety degrees." Again, Benn was using a right angle as a benchmark.

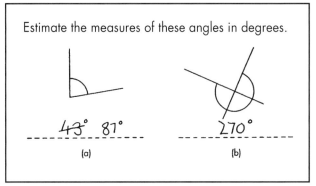

Fig. 2. Benn's estimates of the measures of angles presented on line drawings

The half-turn as a mental image

Joel was asked to estimate rotations of 40 degrees, 160 degrees, and 350 degrees. For each example, he drew on mental images of amounts of turn to reach his estimate. For example, for the angle of 160 degrees Joel's estimate was 175 degrees because "I know half a revolution is one hundred eighty degrees. That's nearly one hundred eighty degrees." For the 350-degree angle, Joel estimated 355 degrees "because it's nearly a revolution and that's three hundred and sixty degrees. . . . I knew it was three hundred and something."

The angles of a polygon as a mental image

With the 60-degree sector, Philippa placed the cardboard sector on the table, as shown in **figure 3**, and

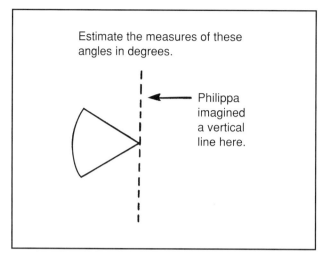

Fig. 3. Philippa's estimates of the measure of the central angle of a cardboard sector

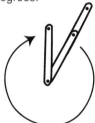

Fig. 4. Philippa's estimates of the measures of angles swept out by geostrips

imagined a vertical line going through the apex of the sector. She recognized the sector as forming one-third of half a circle, "so it's like a hexagon shape." With the geostrips, Philippa again thought of the 60-degree turn in terms of one of the central angles of a hexagon. She also thought of a 45-degree angle as one of the central angles of an octagon and a 30-degree angle in terms of a twelve-sided figure. When she was not able to draw on her mental images of the central angles of familiar polygons, she compared angles with a mental image of a protractor (**fig. 4**).

Conclusion

In a number of our observations, we have noted that students have tended to use their imagined unit as a benchmark so that they can estimate the measure of a given angle as a fraction or multiple of their devised unit. We suspect that the task of estimation, in the example of angles, can be made easier for some students if an appropriate and readily usable mental image is constructed when they are introduced to the concept of angle. For example, it might be helpful for some students to experiment at an early stage with sectors that they cut out and use as units for comparison and measurement of angles. Included in these sectors might be right angles and angles that are multiples of 10 degrees. Additionally, it might be useful to follow up this activity with an introduction to familiar figures that contain specific angles for comparison with their sector units.

A further possibility would be to consider angles as amount of turn so that simple geostrip models can be made and used by students to sweep out an angle that can be compared with the angles made by the spokes on a bicycle wheel or ferris wheel. In other situations, the geostrips can be used to assess the angles made by the components of such familiar structures as bridges and buildings.

Mental images in mathematics may not be accurate, but they need to be functional. The student who meets the concept of angle for the first time may well benefit from opportunities to construct an image in the same way that scientists and engineers construct mental models to serve as useful representations of the phenomena to be understood. These models seem to work for many students. In geometry, teachers have numerous opportunities to facilitate mathematics learning by encouraging imagery. In this article, we have briefly discussed ways in which some students have built their own imagery when working with angles and ways in which the classroom tasks might facilitate the development and use of this imagery.

References

Australian Education Council. *A National Statement on Mathematics for Australian Schools.* Carlton, Victoria: Curriculum Corporation, 1990.

Bright, George W. "Estimation as Part of Learning to Measure." In *Measurement in School Mathematics,* 1976 Yearbook of the National Council of Teachers of Mathematics (NCTM), edited by Doyal Nelson and Robert E. Reys, pp. 87–104. Reston, Va.: NCTM, 1976.

Coburn, Terrence G., and Albert P. Shulte. "Estimation in Measurement." In *Estimation and Mental Computation,* 1986 Yearbook of the National Council of Teachers of Mathematics (NCTM), edited by Harold L. Schoen and Marilyn J. Zweng, pp. 195–203. Reston, Va.: NCTM, 1986.

Gentner, D., and D. R. Gentner. "Flowing Waters or Teeming Crowds: Mental Models of Electricity." In *Mental Models,* edited by D. Gentner and A. L. Stevens. Hillsdale, N.J.: Lawrence Erlbaum Associates, 1983.

National Council of Teachers of Mathematics (NCTM). *Curriculum and Evaluation Standards for School Mathematics.* Reston, Va.: NCTM, 1989.

Siegel, Alexander W., Lynn T. Goldsmith, and Camilla R. Madson. "Skill in Estimation Problems of Extent and Numerosity." *Journal for Research in Mathematics Education* 13 (May 1982): 211–32.

Developing Spatial Sense through Area Measurement

Elizabeth Nitabach and Richard Lehrer

The original sense of geometry, "earth measure," suggests that measurement is an important way in which children can come to understand and develop a mathematics of space. Developing an understanding of measure provides children with opportunities to explore fundamental questions about space, such as "How can shapes look different yet cover the same amount of space?" Traditional instruction in measurement has consisted of helping children gain procedural competence with measuring tools, such as rulers, and teaching children to use formulas to calculate measurements of length, area, and volume. We believe that this traditional instruction in measurement falls short in helping children develop an understanding of space. Consequently, rather than assist children in gaining procedural competence with measuring tools, we aim to help them develop an understanding of measure.

This article focuses on our work with colleagues Carmen Curtis and Jean Gavin and first and second graders to further the children's understanding of space by solving problems involving central concepts of measure. We outline some of the important concepts that serve as the cornerstones of measure and then describe one lesson in which Curtis posed a problem and built on her students' thinking in ways that helped them explore the foundations of area measure. This exploration is only one of many in which children developed fundamental ideas about area measure. At the end of this article, we briefly describe a few others and how they relate to such development. These central concepts have been useful to us in understanding young children's thinking about measure.

Foundations of Measure

A common core of measurement assumptions can guide a teacher in creating problems for children to solve and in making decisions about how to interact with children during problem solving. These assumptions include the following:

1. Units of measure should be adapted to the objects of measure.

We want children to understand the necessity of using a unit that *has* length to measure length, a unit that *has* area to measure area, and so on. This principle is often difficult for children to learn, partly because most of their early experiences with measure deal solely with length, so that units of length are often perceived as being universally applicable. We find that children initially try to measure space by using a length measure, such as a ruler. For example, when children first attempt to measure area, they often resort to measuring the lengths of the sides of a shape, or they assume that they can measure the length of one side of a figure, like a square; record the value; move the ruler over, slightly parallel to the side; and add the current value of length to the previous one. As shown in **figure 1**, they continue this process until they reach the opposite side of the figure (Lehrer, Jenkins, and Osana 1998). This procedure assumes that length has space-filling properties.

A related notion is that children often show a *resemblance bias* when given such manipulatives as

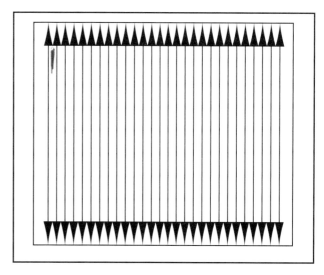

Fig. 1. Children's idea of length as having a space-filling property

cutouts of squares, rectangles, and so on, to measure area. Children believe that the area of squares should be measured with squares, triangles with triangles, and so on. These examples suggest that developing relationships among units of measure and the attribute being measured is not trivial for children.

2. Units of measure should be identical.

We want children to understand that when using repeated units to measure an item, all the units must be identical. Although this axiom is built into commonplace measuring tools like rulers, children's proficiency with these types of cultural tools may impede their consideration of whether, for example, inches on one side of a ruler can be mixed with centimeters on the other. Children often use rulers to mix different units of measure. For example, they might say that an object's length is 17, derived by adding 12, from the inches' side of their ruler, to 5, the measurement of the "extra piece" of the object read from the centimeters' side. We have also observed that some children who use rulers correctly in some situations also use a *mixture* of different-sized paper clips to measure an object's length. These children might tell us that an object is eight paper clips long when they can cover its length with three large, one medium, and four small paper clips.

A related assumption is that units of measure should also be *conventional,* because communal establishment of conventions assists communication. For example, it is much easier to buy lumber measured in feet than in a nonstandard unit, such as paper clips.

3. Measurement involves iteration.

Units of measure for length are *concatenated,* or laid end to end, to produce a quantity. Piaget, Inhelder, and Szeminska (1960) suggested that this process involves mental operations of subdividing a line and moving a standard unit along that line. This concept may seem trivial, in that children often use rulers to count a succession of units of length. However, when the length of an object exceeds the length of the ruler or if a ruler is not available, then the concept of iteration becomes more obvious. Children who can use twelve-inch rulers to measure the length of objects shorter than a foot are sometimes at a loss when using that same ruler to measure longer objects.

4. A scale has a zero point.

The use of measuring tools requires establishing a zero point and recognizing that numbers other than zero can serve in this role. For example, when measuring an eight-inch length of wood, if one end of the wood is aligned with the two-inch mark on the ruler and the other with the ten-inch mark, the two-inch mark serves as the zero point so that the three-inch mark coincides with the first unit of length measure, the four-inch mark with the second, and so on. The conventional use of rulers often obscures this quality of measure.

5. Measurement is characterized by additivity.

Units of measure can be composed so that, for example, the area of a figure can be determined by adding the areas of the subregions of the figure. The areas of different figures can be compared by mentally or physically decomposing one figure into subregions and then rearranging or composing these subregions so that one can decide if they are larger, smaller, or the same size as, or congruent with, a second figure. We have found that children in the primary grades are capable of understanding and using this quality of measure. Children have made such statements as "That square is the same size as that triangle because you could cut the triangle into two little triangles and put them together to make that square." Because many primary-aged children have this understanding, we believe that area measure should not be delayed for later grades when children are presumably ready.

6. Measuring area is based on space filling.

To measure the area of a region, the units of measure should completely tile that region. Our observations suggest that young children often use a boundedness criterion and not a space-filling criterion to measure area. For example, when measuring the area of a square, they might see how many pennies could fit within the square. They tend to be careful to make sure that no penny or part of a penny falls outside the sides of the square, but they do not worry about the gaps between the pennies.

Area Measure

While working with children on area concepts, we think about the foundations described earlier in planning tasks to help children develop their ideas about area. Carmen Curtis also thinks of these foundations and has designed a spiral of tasks, all related to thematic units, to help children develop their ideas about area and its measure. She began with a problem involving three rectangles, designed to help children explore the additivity of areas. This problem initially involved no unit of measure; the children were not asked or told to think of the space in terms of square inches or any other unit. As the lesson proceeded, a unit of measure emerged from the children's explorations and Curtis's skillful guidance.

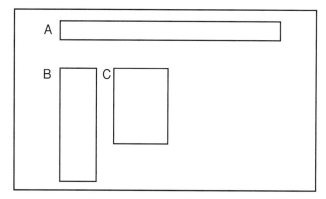

Fig. 2. Three "quilt" shapes presented by Curtis

Three rectangles

Problem description

Curtis presented three paper-strip "quilts," which she fastened to the chalkboard. The problem context of a quilt drew on the children's previous experiences with designing quilts in which they used the transformations of flips, slides, and turns of core squares that were all of the same size to make a series of different quilts. She put this problem in further context by noting that she had been working on three quilts "but [that] when I sewed these patches together, I had three very different-looking shapes. What I want to talk about is how much space each of those shapes covers." Although she did not use the word *area*, Curtis defined area for purposes of this discussion as "how much space would be taken up by a shape."

She chose these shapes so that each quilt was composed of "core squares" in different arrangements (1×12, 2×6, and 4×3 core squares, respectively). The children were not aware that each quilt was composed of core squares, and the core squares were not outlined in any way. The quilts were composed of solid pieces of construction paper without any demarcations. Curtis labeled each different shape with a letter—A, B, and C (see **fig. 2**).

Curtis's rationale for this task's design was firmly anchored in her conceptions of students' thinking. She noted that she expected the children to experience conflict between area conceptions and visual perceptions: "[O]nce they make predictions [about which covers the most space, I expect they might say,] 'Well, I think shape C covers more space. No, no, no. It is A. Look how long it is.' But when I ask them, 'How can we find out?' what are they going to say? Will they suggest covering it? Will they suggest measuring around the outside? Will they suggest folding it in half? [And I will tell them,] you are looking at these three shapes, you have different ideas about which might cover more space, but how are you going to prove to someone what you think might be true?"

She expected the children to think that some shapes looked larger than others and then to challenge their perceptions with their emerging understandings of area. She perceived that these conflicts would open a window into her students' thinking. Curtis also noted that the task lent an opportunity for thinking about invariance and additive composition properties of area. She told us, "By the end of the lesson, some of them will say, 'Well, it looks like it takes up more space, but really you could just push that space around and make it fit.'"

Students' thinking

The students began by claiming that shape B or shape C would cover the greatest amount of space because they were "fatter" or "look bigger." However, some students disagreed and thought that perhaps shapes B and C were really the same size. They sought to convince their classmates by appealing to additive congruence, in other words, by breaking up and rearranging the space of one figure so that it would fit exactly in the other. A picture of Michelle's work is shown in **figure 3**. She looked at B and C and thought that if she folded B in half and rotated it, it would cover exactly half of C. She noted that if the one-half B was flipped horizontally, it would cover the top half of C. The children went on to explore other partitions of B and C that might lead to this result.

Satisfied that B and C did indeed cover the same amount of space, the class turned its attention to A.

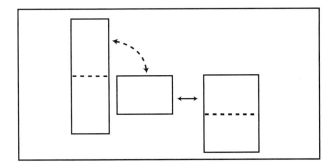

Fig. 3. Michelle's rearrangement of B and C

Michael's work is shown in **figure 4**. He claimed, "You can make C into A," and demonstrated by folding C into four equal parts, each the width of A. He placed this "strip" at the top edge of A and used his finger to demark the bottom edge. Then he moved the entire strip down to his finger and iteratively marked off four equal segments of A. Curtis said, "Michael just showed us that if we divided C into four equal parts, we could make A. Tim thinks he has a different way of turning C into A." Tim folded C into three long strips instead of four short strips (see **fig. 4**). When Tim finished his demonstration of additive congruence and C was unfolded, the fold lines clearly divided it into a four-by-three array of squares. This resulting design was not noticed by the class until Curtis raised the following question: "How come it took Michael four strips and only took Tim three?"

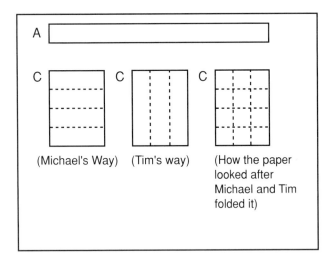

Fig. 4. Various folding techniques used to show that C covers the same space as A

Tim: [Was counting] One, two, three, . . . , twelve.

Curtis: Twelve what?

Tim: Twelve squares! That makes a quilt!

A second student pointed excitedly to A and said, "Then it takes twelve squares to make *that*." In part because of her question, Curtis had instigated a transition from composition-additive-region strategies to units of measure. The children had been talking in terms of cutting up shapes and rearranging figures and later began talking about the number of core squares in each rectangle. The children went on to verify that each of the three forms could be composed of exactly twelve square units, or core squares, the unit of measure they used in their previous quilt lesson. Curtis then asked the children to design "as many shapes as you can" with twelve core squares. She wanted to let the children freely explore the idea that appearances can be deceiving—different-looking forms can cover the same amount of space.

This lesson started the children on the road to exploring the foundations of measure and the relationship between area and shape. In the next series of lessons, the children explored the nature of area measure for irregular forms. In one lesson, the children attempted to rank order how much space was covered by each child's hand. In another, they ranked in order the areas of cutouts of "islands" they had drawn. These problems posed questions about the invention of appropriate units of area measure and raised the issue of "what to do with the leftovers," since the measure of these figures was not a whole number. Toward year's end, Curtis introduced the notion that area could be thought of as being a multiplication of lengths. Curtis's guidance helped the children develop a formula for the area of rectangular "zoo cages." The last lesson in the series also focused on area of zoo cages, but in this instance, issues of scale and use of symbolic diagrams were posed to the children. Although different problems and problem sequences could be posed to accomplish many of the same ends, the important lesson for us was the gradual and progressive emergence of students' understanding when instruction focused on helping them develop a strong foundation concerning ideas about measure.

Action Research Ideas

1. Try Curtis's lesson with students. Do not give students rulers to use when deciding which rectangle has the greatest area. If you have not used the term

area with your students, you could ask them, "Which of these rectangles covers the most space?" This exercise might be particularly interesting for those who have practiced finding the area of figures with rulers and standard formulas. Do they have other ways of reasoning about the amount of space in a region?

2. If this activity is too easy for a particular class, change the figures to a 4 × 4 square, an isosceles right triangle with legs of 4 × 4 oriented so that the hypotenuse is horizontal, and a 2 × 6 rectangle. How do students reason about the relative size of these figures? Do they have any ways of describing how much bigger the square is than the triangle or how much smaller the square is than the rectangle?

3. Students can also practice this type of reasoning with irregular forms. Ask students to find out how they could rank order the area of each person's hand in the class. Look at the list of foundations of measure to see if this activity can reveal anything about students' understandings of measure. How do students account for leftover units when figuring the area of the hand?

References

Lehrer, Richard, M. Jenkins, and H. Osana. "Longitudinal Study of Geometric Conceptions." In *Designing Learning Environments for Developing Understanding of Geometry and Space,* edited by R. Lehrer and D. Chazan. Hillsdale, N.J.: Lawrence Erlbaum Associates, 1998.

Piaget. Jean, Bärbel Inhelder, and Alina Szeminska. *The Child's Conception of Geometry.* New York: Basic Books, 1960.

Spatial Sense and the Construction of Abstract Units in Tiling

Grayson H. Wheatley

Although tiling has been a supplementary topic in elementary school mathematics for many years, its value has not been fully appreciated. This article discusses a variation on tiling that offers rich opportunities for the construction of fundamental mathematical relationships. In particular, *unitizing* as a mental operation is identified as a basis for much mathematical activity in both geometric and numerical settings.

NCTM's *Curriculum and Evaluation Standards* (1989) calls for a reconceptualization of mathematics as more than operations on whole numbers, decimals, and fractions. As we begin to teach for relational understanding, we will need to design tasks to identify the potential for substantial mathematics learning. We will find it important to recognize the mental operations students use in constructing their mathematics. One topic that has great potential for mathematical activity is tiling in the plane.

Creating a tiling pattern is more than just drawing a design—it is the manifestation of powerful mental operations. Consider the tiling shown in **figure 1**. Two "L" shapes are put together to form a rectangle, which becomes a new unit for tiling. Making a tiling pattern requires the mental rotation of a constructed image, determining the feasibility of an anticipated placement, the conceptualization and imagination of a regular pattern of tiles, and the production of the drawing. This activity frequently involves the construction of *abstract units*. For example, **figure 2** shows new tiling units made with a chevron shape. Six chevrons are arranged to form a hexagon, and this shape is taken as a new unit for tiling the plane with hexagons, resulting in a tiling of the plane with chevrons.

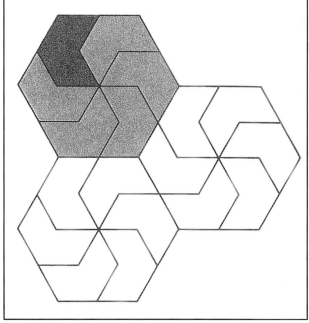

Fig. 2. Tiling units made with a chevron shape

Few activities embody the spirit of mathematics as patterns the way tiling does (Steen 1990). Important mathematical patterns can be constructed by exploring the many ways in which shapes can be packed together. In the act of creating tiling patterns, students build relationships that transcend geometry and involve mathematics in general. The same unitizing seen in the tiling activity is also fundamental in place-value activities, that is, constructing units of ten, hundred, and thousand and coordinating them is a critical construction for students developing their number sense.

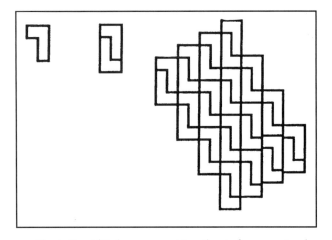

Fig. 1. Two "L" shapes are put together to form a rectangle, which becomes a new unit for tiling.

Senechal (1990) makes a compelling case for tiling, stating that "the study of shape can form a central component of mathematics education, a component that draws on and contributes to not only mathematics but also the sciences and the arts" (p. 140).

Imagery is clearly at the heart of tiling. To create a tiling on dot paper, the student must take the given shape as a unit, construct it as a mathematical object, develop a plan for drawing it, imagine the placement of a tile in a new position, anticipate possible difficulties, and visualize a patterning with the given shape. Few activities offer the opportunities for using and developing imagery that does tiling.

In studies with third-grade students (Wheatley and Reynolds 1991a; Wheatley and Reynolds 1991b), researchers noted that students naturally construct abstract mathematical units as they are engaged in tiling activities. The construction of composite units—that is, taking a pattern of shapes as a new unit with which to tile—signals a cognitive advance in mathematical thought just as thinking in hours, for certain purposes, gives us great advantage over thinking in just minutes. However, the student who can think of an hour as one unit and simultaneously think of it as sixty minutes has greater mathematical power, which will be useful in many situations. Thinking about the length of a trip across the state by car in minutes (e.g., 300 minutes) may not be as helpful as thinking in hours (5 hours), whereas at times the ability to switch to thinking in minutes will be critical. Thus in the tiling activity, some students found it facilitative to think in composite units, such as a hexagon composed of six chevrons as shown in **figure 2,** rather than try to position each basic shape individually over the entire page.

Creating a tiling with the chevron shape on triangular dot paper is a particularly rich mathematical activity. It is rich in potential learning opportunities for several reasons:

1. Chevrons can be put together in many interesting ways to form a tiling.

2. The visualization of the shape in a different orientation, as is necessary in deciding on a new placement, is challenging.

3. The triangular dot paper causes perturbations because horizontal and vertical frames of references do not exist.

4. Many opportunities arise to create abstract units.

When this challenge was put to fifth-grade students, they generated the patterns shown in **figure 3**. Their creations reveal not only evidence of the construction of units but also elaborate color patterning. The students took great pleasure in creating such aesthetically appealing mathematical works of art. Note the use of symmetry in several of the patterns.

The results of these studies suggest that tiling presents many opportunities for students to construct abstract units. The parallel use of units in students' activity involving numbers suggests that the unitizing operation is a general mental operation that is fundamental to many mathematical concepts. The tile itself is taken by the students as a unit and can be considered a unit of area in measurement. Students come to know a hexagon in a deeper sense when they have explored its space-filling quality through tiling; it becomes more than just "a six-sided figure."

As we rethink school mathematics using the curriculum standards as a framework, let us consider the role of imagery and the construction of units in mathematics learning. Tiling is a rich setting for developing spatial sense and coming to think of mathematics as constructing patterns and relationships.

References

National Council of Teachers of Mathematics (NCTM). *Curriculum and Evaluation Standards for School Mathematics.* Reston, Va.: NCTM, 1989.

Senechal, Marjorie, "Shape," In *On the Shoulders of Giants: New Approaches to Numeracy,* edited by Lynn A. Steen, 139–81. Washington, D.C.: National Academy Press, 1990.

Steen, Lynn A., ed. *On the Shoulders of Giants: New Approaches to Numeracy.* Washington, D.C.: National Academy Press, 1990.

Wheatley, Grayson H. "Research into Practice: Enhancing Mathematics Learning through Imagery." *Arithmetic Teacher* 39 (September 1991): 34–36.

Wheatley, Grayson, and Anne Reynolds. "The Construction of Abstract Geometric and Numerical Units." In *Proceedings of the Thirteenth Annual Meeting of the North American Chapter of the Psychology of Mathematics Education,* edited by Robert Underhill. Blacksburg, Va.: October 1991a.

———. "The Potential for Mathematical Activity in Tiling: Constructing Abstract Units." In *Proceedings of the Fifteenth Annual Meeting of the International Group for the Psychology of Mathematics Education,* edited by F. Furinghetti. Assisi, Italy: July 1991b.

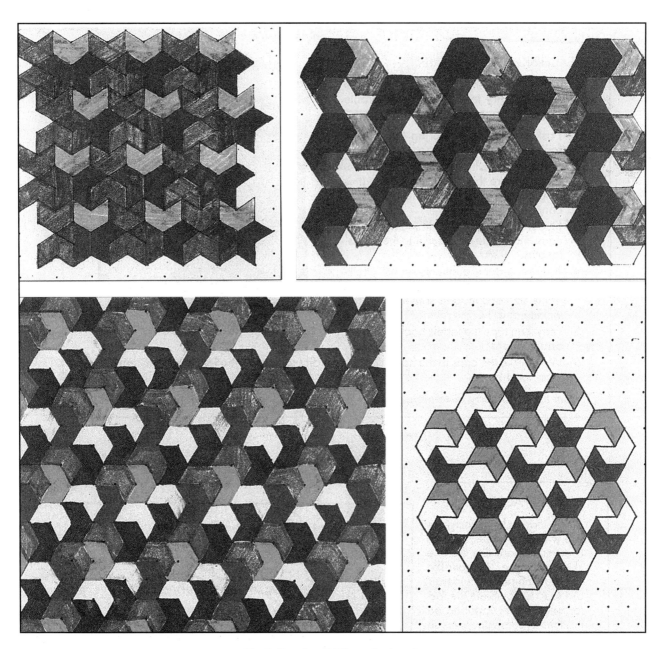

Fig. 3. Samples of fifth graders' work

Similarity in the Middle Grades

Glenda Lappan and Ruhama Even

Similarity is an important topic in geometry, basic to understanding the geometry of indirect measurement, proportional reasoning, scale drawing and modeling, and the nature of growing. When United States' teachers were asked to rate the importance of this topic for the Second International Study of Mathematics, they rated similarity of plane figures as being important for all students in grade 8.

Donald Kerr Jr. (1981) analyzed three exercises from the 1977–78 National Assessment of Educational Progress (NCTM 1981) that dealt with the concept of similarity. He found that 94 percent of thirteen-year-olds were able to recognize shapes similar to a given shape when given the question shown in **figure 1**. But being able to identify similar figures does not mean that students are familiar with most basic properties of similarity. For example, thirteen-year-olds were given the following problem in the 1977–78 NAEP:

> Two triangles are SIMILAR. Indicate if each of the following statements is true or false.

Kerr reports the results given in **table 1**: Results of United States' eighth graders on a similar item in the Second International Mathematics Study (SIMS) were even worse (IAEEA 1985). Less than 30 percent of eighth graders chose correctly the true statement about similar triangles from a set of five statements.

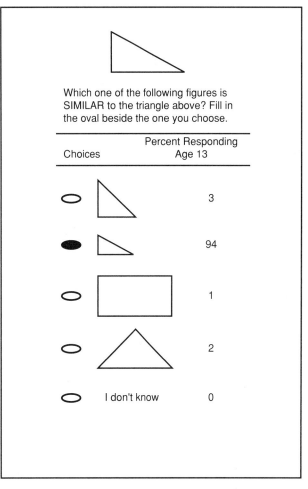

Fig. 1

Table 1

Statement	Response	Age 13
a. Their corresponding sides MUST be congruent	True False* I don't know	54 32 14
b. Their corresponding angles MUST be congruent	True* False I don't know	57 30 12
c. They MUST have the same area	True False* I don't know	51 45 4
d. They MUST have the same shape	True* False I don't know	62 35 4

*Indicates correct response

A common application of similarity in school is using a short object to find the height of a tall object, such as in the problem (see **fig. 2**) that was given to eighth graders (SIMS). A similar problem was given in the NAEP. In both tests only about 40 percent of the students were able to find the correct answer.

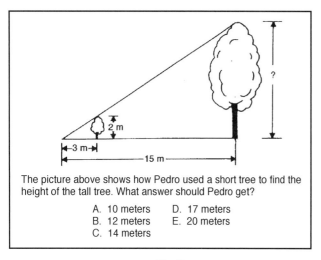

Fig. 2

A past report shows a decline in students' performance. The following problem was given to eighth graders on the First International Mathematics Study in 1964 and on the Second International Mathematics Study in 1982:

On level ground, a boy 5 units tall casts a shadow 3 units long. At the same time a nearby telephone pole 45 units high casts a shadow the length of which, in the same units, is

A. 24   B. 27   C. 30   D. 60   E. 75

Fifty-six percent of the students chose the correct answer at the end of the 1963–64 school year, whereas only 41 percent did so at the end of the 1981–82 school year!

Similarity is a very important topic, and yet it is clear that students do not understand the concept of similarity and cannot use its properties. One reason for this situation could be a lack of opportunity to learn. Although teachers rated similarity as an important topic for eighth grade, half of them did not teach the mathematics needed to respond to the items described.

Providing opportunities to learn is necessary but not sufficient. Similarity is not an easy concept. It has close connections with equivalent fractions, ratio, and proportional reasoning. Behr et al. (1984) found that teaching fraction equivalence requires an extended period of time, and even then children have difficulties with applications of fractions to new situations. Streefland (1985) got the same results about ratio and suggests that visual models will support the long-term learning process. Tourniaire and Pulos (1985) did a comprehensive review of the research on proportional reasoning. They conclude that the passage from additive strategies to multiplicative strategies is slow and complex while fractional ratios are much harder than integer ratios. Hart (1984) did a comprehensive study of children's strategies and errors in ratio in Britain and reports similar findings. Piagetian research (Piaget and Inhelder 1967) shows that similarity tasks are easier than other tasks involving proportion. Therefore, as Friedlander (1984) suggests, the concept of geometrical similarity may be a step toward an understanding of proportionality.

As is usually the approach with difficult concepts, Tourniaire and Pulos recommend the use of manipulative materials, that sufficient time be given to acquire the concept, and that a variety of teaching methods be used.

When working with our students on similarity, we have to keep in mind that most middle school students are at the first two van Hiele levels of development in geometry (Burger and Shaughnessy 1986; Wirszup 1976). They need rich opportunities to investigate similarity using concrete objects to help develop a robust understanding of this concept.

1. The following activity can help students move from an intuitive understanding of similar figures as "having the same shape" to a more analytic and powerful understanding of similarity. They will learn that in similar figures angle size is preserved, all lengths are multiplied by a constant, and ratios of corresponding sides are equal.

   Give the students points to plot that form some simple picture, such as that of a house. Ask the students to transform their houses by (1) multiplying all the y-coordinates by the same number while leaving the x-coordinates untouched, (2) multiplying all the x-coordinates by the same number while not changing the y-coordinates, and (3) multiplying both coordinates by the same constant. After each step the students should record the new pairs of numbers (see **table 2**) and plot the corresponding points.

Table 2

Connect within Each Set	House 1 (x, y)	House 2 (x, 3y)	House 3 (2x, y)	House 4 (2x, 2y)
Set 1	(2,2)			
	(6,2)			
	(6,6)			
	(4,8)			
	(2,6)			
Set 2	(3,4)			
	(5,4)			
	(5,5)			
	(3,5)			

[Graph showing a house shape on coordinate axes from 1–6 horizontally and 1–8 vertically]

Discuss with our students which of the houses are similar. What has remained the same for the similar houses? (shape, corresponding angles, ratio of corresponding sides) What has been changed? (size) What has remained the same for the houses that are not similar? (corresponding right angles, some of the corresponding sides) What has been changed? (shape, corresponding angles that are not right angles, ratio of corresponding sides) Transparencies can be superimposed to help in comparisons of corresponding angles and corresponding sides of different houses.

2. Students often use correct multiplicative strategies in solving similarity problems that deal with integral ratios but use wrong additive strategies when dealing with fractional ratios. In this activity, by using concrete manipulatives, the students can actually find their own mistakes and begin to see the misuse of additive strategies for similarity tasks.

Ask the students to draw a single straight vertical line across a given strip of paper and cut on the line to make a rectangle similar to a given one. Four categories of situations may be considered: situations such as

[rectangle 2 by 4, shaded] and [rectangle 6 by blank],

where the numbers are nice, 2 divides 4 and 6; situations such as

[rectangle 2 by 3] and [rectangle 8 by blank],

where 2 does not divide 3 but divides 8; situations such as

[rectangle 2 by 6] and [rectangle 5 by blank],

where 2 divides 6 but does not divide 5; and finally the hardest situations, those such as

[rectangle 4 by 6] and [rectangle 7 by blank],

where 4 does not divide either 6 or 7. Give the students examples of each type without the dimensions supplied. Let the students make the measurements themselves. Their understandings and their weaknesses are likely to be revealed as you observe and talk to students during this activity.

3. Use the environment for making indirect measures using similarity concepts. The height of an inaccessible object can be found by using similar right triangles that are formed by shadows or by angles of reflections in a mirror (see **fig. 3**).

Ask your students to find the height of their school and other buildings by using the shadow method (on a sunny day) or the mirror method. They should verify their findings by checking the actual heights with the principal or a building maintenance person.

4. The next activity lets students examine properties of similarity using concrete objects. It introduces the relationship between scale factor and the area of the new figure.

Ask your students to put together congruent triangles to make another triangle that is similar to the smaller ones (see **fig. 4**). Ask them to explain why the big triangle is similar to the small ones. The students may use a small triangle to check cor-

Similarity in the Middle Grades

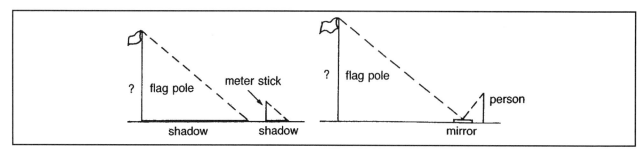

Fig. 3

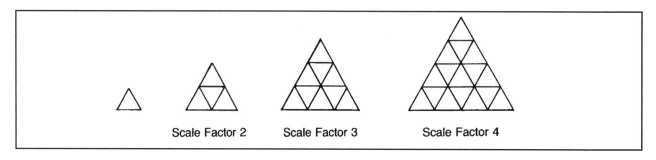

Fig. 4

responding angles and the ratio of corresponding sides. This is a good situation in which to talk about scale factors. Have the students make several similar triangles. Discuss with your students the relationship between the scale factor and the area growth (the scale factor squared). Conclude this activity by going the other way. Ask the students to subdivide a large triangle into smaller congruent triangles that are similar to the large one. How many small triangles will result at the end? (any square number) Can you subdivide any triangle in this way? (yes)

5. This activity emphasizes the complex concept of area growth in an everyday context.

   Mr. Jones has two garden plots. One is shown. His other plot is 3 times as large in length and width. Black plastic to cover his small plot costs $7. How much will it cost for plastic to cover the large plot?

   Your students are likely to answer $21 (3 × 7). If they do, have them cut out several copies of the small plot and actually make a model of the large plot (see **fig. 5**). Then ask the question again. The concrete model shows that the large plot is nine times as big in area and hence will cost $63 to cover.

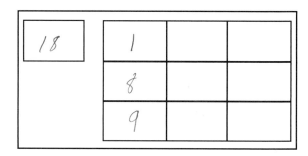

Fig. 5

References

Behr, Merlyn J., Ipke Wachsmuth, Thomas R. Post, and Richard Lesh. "Order and Equivalence of Rational Numbers: A Clinical Teaching Experiment." *Journal for Research in Mathematics Education* 15 (November 1984): 323–41.

Burger, William, and J. Michael Shaughnessy. "Characterizing the van Hiele Levels of Development in Geometry." *Journal for Research in Mathematics Education* 17 (January 1986): 31–48.

Carpenter, Thomas P., Mary Kay Corbitt, Henry S. Kepner Jr., Mary Montgomery Lindquist, and Robert E. Reys. *Results from the Second Mathematics Assessment of the National Assessment of Educational Progress.*

Reston, Va.: National Council of Teachers of Mathematics, 1981.

Crowley, Mary L. "The van Hiele Model of the Development of Geometric Thought." In *Learning and Teaching Geometry, K–12,* 1987 Yearbook of the National Council of Teachers of Mathematics (NCTM), edited by Mary Montgomery Lindquist and Albert P. Shulte, pp. 1–16. Reston, Va.: NCTM, 1987.

Friedlander, Alex. "Achievement in Similarity Tasks: Effect of Instruction and Relationship with Achievement in Spatial Visualization at the Middle Grades Level." Ph.D. diss., Michigan State University, 1984.

Hart, M. Kathleen. *Ratio: Children's Strategies and Errors.* Windsor, Berkshire: NFER Nelson, 1984.

International Association for the Evaluation of Educational Achievement. *Second Study of Mathematics: Technical Report IV.* United States, 1985.

Kerr, Donald R. Jr. "A Geometry Lesson from National Assessment." *Mathematics Teacher* 74 (January 1981): 27–32.

Piaget, Jean, and Bärbel Inhelder. *The Child's Conception of Space.* New York: W.W. Norton & Co.. 1967.

Streefland, Leen. "Search for the Roots of Ratio: Some Thoughts on the Long-Term Learning Process (Towards . . . a Theory) II: The Outline of the Long-Term Learning Process." *Educational Studies in Mathematics* 16 (February 1985): 75–94.

Tourniaire, Francoise, and Steven Pulos. "Proportional Reasoning: A Review of the Literature." *Educational Studies in Mathematics* 16 (May 1985): 181–204.

Wirszup, Izaak. "Breakthroughs in the Psychology of Learning and Teaching Geometry." In *Space and Geometry: Papers from a Research Workshop,* edited by J. Martin. Columbus, Ohio: ERIC/SMEAC, 1976.

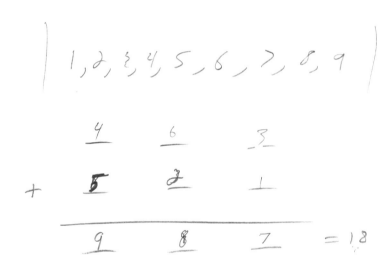

Section 8

Algebra and Algebraic Reasoning

A Foundation for Algebraic Reasoning in the Early Grades

Erna Yackel

For many adults, *algebra* means solving systems of equations; finding the value of an unknown; using the quadratic formula; or otherwise working within a system of formulas, equations, and literal symbols. From this perspective, the suggestion that algebra should permeate the K–12 mathematics curriculum seems unreasonable and certainly indefensible if we take seriously that children should learn mathematics by making sense of things of the basis of their current mathematical understandings. Do we expect children in kindergarten and first grade to solve algebraic equations? If not, then what might it mean to suggest that algebraic thinking should be part of the mathematics curriculum for the elementary grades?

In taking the position that algebra is for all, the NCTM is calling for a complete rethinking of what we might mean by algebra (1994). In effect, the NCTM is advocating that the notion of algebra be expanded to include a range of mathematical activity. To assist in this rethinking process, in 1994 the NCTM appointed an Algebra Working Group charged with developing and elaborating a vision of K–12 algebra that would help teachers and school systems as they grapple with the process of change. Significantly, the working group deliberately chose not to begin by defining algebra or setting forth standards for algebra (NCTM 1995). Such an approach would be static, narrow, and limited and be bounded by historical views and perspectives of algebra. Rather, the group chose to take an emerging view of algebra. This view acknowledges the dynamic nature of mathematics in general and of algebra in particular, treats mathematics as a human activity (Davis and Hersh 1981), and puts students' thinking at the forefront. In this view, we develop our vision of algebra as we consider the mathematical activity and thinking of students. This position is consistent with the views of Smith and Thompson (in press), who argue, "We believe it is possible to prepare children for different views of algebra—algebra as modeling, as pattern finding, or as the study of structure—by having them *build ways of knowing and reasoning which make those mathematical practices appear as different aspects of a central and fundamental way of thinking*" (emphasis added). Thus, the emphasis is not on whether an activity should qualify as being algebraic but on the underlying thinking and reasoning of the students. This view is particularly helpful at the elementary school level because it eliminates the need to focus on what algebra "content" should be included in the elementary grades. Instead, the crucial issue is the nature of the children's reasoning and thinking.

The purpose of this article is twofold. The first is to explore children's thinking that might be foundational to algebraic reasoning. In keeping with the spirit of the Algebra Working Group, no definitive claims are made about what does and does not consititute algebraic reasoning. Instead, readers are invited to consider ways of reasoning and thinking that in their view might be a foundation on which to develop algebraic thinking and reasoning. The second is to describe some instructional activities that can potentially engender such reasoning and thinking. Here again, readers are invited to be active participants by asking themselves what possibilities these instructional activities might have in their own classrooms.

Thinking That Is Foundational to Algebraic Reasoning

To explore the nature of children's thinking, consider the following example that comes from a first-grade classroom-research study conducted by Cobb, Whitenack, and McClain (Cobb et al. 1997) and cited by Smith (in press) as illustrating one kind of algebraic thinking. In the example, students were shown a picture of one large and one small tree and five monkeys (see **fig. 1**). The teacher explained that all the monkeys want to play in the trees, and she asked the students to think about the different ways that the five monkeys could play in the two trees. The children began to generate responses, such as that three could be in the little tree and two in the big tree or that five could be in the big tree and none in the little tree. The children's reasoning in generating these responses might be described as primarily numerical. Their activity involved thinking of, and figuring out, specific instances of how many monkeys might be in each of the two trees. From the observer's perspective, we might say that the children found various ways to partition the number 5 into two parts. The teacher recorded the children's suggestions by drawing a vertical line between the trees and writing the number of monkeys in each tree on the corresponding side of the line, creating a table in the process (see **fig. 1**). As the discussion progressed, the teacher asked if all the possibilities were already recorded and if a way could be found to ensure that they had them all.

Cobb and others note that a shift in the discourse occurred when the teacher asked this question. Jordan explained, "See, if you had four in this [big] tree and one in this [small] tree in here, and one in this [big] tree and four in the [small] tree, couldn't be that no more." He explained how every other partition of 5 into two numbers could yield two possible ways the monkeys could be in the two trees. The various possibilities that had been suggested previously by the children and recorded in the table by the teacher emerged "as explicit objects of discourse that could themselves be related to each other" (Cobb et al. 1997). Cobb and his colleagues use the example to explicate shifts in discourse, but Smith uses the example to focus on the nature of Jordan's thinking. Jordan was no longer thinking about generating specific par-

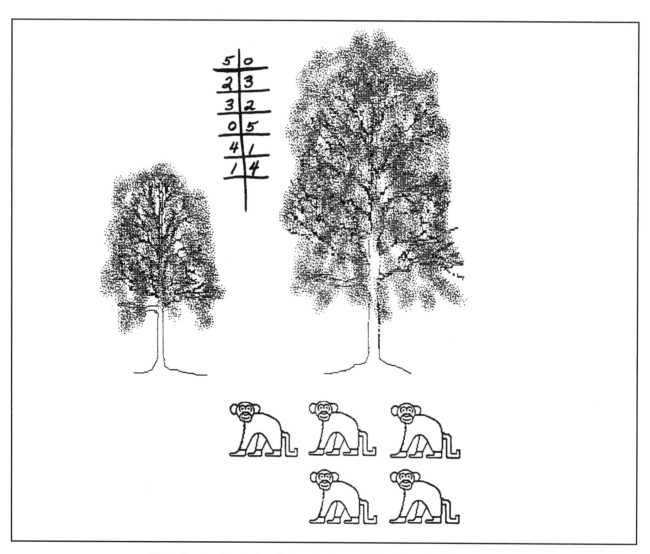

Fig.1. Overhead-projector display showing a table with students' responses

titions of 5. Neither was he checking empirically to see if all possibilities were in the table. As Cobb and others and Smith point out, Jordan was building an understanding of the relationship between the possible partitionings of the monkeys and the possible entries in the table. Because of the focus on relationships, some would refer to Jordan's reasoning as representational or algebraic. Whether the reader concurs with calling Jordan's thinking "algebraic" is unimportant. The crucial point is that Jordan's thinking is qualitatively different from the numerical thinking in which he and the other children engaged initially when they generated the individual instances that first made up the table.

Some—but not all—students in Jordan's class made sense of his explanation. Some children might have been unable to reason about relationships as Jordan did, and this problem might not have engendered higher-level thinking that some would call algebraic. This point is significant. It once again highlights the mathematical activity of the students and indicates that qualitative differences can occur in children's thinking as they attempt to solve a problem. A task in and of itself does not elicit a particular type of thinking. Nevertheless, as the previous example demonstrates, opportunities for various possibilities can be generated by the careful selection of tasks and by the way they are developed in the classroom. The teacher's question "How can you be sure you have all the possibilities?" apparently was the impetus for Jordan's higher-level thinking. By asking the question, the teacher initiated a change in the focus from thinking about the various ways the monkeys might be in the trees to reasoning about the relationship between the various ways and the records in the table. Jordan's reasoning is evidence that the shift in focus was productive for at least one child in the class.

Promising Instructional Activities

The remainder of this article describes several problems that have been proposed by the NCTM Algebra Working Group (1995) as potentially useful in fostering elementary school students' development of a conceptual basis for algebraic reasoning. Although activities from arithmetic or data measurement could have been chosen, these problems are set within the context of measurement. In one sense they can be thought of as focusing on dimension, perimeter, and area and the relationships among them. The purpose is to focus on how they might foster thinking that is foundational to algebraic reasoning. For this reason, possible ways that students might think about and solve the problems are included. These possible solutions and interpretations are representative of the working group members' classroom experiences with the problems.

The first problem described is intended for the early primary grades and is designed to encourage students to investigate various rectangles and their areas where one dimension of the rectangle is fixed.

> Building-rectangles problem: Use some of your (square) tiles to make a rectangle with a base of 2.

The teacher might pose these questions:

- Did you all make the same rectangle?
- How many tiles did you use?
- How did you figure it out?
- What is the height of your rectangle?
- How would you build a rectangle that uses twenty-two tiles?
- Can you figure out how high it will be without building it?
- If you know how many tiles are in a rectangle with a base of 2, can you figure out high it is?
- If you know how high you want a rectangle to be, can you figure out how many tiles you will need?

To find the number of tiles used, which represents the area, some children count them one by one. Others count by twos. Still others count the number of tiles in one column and double the number. These strategies reflect the children's own understanding of, and facility with, number. To answer the follow-up questions, some children begin to reason in a more general way about the relationship between height and total number of tiles, or area. These questions shift the focus from making the rectangle to reasoning about how the dimensions are relevant.

For our purposes, the crucial feature of the instructional activity is the set of follow-up questions that encourage the children to go beyond building rectangles and even beyond producing numerical answers to questions about the dimensions to reasoning about the dimensions and about the relationships between them. The teacher can facilitate the discussion by introducing various ways of recording children's rectangles, such as by making a table or by using grid paper.

The next problem, which is intended for the intermediate grades, involves working with relationships and constraints. Like the first problem, students are asked to construct rectangular shapes but with a fixed perimeter.

> Rectangular-pen problem: Charmaine wants to build a rectangular pen for her pet using 26 feet of fence. Help her figure out some possible pens she might build. Record your results so that someone else can figure out how you thought about the problem. String, pipe cleaners, and grid paper are materials that are available for you to use if you choose.

These questions might be posed:

- What different rectangular pens did you find?
- How did you record your results?
- How does the shape of the pen change by making one side of the pen longer or shorter?
- Have you found all the possibilities?
- How can you be sure?

To find some of the possible rectangular pens, some students make a twenty-six-inch length from pipe cleaners and bend it to form various rectangles. Some record their results by tracing around each rectangle on a piece of paper. After making several rectangles, some of these students make additional rectangles by first folding the pipe-cleaner length in half, then partitioning and folding the first half, and finally folding the other half to complete the rectangle (see **fig. 2**). Some students think about how many pens are possible by reasoning about possible ways to partition the pipe-cleaner length. Other students use grid paper to draw rectangles with perimeter of twenty-six units. The thinking of some of these students is constrained by the lines on the grid paper. As a result they use only integral values for the length and width. Some students record their results only with their drawings on the grid paper. Others make lists of the pairs of dimensions. Very young students use square tiles, arranging them in rectangles and verifying the perimeter by counting and making adjustments as needed.

By using such materials as string and pipe cleaners, students can reason about partitioning a length without thinking numerically. By bending pipe cleaners and folding string, students can reason about how

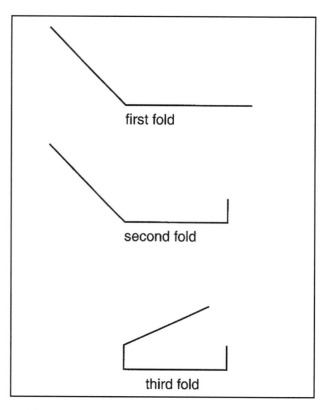

Fig. 2. Diagram depicting reasoning by partitioning and folding a length of pipe cleaner

the length and width of the rectangles are related. Reasoning in this way about quantities, such as length and width, without having specific numerical values in mind, is foundational to algebraic reasoning. As students consider the question of all possible rectangles, they may abstract patterns in their own reasoning and use tables, graphs, and literal symbols to express their thinking. The teacher plays a crucial role in assisting students to develop ways to notate and record their thinking that are consistent with conventional methods. For example, students may reason that the width and the length must add to 13 or that the width can be determined by subtracting the length from 13. With the teacher's help, they can develop ways to notate this reasoning, including the use of such standard symbolic means as $W + L = 13$ or $W = 13 - L$. By introducing methods of recording and organizing as a topic of discussion, the teacher helps students focus explicitly on these means of representation. In this way, representing, notating, and symbolizing, all of which are foundational to algebra, can emerge as the instructional activity unfolds in the classroom.

The problem can be extended by asking which pen provides the most area for the pet. Discussing the maximum area encourages students to reflect on the range of possible rectangles. In some classes, questions of nonintegral values for the length and width will arise. Students can be encouraged to draw graphs and to reason from their drawings in ways appropriate to their grade levels. Some students will reason algebraically using a diagram of a rectangle as the basis of their reasoning.

This discussion illustrates that the initial choice of problem is only one of the relevant factors in the instructional activity that develops. Teacher questions, student solutions, and attempts to follow up on either or both are central to how the instructional activity is realized in action.

Conclusions

As we have illustrated, developing the foundations for algebraic reasoning in the elementary grades can be accomplished through activities that encourage children to move beyond numerical reasoning to more general reasoning about relationships, quantity, and ways of notating and symbolizing, to name but a few. In this way teachers can contribute to the emerging view of algebra for all, K–12, and to the preparation of our students for the twenty-first century.

References

Cobb, Paul, Ada Boufi, Kay McClain, and Joy Whitenack. "Reflective Discourse and Collective Reflection." *Journal for Research in Mathematics Education* 28 (May 1997): 258–77.

Davis, Philip J., and Reuben Hersh. *The Mathematical Experience.* Boston; Houghton Mifflin Co., 1981.

National Council of Teachers of Mathematics (NCTM). "Algebra for Everyone: More Than a Change in Enrollment Patterns." Board of Director's Statement. Reston, Va.: NCTM, 1994.

National Council of Teachers of Mathematics Algebra Working Group. *Algebra in the K–12 Curriculum: Dilemmas and Possibilities.* Final report to the Board of Directors. East Lansing: Michigan State University, 1995.

Smith, Erick. "Algebraic Thinking as a Framework for Introducing Functions in the Elementary Curriculum." In *Employing Children's Natural Powers to Build Algebraic Reasoning in the Context of Elementary Mathematics,* edited by Jim Kaput, in press. Available at http://www.simcalc.umassd.edu/NewWebsite/EABookChapters.html

Smith, John P. III, and Patrick W. Thompson. "Additive Quantitative Reasoning and the Development of Algebraic Reasoning." In *Employing Children's Natural Powers to Build Algebraic Reasoning in the Context of Elementary Mathematics,* edited by Jim Kaput, in press. Available at http://www.simcalc.umassd.edu/NewWebsite/EABookChapters.html

Children's Understanding of Equality: A Foundation for Algebra

Karen P. Falkner, Linda Levi, and Thomas P. Carpenter

Many states and school districts, as well as *Principles and Standards for School Mathematics: Discussion Draft* (NCTM 1998), recommend that algebra be taught in the early childhood years. Although young children often understand much more than traditionally thought, adults can have difficulty conceptualizing what would constitute appropriate algebra for the early childhood years. Fifteen teachers and three university researchers are currently involved in a project to define what algebra instruction can and should be for young children. In this article, we discuss the concept of equality, which is a crucial idea for developing algebraic reasoning in young children.

Misconceptions about the Equals Sign

Even though teachers frequently use the equals sign with their students, it is interesting to explore what children understand about equality and the equals sign. At the start of this project, many teachers asked their students to solve the following problem:

$$8 + 4 = \Box + 5$$

At first, this problem looked trivial to many teachers. One sixth-grade teacher, for example, said, "Sure, I will help you out and give this problem to my students, but I have no idea why this will be of interest to you." This teacher found that all twenty-four of her students thought that 12 was the answer that should go in the box. She found this result so interesting that before we had a chance to check back with her, she had the other sixth-grade teachers at her school give this problem to their students. As shown in **table 1**, all 145 sixth-grade students given this problem thought that either 12 or 17 should go in the box.

Why did so many children have trouble with this problem? Clearly, children have a limited understanding of equality and the equals sign if they think that 12 or 17 is the answer that goes in the box. Many young children do, however, understand how to model a situation that involves making things equal. For example, Mary Jo Yttri, a kindergarten teacher, gave her students the problem $4 + 5 = \Box + 6$. All the children thought that 9 should go in the box. Yttri then modeled this situation with the children. Together, they made a stack of four cubes, then a stack of five cubes. In another space, they made stacks of nine and six cubes. Yttri asked the children if each arrangement had the same number of cubes. The children knew that the groupings did not have the same number of cubes and were able to tell her which one had more. Several children were able to tell the teacher how they could make both groupings have the same number of cubes. Even after doing this activity, however, the children still thought that 9 should go in the box in the equation.

This incident surprised Yttri and the researchers. We had assumed that kindergarten children would have little experience with the equals sign and would not yet have formed the misconceptions about equality demonstrated by older children. Even kindergarten children, however, appear to have enduring misconceptions about the meaning of the equals sign that are not eliminated with one or two examples or a simple explanation. This incident also illustrates that children as young as kindergarten age may have an appropriate understanding of equality relations involving collections of objects but have difficulty relating this understanding to symbolic representa-

---

The research reported in this article was supported in part by a grant from the U.S. Department of Education, Office of Education Research and Improvement, to the National Center for Improving Student Learning and Achievement in Mathematics and Science (R305A60007-98). The opinions expressed do not necessarily reflect the position, policy, or endorsement of the supporting agency. The examples in this article were transcribed from videotaped interactions between teachers and students. Most are available as part of the Cognitively Guided Instruction Project videotape series produced by the Wisconsin Center for Education Research, © 1995. Some episodes were edited for the presentation described here.

Table 1
Percent of Children Offering Various Solutions to 8 + 4 = ☐ + 5

Grade	Answers Given					Number of Children
	7	12	17	12 and 17	Other	
1	0	19	7	0	14	14
1 and 2	6	54	20	0	20	84
2	6	55	10	14	15	174
3	10	60	20	5	5	208
4	7	9	44	30	11	57
5	7	48	45	0	0	42
6	0	84	14	2	0	145

tions involving the equals sign. A concerted effort over an extended period of time is required to establish appropriate notions of equality. Teachers should also be concerned about children's conceptions of equality as soon as symbols for representing number operations are introduced. Otherwise, misconceptions about equality can become more firmly entrenched. (See "About the Mathematics" on p. 206.)

As Behr, Erlwanger, and Nichols (1975); Erlwanger and Berlanger (1983); and Anenz-Ludlow and Walgamuth (1998) have documented, children in the elementary grades generally think that the equals sign means that they should carry out the calculation that precedes it and that the number after the equals sign is the answer to the calculation. Elementary school children generally do not see the equals sign as a symbol that expresses the relationship "is the same as."

Not much variety is evident in how the equals sign is typically used in the elementary school. Usually, the equals sign comes at the end of an equation and only one number comes after it. With number sentences, such as 4 + 6 = 10 or 67 − 10 − 3 = 54, the children are correct to think of the equals sign as a signal to compute.

First and Second Grades

Karen Falkner is currently teaching a first- and second-grade class. Children typically stay in the class for two years. The remainder of this article shows how the children in this class have progressed in their understanding of equality over the past year-and-a-half.

For some time, solving story problems has been an integral part of mathematics instruction in Falkner's class. Students are regularly asked to write number sentences that show how they solved story problems. Falkner expected her students to be successful, therefore, when she first asked them to solve the number sentence 8 + 4 = ☐ + 5. To her surprise, the students answered the problem just as research indicated that they would. Most put 12 in the box, and some extended the sentence by adding = 17. The discussion that followed was interesting. Most said that 12 should go in the box because "eight plus four equals twelve." The following excerpt illustrates the class discussion that took place after students had worked on the problem.

Falkner: Is 8 + 4 the same as 12 + 5?

Anna: No.

Falkner: Then why did you put 12 in the box?

Anna: Because 8 + 4 equals 12. See? [Counting on her fingers] It's 8, 9, 10, 11, 12. [Many children nod their heads in agreement.]

Falkner: Did anyone get another answer?

Adam: It is 7.

Falkner: Why?

Adam: Because you have to have the same amount on each side of the equals sign. That's what the equals sign means.

Falkner: I see. Adam, would you say that again? [Adam repeats his explanation. Other children, considering Adam a class leader, listen attentively.]

Falkner: [Gesturing at the number sentence on the chalkboard.] So, Adam, you say that the equals sign means that however much something is on one side of the equals sign, the same amount has to be on the other side of the equals sign. [Looking at the rest of the class] What do you think about what Adam said?

Anna: Yes, but it has to be 12, because that is what 8 + 4 equals.

Dan: No, Adam is right. Whatever is one side of the equals sign has to equal what is on the other side: 8 + 4 = 12 and 7 + 5 = 12, so 7 goes in the box.

The class wrestled with this problem for some time. The equals sign is a convention, the symbol chosen by mathematicians to represent the notion of equality. Because no logical reason exists that the equals sign does not mean "compute," Falkner thought that it was appropriate to tell the class that she agreed with Adam and Dan. Telling the class what the equals sign meant was not, however, sufficient for many children to be able to adopt the standard use of the sign.

Falkner then chose to develop her students' understanding of the equals sign through discussion of true and false number sentences; this discussion builds on the work of Robert Davis (1964). Falkner presented number sentences, similar to the following, to her students and asked whether the number sentences were true or false.

$$4 + 5 = 9$$
$$12 - 5 = 9$$
$$7 = 3 + 4$$
$$8 + 2 = 10 + 4$$
$$7 + 4 = 15 - 4$$
$$8 = 8$$

The children's reactions were interesting. All agreed that the first sentence was true and that the second was false. They could prove these assertions by a number of means. They were less sure about the remaining sentences.

Falkner: What about this sentence? 7 = 3 + 4. Is it ture or false? [Lots of squirming around, distressed faces, and muttering from the class]

Gretchen: Yes, 3 + 4 does equal 7.

Ned: But the sentence is wrong.

Anna: It's backward.

Falkner: But Adam has told us that the equals sign means that the quantity on each side of it has to be equal. Is that true here?

Anna: Yes, but it's the wrong way.

Falkner: Let's try this. [She models the problem, giving one child seven Unifix cubes in a stack and asking him to stand on one side of her. She gives another child a stack of four Unifix cubes for one hand and a stack of three for the other hand. That child stands on the other side of her.] Now, do these two children have the same number of cubes?

Class: Yes.

Falkner: Does it make any difference which side of me they stand on? [She asks them to change places, which they do.]

Class: No, but....

As you can imagine, the fourth number sentence caused confusion for many children. Some children thought the number sentence was true because 8 + 2 does equal 10. Children who had a firm understanding of equality were able to explain that this number sentence was not true because 8 + 2 is 10 and 10 + 4 is 14 and 10 is not the same as 14.

When Falkner came to the final sentence, 8 = 8, the class was quite disturbed. Anna spoke for the students when she said, "Well, yes, eight equals eight, but you just shouldn't write it that way." In the few remaining weeks of school, Falkner continued to give problems to her students with the equals sign in various locations.

The Next Year

In the fall, Falkner posed the same problem, 8 + 4 = □ + 5, to her class. A few, but not all, of the children who had been in the room the previous spring correctly solved the problem. Many new first graders proudly put 12 in the box; others looked at the sentence in confusion and asked for help. A discussion similar to the one in the spring ensued. This time, however, a few children understood the notion of equality and enthusiastically explained why the number 7 belonged in the box. Lillie gave the most spirited explanation. "The equals sign means that it has to be even. The amount has to be the same on each side of the equals sign. [Gesturing with her hands] It is just like a teeter-totter. It has to be level."

This class discussion was the first of several about similar open number sentences. Each discussion had its doubters, as well as children who once again explained the idea that each side of the equals sign had to "equal" the same amount. As Falkner listened to the discussions, noted who was talking, and looked at facial expressions, it appeared that the children were beginning to grasp this notion of equality but that the concept was not easily or quickly understood. Falkner was convinced that the notion of equality would take time for all the children to understand, and she returned to it often as the year progressed.

Falkner integrated discussion of equality throughout the school year in two ways. First, she continued to present open number sentences in which she varied the location of the unknown. Some examples of these open number sentences included the following: $\square = 9 + 5$, $7 + 8 = \square + 10$, and $7 + \square = 6 + 4$. Second, she presented true and false number sentences, such as those in the examples, to encourage children to reflect on the meaning of the equals sign. She also had the children write their own true and false number sentences. The tasks that Falkner used to build children's understanding of equality were also tasks that build their understanding of number operations.

As the year progressed, more and more children began to understand equality. In March, the class had the following discussion:

Falkner: Look at this number sentence: $8 + 9 = \square + 10$. What should go in the box?

Carrie: It should be 17.

Skip: But $8 + 9$ would equal 17, so $17 + 10$ would equal 27, so 17 isn't OK to put in the box.

Myra: Right; $17 + 10$ does not equal 17.

Ned: I think that 7 goes in the box; $7 + 10$ is 17 and $8 + 9$ is 17. Both sides are even. [The class generally agrees, although Carrie is not yet convinced.]

Falkner: Think about what we know about the equals sign. Look at this number sentence: $4898 + 3 = 4897 + \square$. Can you figure this one out without even doing the addition?

Larry: I think that 4 goes in the box; 4897 is 1 down from 4898, so you need to add 1 more to 3.

Falkner: Did anyone do it a different way? [Children shake their heads. In general, the class agrees that Larry's way gives the right answer and is easy.]

Such discussions about number sentences gave the children an important context for discussing equality throughout the school year. As the year progressed, discussions about equality became integrated with discussions about other algebraic arithmetic concepts. In the following example, the children discuss a much more sophisticated problem that involves an understanding of variables and operations, as well as equality.

Falkner asked the class to look at the sentence $a = b + 2$. She said that the sentence was true and asked the class which was larger, a or b? Children who think of the equals sign as a signal to do something would have trouble with this problem. Because 2 is added to b and nothing is added to a, they might think that b is larger. The class first agreed that a and b were symbols for variables, just as a box or triangle were. The class then quickly agreed that a was larger, and their arguments for that position clearly indicate a sophisticated understanding of equality.

Falkner: Why do you think that a is larger?

Anna: They split the b and 2 apart; a brings them together.

Jerry: I think a [is larger]. That plus 2 is part of a.

Myra: Yes, a has to be bigger because whatever $b + 2$ is has to be higher than b, because you combine them.

Anna: Right; a has the $+ 2$ in it and b doesn't.

Lillie: Together they have to be the same; $b + 2$ has to be the same as a.

Conclusion

Discussions such as these, which involved an ever-growing number of children, indicate that the children have learned to see the equals sign as a symbol describing a relationship rather than as a "do it" sign. Because this article was written before the end of the school year, we have not collected summary data on children's understanding of the problem $8 + 4 = \square + 5$ in this class. In a pilot study involving a similar first- and second-grade classroom in the same town, however, we found that at the end of the year, fourteen out of sixteen children correctly answered that 7 should go in the box.

As we reflect on our introduction of the notion of equality and the equals sign to this class and others, we continue to be amazed at the interest and excitement that the children bring to the discussions. Lillie uses her teeter-totter metaphor with the enthusiasm of a child ready to play on one. Skip is genuinely outraged that anyone should fill in a blank so that an equation reads $17 = 27$. These are not the bored comments of children looking forward to recess but the excited contributions of children who are exploring a new world of thinking and communicating mathematically and who are enjoying the power of that new knowledge. These children are developing an understanding of equality as they learn about numbers and operations. This understanding will allow them to reflect on equations and will lay a firm foundation for later learning of algebra.

About the Mathematics

Children must understand that equality is a relationship that expresses the idea that two mathematical expressions hold the same value. It is important for children to understand this idea for two reasons. First, children need this understanding to think about relationships expressed by number sentences. For example, the number sentence $7 + 8 = 7 + 7 + 1$ expresses a mathematical relationship that is central to arithmetic. When a child says, "I don't remember what 7 plus 8 is, but I do know that 7 plus 7 is 14 and then 1 more would make 15," he or she is explaining a very important relationship that is expressed by that number sentence. Children who understand equality will have a way of representing such arithmetic ideas; thus they will be able to communicate and further reflect on these ideas. A child who has many opportunities to express and reflect on such number sentences as $17 - 9 = 17 - 10 + 1$ might be able to use the same mathematical principle to solve more difficult problems, such as $45 - 18$, by expressing $45 - 18 = 45 - 20 + 2$. This example shows the advantages of integrating the teaching of arithmetic with the teaching of algebra. By doing so, teachers can help children increase their understanding of arithmetic at the same time that they learn algebraic concepts.

A second reason that understanding equality as a relationship is important is that a lack of such understanding is one of the major stumbling blocks for students when they move from arithmetic to algebra (Kieran 1981; Matz 1982). Consider, for example, the equation $4x + 27 = 87$. How do you start solving this equation? Your first step probably involves subtracting 27 from 87. Why may we do so? We may do so because we subtract 27 from *both sides* of the equation. If the equals sign signifies a relationship between two expressions, it makes sense that if two quantities are equal, then 27 less of the first quantity must equal 27 less of the second quantity. What about children who think that the equals sign means that they should do something? What chance do they have of being able to understand the reason that subtracting 27 from both sides of an equation maintains the equality relationship? These students can only try to memorize a series of rules for solving equations. Because such rules are not embedded in understanding, students are highly likely to remember them incorrectly and not be able to apply them flexibly. For these reasons, children must understand that equality is a relationship rather than a signal to do something.

References

Anenz-Ludlow, Adalira, and Catherine Walgamuth. "Third Graders' Interpretations of Equality and the Equal Symbol." *Educational Studies in Mathematics* 35 (1998); 153–87.

Behr, Merlyn, Stanley Erlwanger, and Eugene Nichols. *How Children View Equality Sentences.* PMDC Technical Report, no. 3. Tallahassee, Fla.: Florida State University, 1975. ERIC No. ED 144 802.

Davis, Robert B. *Discovery in Mathematics: A Text for Teachers.* Reading, Mass.: Addison-Wesley Publishing Co., 1994.

Erlwanger, Stanley, and Maurice Berlanger. "Interpretations of the Equal Sign among Elementary School Children." In *Proceedings of the North American Chapter of the International Group for the Psychology of Mathematics Education.* Montreal: 1983.

Kieran, Carolyn. "Concepts Associated with the Equality Symbol." *Educational Studies in Mathematics* 12 (August 1981): 317–26.

Matz, Marilyn. "Towards a Process Model for School Algebra Errors." In *Intelligent Tutoring Systems,* edited by Derick Sleeman and John Seeley Brown, pp. 25–50. New York: Academic Press, 1982.

National Council of Teachers of Mathematics (NCTM). *Principles and Standards for School Mathematics: Discussion Draft.* Reston, Va.: NCTM 1998.

Experiences with Patterning

Joan Ferrini-Mundy, Glenda Lappan, and Elizabeth Phillips

Over the past decade we have learned that children are capable of mathematical insights and mathematical invention that exceed our expectations. We have also learned that we, as teachers, contribute to—or suppress—this insight and inventiveness in our students by the choices we make. We choose the mathematical tasks, the questions, and the expectations for how students are to interact with those tasks and with one another around those tasks. The question of the expectations we knowingly or unknowingly set for our students is nowhere more crucial than in the gatekeeper area called algebra. In this article we will share an example of how algebraic thinking and reasoning might be extended over grades K–6. We hope to stimulate readers to think with us about how we can search for ways to foster algebraic thinking and reasoning by the questions we regularly ask our students.

Although we have no easy answer as to what constitutes algebra or algebraic thinking and reasoning, we can be guided by the view of algebra that emerges from an examination of the NCTM's *Curriculum and Evaluation Standards for School Mathematics* (1989). Standard 13 in the K–4 section, titled Patterns and Relationships, and Standards 8 and 9 in the 5–8 section, titled Patterns and Functions, and Algebra, respectively, suggest that the study of patterns is a productive way of developing algebraic reasoning in the elementary grades. Current curriculum reform efforts and research in learning contend that observations of patterns and relationships lie at the heart of acquiring deep understanding in many areas of mathematics—algebra and function in particular (Steen 1988). When students are presented with interesting problems in context, they observe patterns and relationships: they conjecture, test, discuss, verbalize, generalize, and represent these patterns and relationships. Generalizing and representing patterns are reflected in the example that follows.

An Example for Developing Algebraic Thinking

The following situation was adapted from the NCTM's Algebra Working Group (1995). This situation offers algebraic explorations in grades K–8 or beyond.

Tat Ming is designing square swimming pools. Each pool has a square center that is the area of the water. Tat Ming uses black tiles to represent the water. Around each pool there is a border of white tiles. Here are pictures of the three smallest square pools that he can design with black tiles for the interior and white tiles for the border. (See **fig. 1**.)

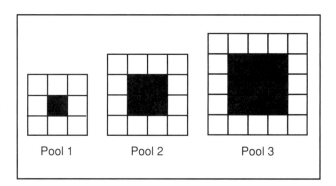

Fig. 1. Swimming pools with borders

What patterns, conjectures, and questions will children find as they explore this situation? Where is the algebra? Let us think about tasks and questions that would fit the various grade levels K–6. The intent of each question is to prompt students to look for patterns among the variables, make conjectures, provide reasons for their conjectures, and represent their patterns and reasoning. The questions and grade levels are only suggestions. Children will generate their own ideas and will pursue their own interests, using this situation as a starting point.

Grades K–2

Kindergarten children will be interested in the colors and in counting the tiles. Two kinds of tiles are used, and the number of each is not necessarily the same. Let us focus on beginning relationships between the numbers of black and white tiles.

- For each square pool, sort the tiles into black tiles for the water and white tiles for the border.
- Count how many tiles are in each pile.
- Are there more black tiles than white tiles?

Next we take the problem a bit further by looking at the pattern in the black-tile squares alone.

Here are pictures of the three smallest black squares that Tat Ming has designed for the water. (See **fig. 2**.)

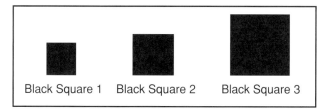

Black Square 1   Black Square 2   Black Square 3

Fig. 2. Pool without borders

- Build each of the three black squares. How many black tiles are in each square?
- Build the next-biggest square that you can make out of the black tiles. Then build the next. Count the squares in each.
- What patterns do you see?
- What is a square?

Here we can return to the original setting and look at the patterns in the two kinds of tiles in **figure 1**.

- Build the three pools using black and white tiles to show the water and the border tiles. Record the information in a table. (See **table 1**.)

Table 1
Organizing the Data

Pool Number	Number of Black Tiles	Number of White Tiles	Total Number of Black and White Tiles

- How many tiles will be in the next-largest pool? Check your answer by building the square.
- Describe your methods for counting the different tiles.
- What patterns do you see?

Looking into the classroom for grades K–2

The beginning questions engage students in sorting and counting the black and white tiles. This activity helps them to look at the relationship between the numbers of black and of white tiles. Students might observe that the first three pools have more white tiles than black tiles. The teacher may ask if this situation is always true and encourage the students to build the next-larger pool. This pool contains more black tiles than white tiles. At this level, students represent their thinking and conjectures with objects that are concrete in nature.

In a first-grade class, some students focus on the black tiles and what it means to be square. They may notice that there are as many rows as columns in the figures. Some may find convenient ways to step-count to find the number of tiles: "two, four" or "three, six, nine." Some students may begin to guess the number of tiles in the next-larger black square. The teacher can follow these observations by asking students how many tiles are on each edge of the black square. Students are beginning to see a connection between the number of tiles needed to build a square and the length of its edge. Their observations are made and checked using the tiles.

In a second-grade class, students begin to organize their data into a table. They use newly developed computational skills to find ways to multiply and add.

Grace, a second grader, at first attended only to the overall size of the squares, not to the differences in color. In exploring this situation, Grace worked with the teacher to reach a definition of a square. This transcription illustrates the tentativeness of the student's concept of square and her need to work with concrete materials to help herself think about the concept. Working in such a relatively open-ended setting can often reveal unexpected student thinking about concepts that we assume students understand thoroughly.

T: Why do you call it a square? What's a square to you?

G: A block.

Experiences with Patterning   **209**

T: If it were longer down like this, would it still be a square?

G: No, it would turn into a rectangle.

T: So what makes it a square?

G: That it's not as far down as a rectangle.

T: Is there anything else about the sides? How long is this side?

G: Three squares.

T: How long is this side?

G: Three squares.

T: So what is a square?

G: Can I try something? I'm putting out three to see if I can scramble them around and make a square. [Grace is working with three unit squares and trying to build a square. Notice that the teacher was trying to draw Grace's attention to the equality of the sides. But, not unexpectedly, she became interested instead in the number 3 and its relationship to the square.]

T: A square out of three? [Grace notices that she will need four squares to make a square and builds it.]

T: How can you be sure it's a square?

G: You can, because all the sides are the same length.

She was also very interested in counting the number of small tiles in each pool. She counted by threes for pool 1, by fours for pool 2, by fives for pool 3, and was able to predict the total number of squares in the fourth pool by intuitively applying the associative property of addition to compute 6 × 6 as follows:

$$6 \times 6 = [(6+6)+(6+6)] + (2 \times 6)$$
$$= (12+12) + 12$$
$$= 24 + 12 = 36$$

Grace was quite intrigued by the prediction and computed the number of tiles in the seventh pool.

$$7 \times 7 = [(7+7)+(7+7)+(7+7)] + 7$$
$$= (14+14+14) + 7$$

Grace was searching for a way to find the total number of squares in a pool. This example illustrates an aspect of algebra that involves developing and generalizing algorithms.

She also filled out the table (see **fig. 3**). She noticed that the four corners would always be present, which seemed to help her in figuring out how to count the border. She physically moved the corner squares away.

Photograph by Joan Ferrini-Mundy; all rights reserved

Fig. 3. Grace builds a square

Grades 3–4

Using the same basic situation, we can begin to ask questions that encourage students to reason about the patterns in the number of black and white tiles for a *given pool* and to reason about the number of border tiles *given the number of black tiles* and the number of black tiles *given the number of border tiles*. (See **fig 1**.)

- Build the first 3 pools, and record the data in a table. (See **table 1**.)

- Continue the table for the next 2 squares. How do you know your answers are correct?

- If there are 32 white tiles in the border, how many black tiles are there? Explain how you got your answer.

- If there are 36 black tiles, how many white tiles are there? Explain how you got your answer.

- Can you make a square with 49 black tiles? Explain why or why not.
- Can you make a square with 12 black tiles? Explain why or why not.

By grade 4, students are learning to make comparisons by looking at the fraction or proportion that a part is of the whole. We can use a version of our problem to give students a new context for using fractions by drawing on patterns. (See **fig. 1**.)

- In each of the first three square pools, decide what fraction of the square's area is black for the water and what fraction is white for the border.
- What patterns do you see?
- What fractions will occur in the next two rows of the table? How do you know that your answers are correct? (See **table 2**.)

Table 2
Looking for Fraction Patterns

Pool Number	Total Number of Black and White Tiles	Fraction of Black Tiles for the Water	Fraction of White Tiles for the Border

Looking into the classroom for grades 3–4

In grade 3, the relationship between the number of black tiles and that of white tiles comes back into play. The teacher can foster the habit of looking for patterns and relationships between the variables by asking, "As the pools get larger and larger, what happens to the number of white tiles and the number of black tiles?" Students may observe that both numbers are increasing but that for the first three squares, more white tiles are found than black tiles. Starting with the fourth square, more black tiles are seen. These observations lead to some beginning insights into different kinds of growth patterns. As the students look for patterns in the table, some may observe that to get the number of white tiles for the next square pool, you always add four.

Ryan, a fourth-grade student, got interested in the patterns he could see in the table and noted, while looking at the squares rather than the table, that "it goes up by fours" in the border (white tiles) column. The teacher asked him why, and he provided a nice geometric explanation.

The following exchange with Ryan illustrates how he used the physical representation to account for the pattern of "going up by fours" that he noted in the border column of the table. At first he associates the four corners with this increment of four. He then finds a more satisfying explanation.

T: Why does it go up by fours?

R: I think it goes up by fours because there's four corners in each. So if you take them out, there's one square on each border [looking at first square], so this would be four, this would be eight. . . . [He decides that this explanation is not adequate.]

T: Why does it go up by four? When it goes by four, from the first to second. . . .

R: Wait! Wait! Wait! I get it now. You see this is four [points to the four white squares bordering the black in the first pool]. Then this goes up another four [looking at the second pool]. This has two [referring to the side of the black square in the second pool].

T: Why does it go up by four?

R: This is one [referring to the black square in pool 1]. It only has one white square on each side. This has two [referring to the side of the black square in pool 2]. So it multiplies, I mean goes up by four because you add a white one to each [new] side. (See **fig. 4**.)

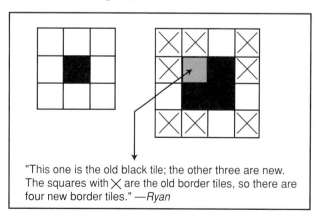

"This one is the old black tile; the other three are new. The squares with X are the old border tiles, so there are four new border tiles." —Ryan

Fig. 4. Ryan explains that the number of border tiles increases by four each time.

Going from one to the next, you pick up four new border squares because you have added four new "outside" edges.

Ryan was intrigued by trying to make the connections between the manipulatives and the table and by trying to justify what he could see in the patterns of the numbers with what he could do with the squares. He had an algebraic definition for squares—a number is square if another number, when multiplied by itself, equals the number—as well as a tentative geometric interpretation. In the following sequence of dialogue with Ryan, we see him making connections between a previous definition of square number and the physical setting.

T: Tell me about the sixth pool.

R: Pool 6 would have 8 across, so that would be 8 times 8, or 64. The total would be 64 [reasoning from the manipulatives].

T: How did you figure that out?

R: There would be 8 across, 8 going down, and 8 times 8 would be 64. The number in the border would be 28.

T: How did you figure that out?

R: 'Cause it goes with the patterns. [Note the ease with which he moves from the physical materials into the table.] Then I'm going to figure this out: 64 minus 28—whatever is left over would be the number of black [again, he is reasoning from the table and using computation to solve the problem]. Okay—36.

T: Are you pretty sure that's right?

R: [He checks by adding 36 and 28.]

T: Would it be a better check to make this square, or are you pretty sure? What if you were going to build it? If you made a square with each side 6, would you get 36 squares in it?

R: Yeah. If you use 6 . . . because 6 times 6 is 36.

T: And that's what it means to make 6 squared. Have you ever heard of 6 squared—6 × 6?

Notice in the following dialogue the tentativeness of Ryan's geometric definition of a square number and how continued exploration in this setting enables him to become more consistent in using his geometric understanding of an algebraic equation.

R: No, but I've heard of square numbers.

T: Is 36 a square number?

R: Yes.

T: Tell me why.

R: Because you can make a square with 36 squares.

T: Show me on the chart which are square numbers.

R: [Ryan looks at the numbers in the border column to see if they are square numbers. He tries to build a square of area 8 and cannot.]

T: Where are the square numbers in the table? You told me a square number was one you could make into a square.

R: Right. So all the squares: 9, 16, 25. . . . Those were the squares!

Ryan's teacher also asked questions about fractions. Even though the fraction questions were new to Ryan, he quickly made a table after the teacher started it. The table shows the fraction of black squares to the total squares in the first few figures. (See **table 3**) When asked if the number of black tiles would be half the total tiles, he said, "Well, it won't be half black until the border and the number of blacks are the same." He looked at the table and noted that it was close in the fifth square (25 out of 49) but not equal. Ryan is beginning to see patterns in equivalent fractions. (See **fig. 5**).

Table 3
Ryan's Fraction Table

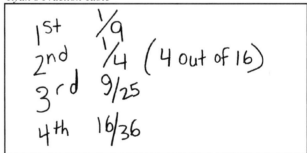

Ryan also is beginning to see that the linear pattern of the white tiles is overtaken by the quadratic pattern of the black tiles even though he does not know the names for these patterns of growth. Both Grace and Ryan used a lot of computation in the process of looking for patterns. Grace added and counted in multiples, and Ryan multiplied and subtracted.

Fig. 5. Ryan completes his table.

Grades 5–6

In grade 5, we can use new ways to represent the relationships between the number of tiles of each color and the number of the square pools. We can begin to make the emphasis on function more explicit. (See **fig. 1**.)

- Make a table showing the numbers of black tiles for water and white tiles for the border for the first six square pools.

- What are the variables in the problem? How are they related? How can you describe this relationship in words?

- Make a graph that shows the number of black tiles in each square pool. Make a graph that shows the number of white tiles in each square pool.

- As the number of the pool increases, how does the number of white tiles change? How does the number of black tiles change? How does this relationship show up in a table and in the graph?

- Use your graph to find the number of black tiles in the seventh square.

- Can there ever be a border for a square pool with exactly twenty-five white tiles? Explain why or why not.

Next we can increase the demand of the problem so that students will look for patterns and make generalizations to help with predicting what will happen in the case of a very large pool. (See **fig 1**.)

- Find the number of black (white) tiles in the 10th pool. The 25th pool. The 100th pool.

- If there are 144 black squares, what is the side length of the square pool including the border? How many white tiles are needed for the border?

Looking into the classroom for grades 5–6

Some students continue to build the squares using tiles, and they notice relationships between the numbers of black and border tiles as they build and then record their data in a table. Some students find that using grid paper is helpful. For some students, the act of building or drawing the pools suggests the relationship between the number of black and the number of white tiles. The number of white tiles is four times the number of black tiles on a side plus four for the four corners. Some students may use a table to find the number of black or white tiles in the 10th pool. But some students begin to reason about the patterns. "In the 10th pool, the square formed by the black tiles is an 8 by 8 square, so there are 64 black tiles. There are 100 tiles, so 100 total tiles minus 64 black tiles equals 36 white tiles." Some students may first reason about the number of white tiles, whereas other students may draw the 10th pool and reason from the picture. As they continue to explore these problems, they begin to notice other patterns. Some notice that the number of white tiles will always be a multiple of 4. Some students question whether every multiple of 4 is a white-tile total (see **fig. 6**).

After discussing the patterns in the table, the teacher suggests that the class explore the graphs of these patterns. The graphs suggest that the relationship between the number of the square pool and the number of white tiles can be represented by a straight line, and that the number of the square pool and the number of black tiles lie on a curved line. In later grades the first pattern is called a *linear function* and the latter pattern is called a *quadratic function*. The students can use the graph to find the number of white tiles (see **fig. 7**) or of black tiles (see **fig. 8**) given the pool number, and, conversely, given the number of white tiles or black tiles, they can find the pool number. By graphing the white- and black-tile patterns on the same grid, students can use the graphs to reason about when the two patterns are equal or when one is greater than the other. (See **fig. 9**.)

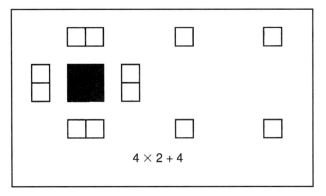

Fig. 6. The number of white tiles is always a multiple of 4.

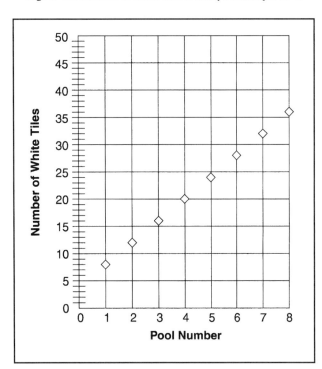

Fig. 7. A graph of the white tiles for each pool

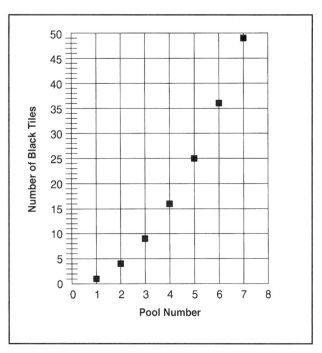

Fig. 8. A graph of the blue tiles for each pool

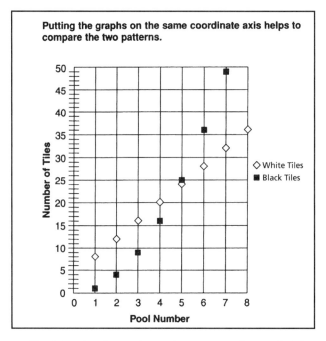

Fig. 9. Putting the graphs on the same coordinate axis helps to compare the two patterns.

Where Is the Algebra?

Throughout the grade-level examples of versions of the pool-design problem, children are being challenged to observe patterns in the growth of the numbers of black tiles and white tiles needed to make the next square pool and to build connections between the physical representations and their verbal descriptions. This activity involves an informal interaction with variables. In the early grades, the focus is on the number of each kind of tile and which is more. Even here the problem has the potential to challenge each child at his or her level of interest and insight. All children can sort and count the black and white tiles, but the teacher can ask questions that push beyond, for example:

> Build your own design out of white and black tiles. How many black tiles and white tiles does your design have?

As the problem moves up the grades in elementary school, the questions asked push toward generalization. The children are challenged to find a way to describe the relationship between the number of black tiles and the number of white tiles and the position in the sequence of pool designs. The ways to represent the change in the variables from one pool to the next also become more varied over time. Verbal descriptions, tables, graphs, and symbolic expressions are all legitimate ways to express the relationships.

Looking for Algebraic Reasoning

Many situations in elementary school mathematics can give teachers an opportunity to generalize and represent mathematical ideas and processes. In this article we offer a geometric setting that illustrates how mathematical ideas can be developed from the study of problems and how algebra emerges as a way to generalize and represent these ideas. Many other settings situated in number, data, and measurement are fruitful sites for developing algebraic reasoning. The following set of questions can serve to organize a classroom discussion in a variety of settings. The wording and choices of representations will vary depending on the experiences of the students.

- What are the variables in this situation? What quantities are changing?

- How are the variables related?

- As one variable increases, what happens to the other variable?

- How can you represent this relationship using words, concrete objects, pictures, tables, graphs, or symbols?

- How can you build connections among representations?

- How can you use this relationship to predict information about the variables?

Bibliography

National Council of Teachers of Mathematics (NCTM). *Curriculum and Evaluation Standards for School Mathematics.* Reston, Va.: NCTM, 1989.

———. *Professional Standards for Teaching Mathematics.* Reston. Va.: NCTM, 1991.

National Council of Teachers of Mathematics Algebra Working Group. *Algebra in the K–12 Curriculum: Dilemmas and Possibilities.* Final Report to the Board of Directors. East Lansing: Michigan State University, 1995.

Steen, Lynn Arthur. "The Science of Patterns." *Science* 240 (29 April 1988): 611–16.

Developing Algebraic Reasoning in the Elementary Grades

Jinfa Cai

Direct modeling with concrete objects can be a powerful problem-solving strategy for young children (Chambers 1996). However, as problem situations become more complex, the value of more powerful strategies becomes apparent. An algebraic approach in which students first describe the problem using an unknown in an equation and then solve for the unknown (Lesh, Post, and Behr 1987) is one such strategy.

The use of algebraic strategies is common among elementary school students in China and Japan but relatively uncommon in the United States. Since mathematics achievement is consistently higher in those countries than in the United States, more emphasis on developing algebraic strategies could promote the twin goals of mathematical power and algebra for all.

Opportunities for Algebraic Reasoning

The Ratio and Proportion Problem (**fig. 1**) assesses students' problem-solving skills and knowledge of ratio and proportion in a map-reading context. Students commonly use three strategies to solve this problem: an arithmetic approach, an algebraic approach, or a measurement approach. In an arithmetic approach, students typically start with known information and work to arrive at unknown quantities. In this problem they might first divide 54 by 3 to find how many miles are represented by a centimeter on the map then multiply by 12 to get the distance represented by the 12 centimeters between Martinsburg and Rivertown. No symbol is used to represent the distance, and no equation expressing the relationships is used.

In an algebraic approach, students first write an equation using ?, __, x, or a short English language phrase to represent the unknown. Students using an

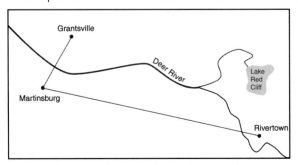

The map below shows the locations of three cities.

The actual distance between Grantsville and Martinsburg is 54 miles. On the map, Grantsville and Martinsburg are 3 centimeters apart. On the map, Martinsburg and Rivertown are 12 centimeters apart.

What is the actual distance between Martinsburg and Rivertown? Show how you found your answer.

Fig. 1. The Ratio and Proportion Problem

algebraic approach to this problem typically use a symbol to represent the unknown distance between Martinsburg and Rivertown and express the distance relationships in a proportion, such as

$$\frac{x}{12} = \frac{54}{3}.$$

This equation can then be solved for x in a variety of ways. **Figure 2a** shows a student's response using an algebraic strategy.

In a measurement strategy, students could use a finger, a paper clip, or a pencil to measure the distance between Martinsburg and Grantsville on the map, then use that measurement unit to measure the distance from Martinsburg to Rivertown. Since one unit represents 54 miles, they multiply the units from Martinsburg to Rivertown by 54 to get the distance in miles. Most students who use this strategy get a good estimate of the actual distance between Martinsburg and Rivertown. See **figure 2b**.

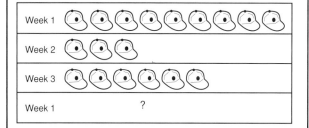

Angela is selling hats for the Mathematics Club. The picture shows the number of hats that Angela sold during the first three weeks.

How many hats must Angela sell in week 4 so that the average number of hats sold is 7? Show how you found your answer

Fig. 3. The Average Problem

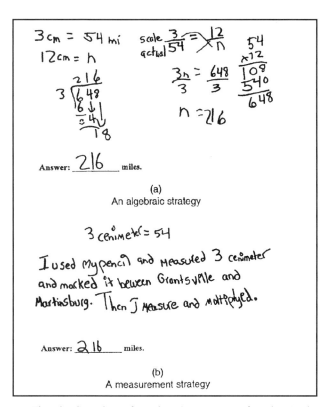

Fig. 2. Samples of students' responses for the Ratio and Proportion Problem

In an international study (Cai 1995), United States and Chinese sixth-grade students had similar mean scores on this problem, but differences were evident in their solution processes. More Chinese students used algebraic approaches, whereas more United States students used a measurement strategy. No Chinese students used the measurement strategy. About half the students from each country used an arithmetic approach.

The Average Problem (see **fig. 3**) also provides an opportunity for students to use algebraic reasoning. To be successful, students must understand the concept of *average*. Chinese students performed significantly better on this problem than United States students, and a significantly larger percent of Chinese students used algebraic approaches. In an algebraic approach, students assume that x hats were sold in week 4:

$$(9 + 3 + 6 + x) \div 4 = 7$$

By solving the equation for x, they determine the answer, which is 10.

A considerable number of United States and Chinese students used an arithmetic approach in which they calculated the total number of hats by multiplying the average number of hats by the number of weeks, that is, $7 \times 4 = 28$. Next they calculated the sum of the hats sold during the first three weeks, which was $9 + 3 + 6 = 18$, and then subtracted to determine the number sold in week 4, which was $28 - 18 = 10$. In this arithmetic approach, the answer is obtained by working backward through a string of inverse operations using the averaging algorithm. In contrast, in the algebraic approach, the answer is obtained by working forward using the averaging algorithm to directly describe the problem situation.

Some United States students solved this problem by adding or subtracting hats to "even off" the number of hats sold each week at the average number of hats sold (7). These students used the average number of hats sold as the basis for lining up the numbers of hats sold in weeks 1, 2, and 3. Since nine hats were sold in week 1, that week has two extra hats. Since three hats were sold in week 2, an additional four hats are needed to line up with the average. Since six hats were sold in week 3, one additional hat is needed to line up with the average. To line up the average number of hats sold over four weeks, ten hats should be sold in week 4.

A few students used a trial-and-error strategy: they first chose a number for week 4, then checked to see whether the average of the numbers of hats sold for the four weeks was 7. If the average was not 7, they chose another number for the week 4 and checked again, until the average was 7.

Developing Algebraic Reasoning in the Elementary Grades

What Can We Learn from International Studies?

Findings from cross-national studies (e.g., Becker [1992]; Cai [1995]) have shown that fourth- and sixth-grade students are able to use algebraic approaches to solve problems. The frequencies of using algebraic strategies among United States and Chinese students were strikingly different. Sixth-grade students in the United States tended to use visual approaches more frequently than Chinese students. Chinese sixth-grade students tended to use algebraic approaches more frequently than United States students (Cai 1995). Japanese fourth-grade students tended to use relatively more sophisticated representations in their solution processes than their United States counterparts (Becker 1992; Silver, Leung, and Cai 1995). In situations in which Japanese fourth-grade students were more likely to use multiplication, United States fourth-grade students were more likely to use addition.

The differences between United States and Asian students' uses of solution strategies might be related to instructional practice. Chinese teachers are encouraged to use concrete examples or materials in mathematics classrooms, but their role is to mediate the understanding of mathematical concepts. The United States students' relatively less frequent use of mathematical expressions and more frequent use of concrete visual representations may suggest that United States teachers less frequently encourage students to move to more abstract representations and strategies. United States teachers tend to believe that young children need concrete experiences to understand mathematics, at times asserting that concrete experiences automatically lead to understanding (Stigler and Perry 1988). Concrete visual representations do help students make sense of mathematics. However, we should expect students to move to an understanding that goes beyond concreteness (Clements and McMillen 1996). To facilitate the transition from concrete to abstract, teachers can capitalize on students' solutions that are more abstract. Class discussions of those solutions help students see their advantages. Promoting the transition from concrete to more abstract ways of thinking increases students' success in later learning (NCTM 1989).

The finding that some sixth-grade students were able to use algebraic approaches to solve open-ended problems indicates that it is possible for young children to develop algebraic thinking. In fact, those United States sixth-grade students who used algebraic approaches were better problem solvers than those who used verbal or visual representations (Cai 1998). Meanwhile, the findings that United States students are less likely than their Chinese peers to use algebraic approaches challenges us to seek better ways to move beyond concrete experience to develop students' algebraic thinking. The mathematics curriculum, even in elementary and middle schools, should include explorations of algebraic ideas and processes so that students can use algebraic thinking to solve a variety of real-world problems. The development of students' algebraic thinking is an important step toward becoming mathematically literate and should not be postponed until middle or high school. Here again, the Chinese experience may be helpful.

In Chinese elementary schools, teachers consistently encourage students to solve a problem both arithmetically and algebraically. Textbooks and teacher reference books in China contain examples showing how a problem can be solved using different strategies and recommend spending several lessons on comparisons of arithmetic and algebraic approaches. They also contain exemplary lesson plans. The objectives of teaching students to solve problems both arithmetically and algebraically are (*a*) to help students attain an in-depth understanding of quantitative relationships by representing them both arithmetically and algebraically, (*b*) to guide students to discover the similarities and differences between arithmetic and algebraic approaches, and (*c*) to develop students' thinking skills and flexibility of using appropriate approaches to solve problems (Division of Mathematics of People's Education Press 1993).

The teacher's guides supplied in Chinese classrooms highlight the importance of encouraging students to represent quantitative relationships using different strategies. The Damaged Receipt Problem, shown in translation in **figure 4,** is one example. Students are given a damaged receipt from the purchase of four chairs and two tables. On the receipt students are able to see the number of chairs and tables purchased, the unit price for chairs, and the total amount paid. However, the part of the receipt with the subtotal for chairs, subtotal for tables, and the unit price for tables was damaged. Students are asked to figure out how much each table costs. Teachers first ask students to solve the problem using both arithmetic and algebraic approaches, then present both approaches on the chalkboard. The presentation of the two approaches allows students to easily compare them.

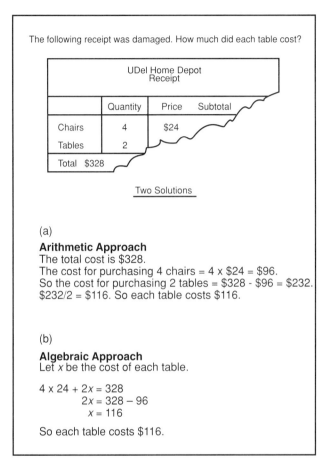

Fig. 4. The Damaged Receipt Problem and its two solutions

After the presentation of the two approaches, teachers and students discuss the similarities and differences and the teacher summarizes them. For example, both approaches involve the following quantitative relationship: the cost for purchasing chairs + the cost for purchasing tables = the total cost. In the algebraic approach, the unknown, or the cost for each table, is represented as x and is directly involved in the solution process. In contrast, the arithmetic approach uses known values until the unknown cost of each table is determined at the end. This comparison gives students an opportunity to experience the advantages of using the algebraic approach to solve the problem.

A rich collection of teaching ideas leading to the development of algebraic reasoning can be found in the "Algebraic Thinking" focus issue of *Teaching Children Mathematics* (NCTM 1997).

Action Research Idea

Pose a problem, such as the Ratio and Proportion Problem; the Average Problem; or the Damaged Receipt Problem, which has been modified to reflect a familiar local context, to your class. As the students work on the problem, watch to see whether both algebraic strategies and arithmetic strategies are being used. Note which students use which strategy. Ask students to present a variety of strategies, and have them discuss the relative advantages and disadvantages of each. When students use arithmetic strategies, ask them to try using algebraic strategies, as well. It may be inappropriate to use the terms *arithmetic* and *algebraic*. You may want to name the strategies after students in the class who use them, such as "Brenda's strategy" or "Miguel's strategy." Note which students tend to use each type of strategy. Continue to discuss the different strategies and their relative advantages over a period of several months. Continue to chart the strategies used by each student, watching to see whether the number of students who use algebraic strategies increases over time.

Some problems that can be used for this purpose are given here. More complex problems, such as problem 4, make the advantages of an algebraic approach even more apparent.

- Problem 1: Uncle John bought 5 pounds of apples. He paid $10.00 and got $6.00 back. How much did each pound of apples cost?
- Problem 2: Jamie's mom spent $7.00 on 4 pounds of bananas and 6 pounds of peaches. Bananas cost $0.40 per pound. How much does each pound of peaches cost?
- Problem 3: Ms. Johnson spent $336.00 to buy 3 tables and 4 chairs. Each chair cost $32.00. How much did each table cost?
- Problem 4: The Del Video Shop offers two rental plans. Plan A costs $18.00 annual membership fee plus $1.50 per video rented. Plan B has no membership fee but costs $2.00 per video rented. For what number of video rentals will these two plans cost exactly the same in a year?

References

Becker, Jerry P., ed. *Report of U.S.-Japan Cross-National Research on Students' Problem Solving Behaviors.* Carbondale. Ill.: Southern Illinois University, 1992.

Cai, Jinfa. "A Cognitive Analysis of U.S. and Chinese Students' Mathematical Performance on Tasks Involving Computation, Simple Problem Solving, and Complex Problem Solving." *Journal for Research in Mathematics Education* Monograph Series No. 7. Reston, Va.: National Council of Teachers of Mathematics. 1995.

———. "Exploring Students' Conceptual Understanding of the Averaging Algorithm." *School Science and Mathematics* 98 (February 1998): 93–98.

Chambers, Donald L. "Research into Practice: Direct Modeling and Invented Procedures: Building on Students' Informal Strategies." *Teaching Children Mathematics* 3 (October 1996): 92–95.

Clements, Douglas H., and Sue McMillen. "Rethinking 'Concrete' Manipulatives. *Teaching Children Mathematics* 2 (January 1996): 270–79.

Division of Mathematics of People's Education Press. *Teachers' Reference Books in Elementary School.* Beijing: People's Education Press, 1993.

Lesh, Richard, Thomas Post, and Merlyn Behr. "Dienes Revisited: Multiple Embodiments in Computer Environments." In *Developments in School Mathematics Education around the World,* edited by Izaak Wirszup and Robert Streit, pp. 647–80. Reston, Va.:

National Council of Teachers of Mathematics, 1987.National Council of Teachers of Mathematics (NCTM). *Curriculum and Evaluation Standards for School Mathematics.* Reston, Va.: NCTM, 1989.

———. "Algebraic Thinking" Focus Issue. *Teaching Children Mathematics* 3 (February 1997).

Silver, Edward A., S. S. Leung, and Jinfa Cai. "Generating Multiple Solutions for a Problem: A Comparison of the Responses of U.S. and Japanese Students." *Educational Studies in Mathematics* 28 (1995): 35–54.

Stigler, James W., and Michelle Perry. "Cross Cultural Studies of Mathematics Teaching and Learning: Recent Findings and New Directions." In *Effective Mathematics Teaching,* edited by Douglas A. Grouws, Thomas J. Cooney, and Douglas Jones, 194–223. Reston, Va.: National Council of Teachers of Mathematics, 1988.

Helping to Make the Transition to Algebra

Carolyn Kieran

> Knowledge is not an entity which can be simply transferred from those who have to those who don't.... Knowledge is something which each individual learner must construct for and by himself. This view of knowledge as an individual construction ... is usually referred to as constructivism. (Lochhead, cited in Blais 1988, p. 624)

One might consider that algebra would be a difficult subject to teach within a constructivist perspective—after all, algebra involves the use of letters, along with formal rules for operating on these letters. Nevertheless, the instructional approaches presented in this article illustrate ways of developing students' understanding of nonnumerical notation that are compatible with a constructivist stance. These approaches have been researched and found to be accessible to both elementary and middle school students.

Letters are used in algebra in several different ways (Küchemann 1981); however, two uses predominate. One of these is the use of letters to represent a range of values, as in the expression of a general solution or the generalization of number patterns (e.g., $3t + 6$); the other is the use of letters as unknowns in, say, equation solving (e.g., $n + 5 = 17$). The NCTM, in its *Curriculum and Evaluation Standards for School Mathematics* (1989), has emphasized both of these uses of letters as goals for the teaching of mathematics in grades 5–8.

Letters to Represent a Range of Values

The use of letters to represent a range of values is far more neglected in the teaching of prealgebra than their use as unknowns. Because students have little experience in using algebraic symbolism as a tool with which to think about and to express general relations, they encounter difficulty with this use of letters. For example, in a study on students' conceptions of generalization and justification, Lee and Wheeler (1987) noted that only 8 percent of the students they interviewed could use algebraic notation for such problems as the following:

A girl multiplies a number by 5 and then adds 12. She then subtracts the original number and divides the result by 4. She notices that the answer she gets is 3 more than the number she started with. She says, "I think that the same thing would happen, whatever number I started with." Using algebra, show that the girl is right.

Most of the students worked out numerical examples and concluded from these examples. Lee and Wheeler point out that students do not appear to see algebra as generalized arithmetic and, furthermore, do not believe that arithmetic can be generalized.

Peck and Jencks (1988) have developed a teaching approach that helps students make explicit links between their arithmetic and the nonnumerical notation of algebra. They describe some lessons on multiplication in a fifth-grade classroom in which students who were taught "to understand parts of arithmetic in depth, found the corresponding algebra completely sensible and a natural product of their own thinking" (p. 85). In this classroom, the teacher was consistently a "question asker and doubter, never an explainer nor judge of right or wrong" (p. 86). Thus, the traditional roles of teacher and student were reversed. The teacher also believed that physical materials were necessary in helping students construct meaningful mathematics internally for themselves.

The teacher did not require the students simply to compute the answers to the various multiplication problems. Instead, she began by asking them to use graph paper and small masking strips to show, say, the idea of "2 times 3" (three squares seen two times; see fig. 1a). Later, special graph paper heavily marked on every tenth line was used to model two-digit multiplications. The students explained shortcuts for multiplying such pairs as 24 3 26 in terms of transposing the rows of squares below the last heavily lined row over to the right-hand vertical end (see fig. 1b). Eventually, the teacher asked what would happen if the tens digits were not the same and if the units digits did not sum to 10. The students relied heavily on the graph-paper model as a means of arriving at their own decisions and ultimately came up with a general model like the one shown in figure 1c. When it became clear that frames of the same shape repre-

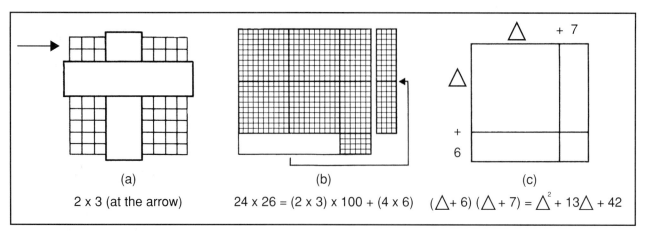

Fig. 1. Linking multiplication in arithmetic and algebra (adapted from Peck and Jencks [1988, pp. 87–88])

sented the same number, the students were able to record such statements as

$$(\Box + 5) \cdot (\Box + 8) = \Box^2 + 13\Box + 40$$

and even

$$(x + 5)(x + 8) = x^2 + 13x + 40.$$

They could handle expressions like

$$(x + 9)(x - 4)$$

or

$$(x + 5)(x - 5)$$

by imagining the graph-paper model and adding on or removing portions of the sides. Note that Peck and Jencks make extensive use of squares and triangles as beginning symbols for general numbers; the reader is urged to refer to their report for details of the teaching approach.

According to Peck and Jencks, one of the bonuses of this approach is that the students were able to "draw together variations of an operation or principle into large conceptual themes and treat them as manifestations of a single concept rather than as separate ideas" (1988, 89). For example, such generalizations as

$$a^2 - b^2 = (a + b)(a - b)$$

and

$$a^2 + 5ab - 14b^2 = (a + 7b)(a - 2b)$$

were viewed as simple variations of the same conceptual theme.

Another way of introducing letters to represent a range of values is the use of tables that represent functional relationships. W. W. Sawyer's Guess My Rule game, described by Davis (1985), is a good means of generating students' interest in this kind of activity. Students are asked the question "Can you make up a rule so that we'll tell you a number, you'll use your rule on that number and tell us the answer, and we'll try to guess what your rule is?" (p. 201). For example, the "rule" might be "Whatever number you tell us, we'll double it and add eight." If $\Box$ is used to represent the number given to the student and $\triangle$ represents their response, the table shown in **figure 2** might be generated, followed by the writing of the rule

$$(\Box \times 2) + 8 = \triangle.$$

It should be noted from all that has been said here that what is meant by a "transition to algebra" is not

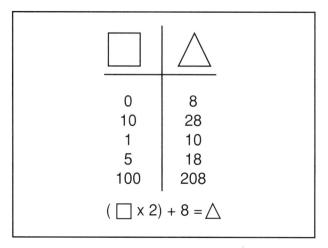

Fig. 2. Using a table to represent a functional relationship (adapted from Davis [1985, p. 202])

a preparation for the traditional ninth-grade algebra course, in which the emphasis is usually on learning to manipulate meaningless symbols by following rules learned by rote. What is meant is an exploration of some key algebraic ideas in which students *(a)* think about the numerical relations of a situation; *(b)* discuss them explicitly in simple, everyday language; and *(c)* eventually learn to represent them with letters or other nonmisleading notation, such as squares and triangles (adapted from Davis [1985]).

Letters as Unknowns

Students encounter unknowns early in elementary school, usually in the context of such open sentences as $\square + 5 = 8$. The ways that sixth and seventh graders think about such equations and attempt to solve them were the focus of a study carried out by Kieran (1988). The students were asked what the letter means in such equations as $5 + a = 12$. One typical answer was "An answer—twelve minus five is seven." Those who responded in this way believed that an equation had to be reversed—that is, undone—for the letter to have meaning. Other students replied that the letter means the whole number that should be added to 5 to equal 12; they also viewed the operations of an equation in the left-to-right sequence in which they were presented. These latter students relied on substituting trial values into equations to solve them, in contrast with the former students, who solved simple equations by transposing numbers and using the inverse operations.

All the students were then shown how equations can be built from arithmetic equalities by "hiding" one of the numbers (see Herscovics and Kieran [1980] for details). Subsequently, they were taught to solve equations by performing the same operation on both sides of the equality symbol. It was found that the students who had begun the study with a tendency to reverse the form of the equation by using undoing operations had a great deal of difficulty in making sense of the solution procedure of doing the same operation on both sides of the equation. This procedure seemed most accessible to those students who had interpreted such equations as $5 + a = 12$ as "the number that should be added to five to equal twelve." This outcome suggests that acquiring meaning for the "same operation on both sides" procedure may require that a good deal of time be spent at a *prior* stage on developing a kind of "forward operations" thinking (i.e., for the foregoing example, "What num-

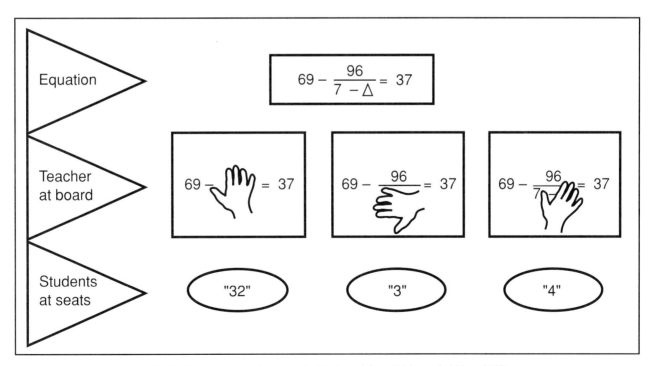

Fig. 3. The cover-up solving method (adapted from Whitman [1982, p. 202])

ber, when added to five, gives twelve?"). The teaching of this kind of thinking was the focus of a study carried out by Whitman (1982).

Whitman began her instructional sequence with equations containing one operation (e.g., $\Box + 17 = 21$), making sure that students learned to interpret equations easily by asking such questions as "What number plus seventeen gives twenty-one?" In other words, the emphasis was on reading the equation as a *question* to be answered with a number. Next, students were given practice with two-operation equations (e.g., $2 \cdot \Box + 5 = 47$, $3 \cdot \triangle - 8 = 31$, $48 - 3 \cdot \S = 6$), and much time was spent verbalizing such questions as, for the first example, "What number plus five is forty-seven?" (42); "Two times what number is forty-two?" (21).

Whitman points out that it is important that students learn to associate the product of a number and a frame as a single number. She emphasizes that the "cover up" procedure—as she called it (illustrated in **fig. 3**—assisted students in overcoming the tendency to separate the number from the frame at the wrong time. This procedure was also of great use in helping students to see the overall structure of equations and to unpack this structure in a systematic way.

Another aspect of Whitman's research was the teaching of the formal solving technique of performing the same operation on both sides of the equation. She found that students who had been taught the cover-up equation-solving procedure followed by the formal technique were more successful at equation solving than the students who had learned only the formal technique.

Concluding Remarks

This brief report has focused on a conceptual theme that traditionally causes difficulty for beginning algebra students—the use of letters. It has attempted to illustrate how students' algebra can arise out of their arithmetic if their arithmetic is completely sensible to them. Peck and Jencks (1988) have pointed out that far too often, students are expected to find meaning in mathematics not through their own reasoning but by reacting to the reasoning of teachers, textbook writers, and so on. Contrast this expectation with a vision of a constructivist classroom where students are encouraged to explore their own thinking and to talk about their approaches with one another and with the teacher. In such a class, the teacher asks questions that are designed to stimulate exploration, investigation, and decision making on the part of the students. By encouraging students to justify their own thinking and to express their justifications by means of nonambiguous notation, teachers foster the view that algebra is a natural way of expressing what is sensible in arithmetic. It is hoped that the activities described in this article will serve as catalysts for thinking about and generating constructivist approaches to the teaching of other prealgebraic concepts.

References

Blais, Donald M. "Constructivism—a Theoretical Revolution for Algebra." *Mathematics Teacher* 81 (November 1988): 624–31.

Davis, Robert B. "ICME-5 Report: Algebraic Thinking in the Early Grades." *Journal of Mathematical Behavior* 4 (October 1985): 195–208.

Herscovics, Nicolas, and Carolyn Kieran. "Constructing Meaning for the Concept of Equation." *Mathematics Teacher* 73 (November 1980): 572–80.

Kieran, Carolyn. "Two Different Approaches among Algebra Learners." In *The Ideas of Algebra, K–12,* 1988 Yearbook of the National Council of Teachers of Mathematics (NCTM), edited by Arthur F. Coxford, pp. 91–96. Reston, Va.: NCTM, 1988.

Kuchemann, Dietmar. "Algebra." In *Children's Understanding of Mathematics: 11–16,* edited by K. M. Hart, pp. 102–19. London: John Murray, 1981.

Lee, Lesley, and David Wheeler. *Algebraic Thinking in High School Students: Their Conceptions of Generalisation and Justification.* Research Report. Montreal: Concordia University, Department of Mathematics, 1987.

National Council of Teachers of Mathematics (NCTM), *Curriculum and Evaluation Standards for School Mathematics.* Reston, Va.: NCTM, 1989.

Peck, Donald M., and Stanley M. Jencks. "Reality, Arithmetic, Algebra." *Journal of Mathematical Behavior* 7 (April 1988): 85–91.

Whitman, Betsey S. "Intuitive Equation-Solving Skills." In *Mathematics for the Middle Grades (5–9),* 1982 Yearbook of the National Council of Teachers of Mathematics (NCTM), edited by Linda Silvey, pp. 199–204. Reston, Va.: NCTM, 1982.

Section 9

Statistics and Statistical Reasoning

What Do Children Understand about Average?

Susan Jo Russell and Jan Mokros

The statistical idea we come across most frequently is the idea of average. Children in fourth grade and beyond learn to apply the algorithm for finding the mean fairly easily, but what do they understand about the mean as a statistical idea?

Many students do not have opportunities to learn about various kinds of averages as statistical concepts. They view an average as a number found by a particular procedure rather than as a number that represents and summarizes a set of data. Students may learn to find a mode, median, or mean—which technically are all averages even though average often refers to the mean—but they do not necessarily know how these statistics relate to the data being represented.

Average as a Statistical Idea

To investigate students' understanding of the idea of average, we use what we call "construction" problems. Instead of asking students to find the average for a given set of numbers, we give students an average and ask them what could be the data set it represents. This kind of problem is similar to situations we often encounter in life; we read about a median or mean in the newspaper or come across it in our work and need to interpret what might be represented by that number. Statistically literate readers can think about such a statement as "the median price of a house is $150 000" or "the average family size is 3.2" in terms of what it tells or does not tell them about the distribution of the data.

Most students are unable to imagine what kind of data an average might represent. One student, who had had plenty of experience calculating the mean, said, "I know how to get an average, but I don't know how to get the numbers to go into an average, from an average."

Asking students to imagine what the data could be for a given average yields interesting insights into students' thinking about the relationship between data and the average of the data. They do not find it easy to use the memorized algorithm to construct data. They have to think about how the average represents the data. You might want to try the following problem, one we gave to students, before reading about how students solve it.

> We took a survey of the prices of nine different brands of potato chips. For the same-sized bag, the typical or usual or average price for all brands was $1.38. What could the prices of the nine different brands be?

We use the language "typical or usual or average" to keep the conversation open to any ways that students have to think about an average. When they show us one way, we ask them for other ways so that we get a view of the range of their thinking. Since these problems are administered in an interview, we are able to interact, ask questions, and probe students' thinking.

Average as mode

In interviews with fourth graders, many students consistently associated the "typical or usual or average" value with the mode. In construction problems, they produced the data set by making all or most of the values the same as the average value. They might have made a few adjustments to the data when pushed, but despite probing for other approaches, these students stuck to a view of the average as the most frequent piece of data. As one fourth grader explained, "Okay, first, not all chips are the same, as you told me, but the lowest chips I ever saw was $1.30 myself, so since the typical price is $1.38, I just put most of them at $1.38, just to make it typical, and highered the prices on a couple of them, just to make it realistic."

The action-research ideas in this article were prepared by Donald Chambers.

Average as median

Another group of students relied more on reasonableness in constructing the data from the average. They drew on what was realistic in their own lives but were also concerned about mathematical reasonableness. These students usually thought about an average as being a middle value and constructed their data sets so that high values were countered by low values. They did not necessarily find a precise middle but rather centered their average value roughly in the middle of their data.

Some students used an even stronger understanding of "middle." These students often developed perfectly symmetrical distributions around their average value. For example, in the potato-chip problem they would put one price at $1.38, then one at $1.37 and one at $1.39, then one at $1.36 and one at $1.40, then perhaps one at $1.30 and one at $1.46, and so forth. If we moved one of the data points on their distribution to a new value, they could easily adjust their data set to keep the same average. However, if we introduced a constraint that did not allow them to make a perfectly symmetrical distribution, they were often stumped. When we asked students to make prices for the potato-chip problem without using the value $1.38, most said that the task could not be done. Others made minor adjustments to their symmetrical distribution, such as changing the $1.38 to $1.37 and arguing that this modification was the best they could do.

To help students learn more about averages, it makes sense to build on their developing understandings. Students often invent the idea of looking at the middle of the data as a way to compare data sets. The median—the value of the middle piece of data—connects clearly to these informal understandings and has increasingly become that statistic of choice in many real data-analysis contexts because unusually high or low values in the data do not greatly affect it. Although developing an understanding of how the median represents the data is not without complexity (see Russell and Crowin [1989, p. 54]), it is a good place for upper elementary students to begin learning about averages.

When finding a median, we can actually point to a value in the data set—the middle value or the midpoint between the two middle values. Finding the middle value is easy if all the data are arranged in order. For example, to find the median of the heights of the students in a classroom, the students could line up in order or height. The height of the middle person, or the midpoint between the heights of the two middle people, is the median value. Half of the data are below the median, and half of the data are above it.

With experience, students can begin envisioning the variety of data sets that might be represented by a median value of, for example, 48 inches. **Figure 1** pictures three such data sets: (a) is fairly symmetrical around 48 but has a small range; (b) is also fairly symmetrical but has a big range; and (c) is rather bimodal, with few data near the median itself.

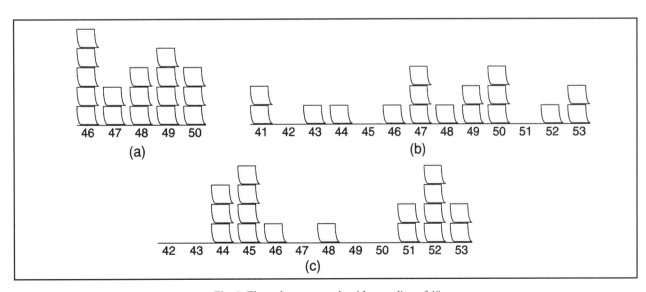

Fig. 1. Three data sets, each with a median of 48

Average as a procedure

Fourth graders who tried to use the algorithm for the mean to solve the potato-chip problem generally got stuck and frustrated. One student divided $1.38 by 9, resulting in a price close to $0.15. We asked her if pricing the bags at $0.15 would result in a typical price of $1.38, and she responded, "Yeah, that's close enough." Another student chose for her prices pairs of numbers that totaled $2.38, such as $1.08 and $1.30. She thought that this method resulted in an average of $1.38. Other students did other meaningless calculations, such as choosing one price then getting the next price by subtracting $0.38.

Through many interviews, we have found that most students' knowledge of the procedure for finding a mean is not at all connected to understanding what the mean represents. However, we have also found that fourth through sixth graders are developing important understandings about the nature of an average, including these:

- A value in the middle of the data can represent a set of data.
- The average is situated in the data in such a way that values higher than the average are countered by values lower than the average.
- An average for a given set of data furnishes a reasonable sense of the values of the data.
- The shape of the data represented by an average should reasonably reflect what we know about the context from which these values are derived, for example, heights or family size.

Learning about the Mean

Students need many experiences with data sets and the median before they can understand how the mean represents the data. Unlike the median, the mean is a mathematical abstraction. We can "see" where the median falls among the data. The mean has no such clear identity within the data themselves; its value may not appear in the data set at all. The average family size may be a number such as 3.6, a number that is not the size of any family. The mean is a mathematical construction that represents certain relationships in the data—a kind of abstraction that may be new to many upper elementary school students.

Over the years mathematics educators have tried to develop models that support students as they build an understanding of the mean. Early in our work with sixth graders, we assumed that students would most effectively learn this relationship by having an opportunity to transform a set of data into "fair shares." If each person has a certain number of pets, the way to find the average is to pool the pets then distribute them equally among the children who pooled their pets. This model reflects exactly what happens when the averaging algorithm is used: the data are pooled then divided evenly, as if each piece of data had the same value. That value is the mean.

We discovered that although this model is effective in teaching children about division, it simply does not help them think about the statistical relationship between the data and the mean. The problem with this model is that in the act of redistributing quantities so that each data point has the same value, the relationship between the actual data and the mean is completely obscured.

For example, consider a problem in which eight children checked the following numbers of books out of the library: 5, 4, 3, 2, 1, 4, 5, and 2. What is the mean number of books checked out by each child? In the fair-shares model all the books are put together then the twenty-four books are redistributed into eight groups (or divide by 8) so that three books are in each group. We then have an image of eight children with three books each. The mean is 3. But what is the relationship of this "3" to the original set of data? In the course of redistributing the data, we have lost the original set of data. It is no wonder, then, that children cannot see the relationship between data and mean using this model. They do not have the opportunity to see them together! Once the total is calculated, the individual data are lost. This scenario is not what happens in real life, where averaging rarely involves making fair shares. In life, and in real statistical situations, the data still exist in their original form. We found that learning the fair-shares model to find the mean did not result in better understanding of the relationship between mean and data. Students were still unable to solve construction problems, such as the potato-chip problem described in the foregoing, in which their job was to create a data distribution with a particular mean value.

Another model emphasizes the mean as the value about which all the data "balance." Rather than have the same number of data points on either side, as with the median, we want the sums of the differences from the mean on either side to be equal. **Table 1** shows two distributions for which the mean value is $1.38. In both distributions, the sum of the differences from the mean is 0. In our research, very few students—and none younger than sixth grade—were developing notions about the mean as a point of balance.

Table 1
Two Distributions for Which the Mean Value Is $1.38

Case 1		Case 2	
Price	Price – $1.38	Price	Price – $1.38
$1.40	+0.02	$1.50	+0.12
$1.39	+0.01	$1.49	+0.11
$1.38	0.00	$1.39	+0.01
$1.37	– 0.01	$1.34	– 0.04
$1.36	– 0.02	$1.18	– 0.20

Balance models are not new (Pollatsek, Lima, and Well 1981) but can be problematic because they depend on an understanding of the relationship between weight and distance on a balance beam. Using one poorly understood set of ideas—the physical relationship of weight and distance—may not help students understand another set of difficult ideas—the numerical relationship between the mean and the data.

The unpacking model

In our work with sixth graders, we use a new model of thinking about the mean that involves neither fair shares nor balances. We use "unpacking" tasks. Like construction problems, unpacking tasks begin with a mean and work backward to the data. For example, if the mean family size is 4, what might the data look like? Students work with a line plot and stick-on notes, starting with all the notes stacked on the value 4. (See **fig. 2**.) Although the data distribution certainly could look like this representation, it is unlikely. Students know from their own experience that some families are the same size as the average but that many are larger and many are smaller. We ask students to make a distribution that is more realistic by asking, "If we knew that one family really had three people instead of four, what could we do to make the average come out to 4?" Students usually suggest moving one data point, represented by a stick-on note, from 4 to 5 so as to balance the 3. The teaching intervention continues by having students move more data points until they are happy that the data look "like real life" and that the average is still 4. We keep the numbers small so that students can easily calculate whether the mean is still 4. (See **fig. 3**.)

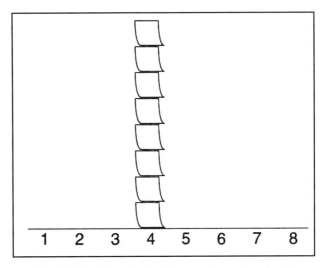

Fig. 2. Line plot with all stick-on notes on the value 4

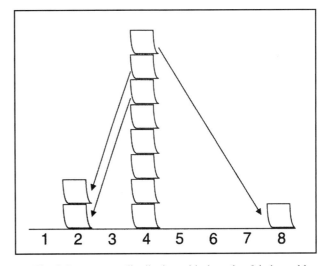

Fig. 3. In-progress distribution with the value 8 balanced by two stick-on notes on the value 2

When solving this problem, many students stick with a symmetrical distribution in which each point above the average is matched by a point the same distance below the average. Once they are comfortable with this approach, we introduce the idea of asymmetrical balancing by asking what would happen if one family has 8 people. In this situation, the move from 4 to 8 cannot be balanced by a comparable move on the left-hand side of the mean because no family can have zero people in it. In our work in classrooms, students have always come up with the idea of moving two different stick-on notes down a total of 4 units so as to balance the upward move of 4.

As students work with this complex idea using a variety of problem contexts, it is important to move slowly and let them share their ideas and strategies as they develop their ideas about balance and distance (see Friel, Mokros, and Russell [1992]).

Conclusions

Understanding averages must be grounded in many experiences with a variety of data sets. As students describe, summarize, and compare data sets, they naturally begin to talk about what is typical of a particular set of data. They develop their own descriptions of the characteristics of a number that summarizes a whole set of data. For example, during a fourth-grade class discussion of plans to compare the heights of first graders in their school with their own heights, one student described his idea saying, "We could find the one number that's sort of in the middle or that the other numbers are crowded around, and then do the same thing for the first grade (Russell and Corwin [1989]).

Once students develop some ideas about how the middle of a data set is important, they can be introduced to the definition and use of the median as a statistical measure. Used in conjunction with the range of the data, the median supplies a great deal of information about the data set. We recommend that work with the median begin in about fourth grade and continue through the upper elementary and middle school years.

Even though the procedure for finding the mean is easy to teach, we recommend delaying the teaching of this procedure until about sixth grade, and only after students have developed their own ideas about balance. Prematurely teaching the averaging algorithm does nothing to help children develop a solid understanding of the relationship between the mean and the data it represents, and may actually interfere with the development of this understanding (see Mokros and Russell [1995]). To understand what the mean represents and how it relates to the data, the concept of distance from the mean is essential.

Some questions that might be useful for teachers to ask before introducing the algorithm are the following: What do my students do if I give them an average and ask them to make up a data set that reflects that average? Do they have some idea about an average as a kind of middle? Do they have a sense that an average is situated in the data such that high values are balanced by low values? Can they begin to think about the significance of the distances of the values from the average?

If students have not developed these ideas, which may not be understood until late in middle school, then teaching the algorithm is pointless. Students need more experiences describing and comparing sets of data. The algorithm is a useful shortcut but does not model the intricate statistical balancing act involved in finding the mean. Far more important is the relationship between the data and the average that summarizes the data. A focus on this relationship should be a top priority in statistical education.

Action Research Ideas

Action-research idea 1

Ask your students to solve the potato-chip problem as it is worded in the article.

1. See which students think of "typical or usual or average" as mode, which as median, and which as mean, by asking them for their numbers and their reason for that choice.

2. Determine which students are able to explain their choice. Do any students change their mind as a result of these explanations?

3. Some students may know an algorithm for finding the mean. Determine whether any of them can explain why their procedure is a good way to find a "typical or usual or average" value.

4. The authors indicate that many fourth graders consistently used the mode as the "typical or usual or average" value. Is this finding true in your class?

Action-research idea 2

Tell your students that the typical or usual or average family size for eight families is 4. Ask them what the data might look like.

1. If the students have trouble getting started, you might show them the distribution in **figure 2** and ask them if that would work. Then ask what other data sets might also work.

2. Ask, "If we knew that one family really had 3 people instead of 4, what could we do to make the average come out to 4?" The authors indicate that students usually suggest moving one data point, a stick-on note on the graph, from 4 to 5 so as to balance the 3. Is this suggestion made in your class? Determine whether the students want to move more data points to make the data look more realistic. After each point is moved, ask, "Is the average still 4?"

3. Introduce asymmetrical balancing by asking what would happen if one family had 8 people. The authors indicate that students always come up with the idea of moving two data points from 4 to 2. Is this suggestion made in your class? Can it be done another way?

4. Do the students have some idea about an average as a kind of middle? Which students?

5. Do the students have a sense that an average is situated in the data such that high values are balanced by low values? Which students?

6. Can the students begin to think about the significance of the distances of the values from the average? Which students?

Continue to pose this kind of "unpacking" task throughout the school year. Note how each student's understanding of the fundamental concepts of average develops over time.

References

Friel, S. N., J. R. Mokros, and S. J. Russell. *Used Numbers: Middles, Means, and In-Betweens.* Palo Alto, Calif.: Dale Seymour Publications, 1992.

Mokros, Jan, and Susan Jo Russell. "Children's Concepts of Average and Representativeness." *Journal for Research in Mathematics Education* 26 (January 1995): 20–39.

Pollatsek, Alexander, Susan D. Lima, and Arnold D. Well. "Concept of Computation: Students' Understanding of the Mean." *Educational Studies in Mathematics* 12 (1981): 191–204.

Russell, Susan J., and Rebecca B. Corwin. *Used Numbers: The Shape of the Data.* Palo Alto, Calif.: Dale Seymour Publications, 1989.

Fourth Graders Invent Ways of Computing Averages

Constance Kamii, Michele Pritchett, and Kristi Nelson

Children easily learn the conventional algorithm of averaging: to find the *average* of a set of numbers, such as 80, 90, 94, and 100, just add the four numbers and divide the total by 4. However, the authors are working on an approach to mathematics based on Piaget's theory, constructivism. For reasons explained later in this article and elsewhere (Kamii 1985, 1989, 1994), our approach is to encourage children to do their own thinking and to refrain from giving rules made by someone else. The purpose of this article is to describe what fourth graders can do when they are encouraged to invent their own ways of getting the average. The article also shows the teacher's active role in constructivist teaching.

Three Methods Invented by Children

We had seen or heard about three ways invented by children to compute averages. All these ways were observed in classrooms of teachers who did not teach any conventional algorithms and, instead, encouraged students to invent their own procedures. None of the children had received any instruction in how to compute averages, but they had all heard of "batting averages" and knew that the letter grades they received were "averages" of many numbers. The children may well have heard that the teacher added many scores and divided the total by the number of scores, but it is impossible to know what the children might have made of this verbal explanation.

The first method. The first example was invented by Nick, a third grader who had gotten 40 points on a test out of a maximum of 100. His teacher announced

---

The authors gratefully acknowledge the assistance of Barbara J. Reys, who read an earlier version of this article and offered many valuable suggestions.

that the students' grade for the week would be the average of this test and the next one. Nick raised his hand and asked if he could still get a C if he made 100 on the next test. "What do *you* think?" the teacher inquired back, and Nick surprised her with a brilliant answer: "I think my average would be 70, and that's a C (see the criteria in **fig. 1**) because half of 40 is 20 and half of 100 is 50. That's a C 'cause 20 plus 50 is 70," he explained.

```
A   90 – 100
B   80 – 89
C   70 – 79
D   60 – 69
F    0 – 59
```

Fig. 1. The criteria for letter grades

Nick's reasoning was difficult for us to understand. We finally figured out that it worked for the following reason: If

$$a = 40$$

and

$$b = 100,$$

then

$$\frac{1}{2}a + \frac{1}{2}b = \frac{a+b}{2},$$

which is the conventional algorithm. We also tried other sets of numbers, such as 80, 90, 94, and 100. The average could be obtained either by adding the results of $80 \div 4$, $90 \div 4$, $94 \div 4$, and $100 \div 4$ or by using the conventional algorithm. Both approaches yield the same outcome: If

$$a = 80,$$

$$b = 90,$$
$$c = 94,$$
and
$$d = 100,$$
then
$$\frac{1}{4}a + \frac{1}{4}b + \frac{1}{4}c + \frac{1}{4}d = \frac{a+b+c+d}{4}.$$

The second method. The second example was furnished by Andy, a fourth grader, when his teacher introduced averaging by asking the class for the average of three bowling scores—150, 125, and 200. Estimating that the average should be a little over 150, Andy worked around this number and wrote as shown in **figure 2**. He explained that he used 25 of the 50 points that he had taken away from the 200 "to give 25 to the 125, to make it 150." After thus making three 150s, he was left with 25 points to distribute among the three scores. He first gave 5 points to each score and was left with 10 more points to distribute. That was enough to give 3 more points to each score, and the average came out to be 150 + 5 + 3 = 158, with a remainder of 1 point.

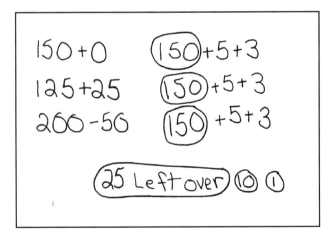

Fig. 2. Andy's way of averaging 150, 125, and 200 and getting the answer of 158 with a remainder of 1

Andy's procedure baffled us, and we could not follow his reasoning even after he explained it three times! But Andy was sure of what he was doing and was willing to repeat his explanation as many times as we requested with many other sets of numbers. Eventually, we understood that Andy was making an estimate, equalizing all the scores, and then distributing the leftover points.

The third method. The third example was invented by another fourth grader in response to a homework assignment given by one of us: Think of a set of numbers you might want to average, and find the average of those numbers. The child explained that the average of 2, 9, 3, and 6 is 5 because the midpoint between 2 and 6 is 4 (see **fig. 3**), and the midpoint between 9 and 3 is 6. Since the midpoint between 4 and 6 is 5, the average of the four numbers is 5, he asserted.

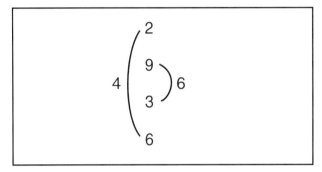

Fig. 3. A fourth grader's procedure for getting the average of 2, 9, 3, and 6

We were impressed by this method of averaging two pairs of numbers first and then averaging the two averages. As another child pointed out, however, this procedure works only with sets of four numbers and is, therefore, of limited utility. (We later realized that this procedure could also work with many other sets of numbers, such as a set of eight numbers.)

On hearing about the preceding child-invented procedures, one of us took her turn to experiment in her class.

The first day

We began by asking the children what "average" meant to them. The following are some of the definitions they gave: "It's not the best and not the worst, but in the middle"; "It's a bunch of numbers put together into one"; "It's the usual amount"; and "If you get an A and an F, your average is a C." We realized how hard it was to define this term without using big words like "typical" and "representative." The children's definitions were neither general nor complete, but the examples they gave were appropriate and reflected their intuitive understanding of average based on life experiences.

The problem we planned for the first day asked for the average of three test scores—85, 95, and 100. No one invented a defensible procedure for getting an exact value, but the good news was that the children were able to make good estimates. The various lines of reasoning that came out during the whole-class discussion are grouped into three categories, as follows:

Qualitative reasoning using letter grades.
All the children thought that for two A's and one B, according to the criteria listed in **figure 1**, the average should be "something like a low A, maybe 91."

Spatial reasoning
Some children drew the line shown in **figure 4** to represent all the numbers from 85 to 100 and said that the average had to be in the middle, around 92 or 93.

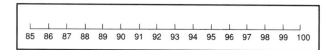

Fig. 4. The line drawn by some children to get the average of 85, 95, 100

Numerical reasoning

Having decided that the average had to be between 85 and 100 and a little above 90, many children tried a variety of numerical methods in an attempt to produce a number close to 90. The following two examples are both incorrect but contain elements of correctness.

Laura wrote:

$$\begin{array}{rl} 10 & \text{(from 85 to 95)} \\ +5 & \text{(from 95 to 100)} \\ \hline 15 & \end{array}$$

Half of 15 is 8. So the average is 85 + 8 = 93.

Chris wrote the following, but Lynn disagreed with him:

$$100 - 95 = 5$$
$$85 + 5 = 90$$

The average had to be higher than 90, Lynn argued, because the score of 95 had to count for one-third of the grade. The idea of one-third indicated her awareness of an important element of averaging.

We thus learned on the first day that most of the children made good estimates and were interested enough to think hard and to struggle. We decided that the next day should begin with Nick's problem, which had only two scores, and that the second problem should involve three scores—80, 80, and 98.

The second day

The average of 40 and 100 proved to be very easy. When this problem was presented, the hands started to go up immediately, without even touching a pencil. Almost all the hands were eventually up, and the children described three similar ways of approaching the problem.

The first way was to work from both ends toward the middle by adding 10 and subtracting 10 many times (40 + 10 = 50, 100 − 10 = 90; 50 + 10 = 60, 90 − 10 = 80; and 60 + 10 = 70, 80 − 10 = 70). The second way was to take the difference between 100 and 40, divide it by 2 (60 ÷ 2 = 30), and add the result to 40 (40 + 30 = 70). The third way was a guess-and-check approach, trying to add a number to 40 and to subtract the same number from 100. Interestingly, no one in the class came up with Nick's method.

The second problem, getting the average of 80, 80, and 98, was much more difficult for the class. The children all computed the difference between 80 and 98 first but then proceeded in different ways. The most common method was this technique:

$$18 \div 2 = 9$$
$$80 + 9 = 89$$

When asked if this method made each of the three tests count equally as a third of the grade, some students divided 9 by 2, thereby getting a fourth of the difference between 80 and 98. Fortunately, these students had enough number sense to then say that they were stuck because their way did not make sense.

One child, Carl, however, did the following and was unbearably puzzled:

$$18 \div 3 = 6$$
$$80 + 6 = 86$$
$$80 + 6 = 86$$
$$98 - 6 = 92$$

We unfortunately had to stop the discussion at this point because it was time to go to physical education. The students were asked to think about Carl's idea because we would come back to this very point the next day. All the way to the gymnasium, Carl kept insisting, "I just *know* it's 86; it just *has* to be 86."

The second day thus resulted in our learning that finding the average of two numbers was very easy. Our task in preparation for the third day was to figure out how to facilitate children's use of what they knew to solve a problem with three scores.

Two things seemed to be in need of change. First, the students were focusing only on the difference between the scores, that is, between 80 and 98, whereas they needed to pay attention to the scores themselves—from 0 to 80 twice and from 0 to 98. Second, all the children except Carl had to think about how to average three scores instead of two. We decided that since some children seemed to be using lines to represent the sequence of numbers, one way to decenter their thinking was with the lines shown in **figures 5** and **6**. **Figures 5a** and **6** represent individual scores, and **figure 5b** shows spatially what the average of 40 and 100 means.

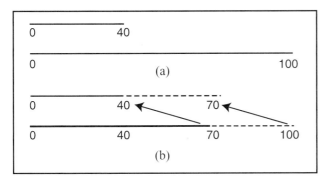

Fig. 5. The lines drawn by the teacher to explain (a) the scores of 40 and 100 and (b) the average of the two scores

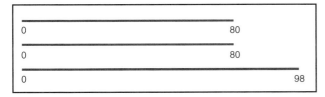

Fig. 6. The lines drawn by the teacher to clarify the problem of getting the average of 80, 80, and 98

As a result of the preceding reflection, we also thought about a better way to define *average:* The average is a number we imagine that gives an overall picture of all the scores, a number you get by balancing out the higher scores against the lower ones.

The third day

The review of the previous day's work with **figures 5** and **6** greatly helped the children's thinking. Almost all the students thought about getting the average of 80, 80, and 98 by dividing 18 by 3, obtaining 6, and adding 6 to 80. Some children even remarked, "I know what Carl did wrong yesterday." Carl was unfortunately out of the room for a chess tournament.

The new problem for the day was to get the average of 100, 91, and 94. After drawing the three lines illustrated in **figure 7a**, the teacher asked what the person *at least made* on all three tests—the minimum that all three tests had in common. On hearing the answer of 91, the teacher drew on the chalkboard the lines illustrated in figure 7b. Most of the children used this hint, added 3 and 9 (because $94 - 91 = 3$ and $100 - 91 = 9$), divided the result by 3 ($12 \div 3 = 4$), and added 4 to 91, which equaled 95. However, two children invented other procedures that made us realize that **figure 7b** was superfluous and inappropriate.

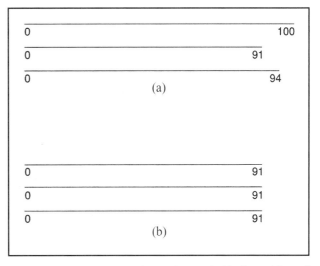

Fig. 7. The lines drawn by the teacher to clarify the problem of getting the average of 100, 91, and 94

Reasoning that the average had to be higher than 94, Lynn decided to find out if 95 was a good estimate. She took 5 points from 100 (since $100 - 5 = 95$) and found out with pleasure that these points could be given to the two other scores ($91 + 4$ and $94 + 1$) to equalize the three scores.

The other unexpected procedure was to try to equalize 91 and 94 first. Chris changed 94 to 93 and 91 to 92 by taking 1 away from 94 and giving it to the 91. He then took 7 from 100 (because $100 - 7 = 93$) and gave 1 of the 7 to the 92 to have 93, 93, and 93. He was left with a remainder of 6, distributed the 6 points among the three scores, and ended up with an average of $93 + 2$.

Once the children knew that averaging meant to equalize the scores by distributing the differences among them, other problems became easy for most of them. When three widely different scores, such as 100, 50, and 62, were later presented, some children solved them by the methods that follow. The first method was based on an estimated average of 70, but the second one began with 62, which is the score in the middle (the *median*).

Some readers may be disturbed by such "equations" as "$100 - 28 = 72 - 2 = 70$." We do not correct this kind of writing because children in these situations are representing their *process* of thinking and using writing in the service of thinking. Correcting children in such a situation would distract them and interfere with their thinking. The convention of how to write equations will be easy to teach at a later date.

One procedure

$$50 + 20 = 70$$
$$\underline{62 + 8 = 70}$$
$$28 \quad \text{(the sum of 20 and 8)}$$
$$100 - 28 = 72 - 2 = 70$$

The average is 70 R. 2.

Another procedure

$$50 + 12 = 62 + 8 = 70$$
$$62 + 0 = 62 + 8 = 70$$
$$\underline{100 - 38 = 62 + 8 = 70}$$
$$-26 \text{ (the sum of 12, 0, and } -38\text{:}$$
$$\text{the difference that still needs}$$
$$\text{to be distributed)}$$

The average is 70 R. 2.

Like Andy, these children both knew that whatever value they added to a low score had to come from a high score. Once the children "demystified" the idea of an average, some even asked if they could go over all their test scores in the teacher's grade book to see if they were averaged correctly.

Conclusion

Children easily learn the conventional algorithm for getting the average, or the arithmetic mean. However, researchers (Mokros and Russell 1992; Strauss and Bichler 1988; Pollatsek, Lima, and Well 1981) have documented children's and university students' inability to reason logically after being taught to use the algorithm. Mokros and Russell (1992, pp. 29–30) even made the following statement within the perspective of how to make sense of large sets of data:

> Our work with adults also leads us to suspect that the arithmetic mean is a mathematical object of unappreciated complexity (belied by the "simple" algorithm for finding it) and that it should only be introduced relatively late in the middle grades, well after students have developed a strong foundation of the idea of representativeness and have a great deal of experience using the median, a measure that is not only extremely useful, but more easily connects to children's informal ideas about average.

We agree that the median is an easier concept for fourth graders to learn than the mean. However, we think that encouraging children to invent their own ways of getting the mean is a good way for them to clarify the idea of representativeness. We also think that this invention of ways to get the mean paves the way for children to reinvent the conventional algorithm.

We found that our fourth graders had three strengths: *(a)* They could easily figure out the average of two scores, *(b)* they could make good estimates, and *(c)* they all knew that they had to do *something* arithmetically with the differences among the scores. By analyzing children's errors and the fact that many of them came up with the idea of using lines to represent scores, we thought that letting them visualize all the scores in their entirety—instead of only the differences among them—might facilitate their thinking.

Children use what they know to construct, or invent, new knowledge. By analyzing what they know and how they think, the teacher can facilitate their construction of higher levels of reasoning. We think that children can reinvent the conventional algorithm for getting averages, but this hypothesis needs to be tested in the future.

References

Kamii, Constance. *Young Children Reinvent Arithmetic.* New York: Teachers College Press, 1985.

———. *Young Children Continue to Reinvent Arithmetic, 2nd Grade.* New York: Teachers College Press, 1989.

———. *Young Children Continue to Reinvent Arithmetic, 3rd Grade.* New York: Teachers College Press, 1994.

Mokros, Jan, and Susan Jo Russell. *Children's Concepts of Average and Representativeness.* Working Paper 4–92. Cambridge, Mass.: TERC, 1992.

Pollatsek, A., S. Lima, and A. D. Well. "Concept or Computation: Students' Understanding of the Mean." *Educational Studies in Mathematics* 12 (May 1981):191–204.

Strauss, Sidney, and Efriam Bichler. "The Development of Children's Concepts of the Arithmetical Average." *Journal for Research in Mathematics Education* 19 (January 1988):64–80.

Teaching Statistics: Mean, Median, and Mode

Judith S. Zawojewski

Research . . .

Statistics is increasingly being recognized as an important and useful branch of mathematics that needs increased emphasis in the elementary school curriculum (e.g., National Council of Teachers of Mathematics [1980]; Conference Board of Mathematical Sciences [1982]; National Commission on Excellence in Education [1983]). Although data from the Second International Mathematics Study (SIMS) indicate that typical eighth-grade teachers allot little time to teaching statistics (i.e., four days per year), statistics has been found to be one of the few areas in which the United States performs as well as, or better than, students in other countries (Travers 1986). Considering the small amount of class time spent on statistics and the relatively promising achievement, statistics has good potential for inclusion in the middle school curriculum. However, a focused look at a traditional junior high school statistics topic—mean, median, and mode—indicates that students' performance may reflect a mastery of memorized procedures rather than a thorough understanding of the underlying concepts.

Most junior high school students can compute an arithmetic average when computations involve whole numbers (Carpenter et al. 1981; Travers 1986). However, the SIMS data indicate that even this skill level of performance has dropped from 84 percent correct in 1964 to 76 percent correct in 1982. Performance is sharply lower in items that require computation with decimals (Travers 1986), items that use the term *mean* (Carpenter et al. 1981), and items that move from simple computational procedures to ones that require a deeper understanding of the mean. For example, only 16 percent of thirteen-year-olds can find a missing piece of data when given the average and all other data entries, and only 7 percent perform successfully on items involving the concept of a weighted mean (Carpenter et al. 1981). Misconceptions and difficulties in understanding the mean apparently are not overcome by the time students reach college age (Pollatsek, Lima, and Well 1981; Hardiman, Well, and Pollatsek 1984; Maraveck 1983). These researchers concur with the Carpenter et al. (1981) conclusion that students apparently only understand an arithmetic average as a totally abstract formula, devoid of any meaning. Pollatsek et al. (1981, p. 191) summed up that "for many students dealing with the mean is a computational rather than a conceptual act."

Less than half the thirteen-year-olds in the United States are able to find the median of a set of data listed in order from least to greatest (Carpenter et al. 1981). Evidence shows that performance may be even worse when the list of data is reported in nonnumerical order or in a frequency table (Barr 1980; Zawojewski 1986). In these instances many students choose the value that is in the center of a table or list (whether ranked or unranked), which seems to indicate the use of an impoverished definition of the median as the "middle of *something*." Unfortunately we have little information on students' ability to interpret the relationship between the mean and the median. For example, do students understand why average housing costs are reported in medians? (Extremes in housing costs affect the mean value, so the median is a better representation of the average.) Given students' poor conceptual understandings of the median and mean, it is doubtful whether they understand the relative effect of extreme data on each.

Over a third of the thirteen-year-olds in the United States do not know what a mode is; only a fourth can correctly identify the mode of a list of data; and only 12 percent are able to identify the appropriate use of the mode in a realistic application (Carpenter et al. 1981). When students have some familiarity with the mode, oversimplified definitions may lead to predictable errors. For example, Zawojewski (1986) found that some students identified the greatest number in a list or table as the mode, using "the most" as their definition, rather than figure out what was the most frequently occurring piece of data.

Few instructional interventions for teaching mean, median, and mode have been systematically assessed. Hardiman et al. (1984) found that a balance-beam analogy is useful in helping some college students perform successfully on weighted-mean problems. Performance on these measures of central tendency by a small group of seventh and eighth graders improved as a result of seven days of instruction emphasizing concepts and problem solving with mean, median, and mode (Zawojewski 1986). Although we clearly need more studies on the teaching and learning of the topic, the information available does offer some guidance in developing instructional practices for classroom use.

... Into Practice

Link concrete experiences to procedures.

When fourth-grade students are first exposed to the arithmetic average, teachers can help them discover the "add-'em-up and then divide" rule by providing concrete experiences. Show students a bundle of eight pencils of varying lengths and ask them to cut a straw to the length that they would estimate as the average length of all the pencils. After the estimates are in, lay the pencils end-to-end and cut a strip of adding machine tape the same length. This action illustrates the "add-'em-up" step. Then fold the strip into eight equal parts to illustrate the division step. This concrete experience can be related to numerical manipulation by measuring and finding the sum of the total length of the pencils to the nearest centimeter and dividing by the number of pencils.

Introduce students to median and mode by physically arranging the children in numerical order depending on the number of pets they own. Identify those who represent the median and mode number of pets owned. Repeat the exercise for shoe sizes. Then without changing their order, ask students to identify the median number of brothers and sisters. Students should realize that this cannot be done until the data are organized in order from least to greatest. These exercises are concrete illustrations that the median is the middle value in a set of numerically ordered data and that the mode is the most frequently occurring value. Then ask students to identify the median hair color. They should note that since no numerical order is possible, the concept of median does not apply, although a mode can be identified.

Give students the opportunity to develop language.

Some students tend to reduce their understandings of mean, median, and mode to oversimplified definitions that result in shallow concepts. Ask students to talk about and write their own definitions for the summary statistics that reveal the relationship between the measure of central tendency and the data set. For example, rather than allow students to say, "the mode is the most," require students to explain that "the mode is the shoe size that occurs most often" or "the mode is the most frequently occurring score."

Give students a variety of problem types.

Students are capable of dealing with problems that go beyond computing averages for lists of numbers. Here are some examples of more challenging types of problems:

- If 4 pounds of ground beef is bought at $2 per pound and 5 pounds of round steak at $3 per pound, what is the average cost per pound for this meat? (Carpenter et al. 1981, pp. 111–12)
- Beth averaged 150 points per game in the bowling tournament. On her first two games she scored 170 and 110. What was her score on the third game?
- Seven girls are at a slumber party. Their shoe sizes range from 5 1/2 to 9 (with half-sizes included). If the median shoe size for the girls is 7, what are some possible combinations of shoe sizes for the girls?
- The mean of five brothers' ages is 4, and the mode is 3. What are some possible ages for the five brothers?

Use problems that represent the data in various forms.

Students have difficulty identifying the mean, median, and mode when the data are in a nonnumerically ordered list or a table. Students may need experiences in both organizing the data into a table or graph and "unpacking" the data from a table or graph back into a list of individual entries.

Have students collect and summarize their own data.

Collecting and organizing one's own data is very motivating. Making graphs and tables of their own data can help students flexibly translate from one form of representation to another. Students searching for solutions to their own real-world questions are involved in applied problem solving. When they decide which summary statistic to use to best represent their data, students engage in a decision-making process that involves integrating knowledge of the context, value judgments, and their ability to interpret the various measures of central tendency.

Use a calculator for weighted-mean problems.

A common error in weighted-mean problems is simply to pull out some numbers and compute a simple mean. However, students who use a calculator can focus less on computations and more on the meaning of the data representation.

References

Barr, G. V. "Some Student Ideas on the Median and Mode." *Teaching Statistics* 2 (May 1980): 38–41.

Carpenter, Thomas P., Mary Kay Corbitt, Henry S. Kepner, Mary Montgomery Lindquist, and Robert E. Reys. *Results from the Second Mathematics Assessment of the National Assessment of Educational Progress.* Reston, Va.: National Council of Teachers of Mathematics, 1981.

Conference Board of Mathematical Sciences. "The Mathematical Sciences Curriculum K–12: What Is Still Fundamental and What Is Not." Report to the NSB Commission on Precollege Education in Mathematics, Science, and Technology, 1982.

Hardiman, Pamela T., Arnold D. Well, and Alexander Pollatsek. "Usefulness of a Balance Model in Understanding the Mean." *Journal of Educational Psychology* 76 (October 1984): 792–801.

Maraveck, Zemira R. "A Deep Structure Model of Students' Statistical Misconceptions." *Educational Studies in Mathematics* 14 (November 1983): 415–29.

National Commission on Excellence in Education. *A Nation at Risk: The Imperative for Educational Reform.* (GPO # 065-000-001772-2). Washington, D.C.: U.S. Government Printing Office, 1983.

National Council of Teachers of Mathematics (NCTM). *An Agenda for Action: Recommendations for School Mathematics of the 1980s.* Reston, Va.: NCTM, 1980.

Pollatsek, A., S. Lima, and A. D. Well. "Concept of Computation: Students' Understanding of the Mean." *Educational Studies in Mathematics* 12 (May 1981): 191–204.

Travers, Kenneth J., ed. *Second International Mathematics Study Detailed Report for the United States.* Champaign, Ill.: Stipes Publishing Co., 1986.

Zawojewski, Judith S. "The Teaching and Learning Processes of Junior High School Students under Alternative Modes of Instruction in the Measures of Central Tendency." Ph.D. diss., Northwestern University, 1986.

Learning Probability Concepts in Elementary School Mathematics

Albert P. Shulte

Many curriculum groups and individuals have recommended the inclusion of topics in probability in the elementary school mathematics program. In "A Case for Probability," an important position statement in this journal, Jones (1970) recommended probability because *(a)* the subject deals with ideas and patterns that grow over time; and *(b)* evidence suggests that appropriate units can be taught in the elementary school.

Piaget's pioneering studies of the ideas of chance (Piaget and Inhelder 1975) are the foundation of research in this area. On the basis of his explorations, he divided the development of probabilistic thinking in children into three age-related stages. In stage 1 (up to age 7 or 8), the student does not understand random phenomena but continually looks for a hidden order. The ability to make predictions is unstable, depending on preference, on expectation of an outcome to "catch up," or on observation of the largest frequency to date. "Miracles" (extremely unlikely events) cause no surprise. In stage 2 (up to age 11 or 12), a global understanding of randomness is achieved, but not of the effects of large numbers. "Miracles" cause a search for an underlying cause. In stage 3, formal thought, the law of large numbers is understood and probabilities are successfully assigned.

Since Piaget's work, a number of studies have dealt with the study of probability in the elementary school. These studies tend to fall into one of two types: (1) What do elementary school students know about probability prior to formal instruction? (2) What topics in probability can elementary school students learn?

1. What do elementary school students know about probability prior to instruction?
Work with younger students using simple probability devices, by Perner (1979) and by Falk, Falk, and Levin (1980), found that the ability to predict the most probable outcomes improves with age, beginning at about age 6. In grades 2 and 4, McLeod (1971) investigated a number of topics, including reasoning from distributions, independence, and the law of large numbers. He found that most students were able to apply the concepts prior to study.

Schroeder (1983) found that students in grades 4–6 had notions of probability, chance, and relative frequency. Doherty (1965) looked at several formal probability topics in the same grade levels and found that older students and those with a higher level of arithmetic achievement scored higher. In contrast, Mullenex (1968), who worked with grades 3–6, concluded that age, general ability, basic vocabulary skill, and work-study skill are not relevant to understanding randomness, assigning probabilities to events, or recognizing the long-range stability of random phenomena.

Students' misconceptions about various aspects of probability are discussed in Falk, Falk, and Levin (1980); Green (1983a, 1983b); Dunlap (1980); Armstrong (1972); and, of course, in Piaget and Inhelder (1975).

Common misconceptions mentioned include counting the number of desirable outcomes rather than considering the ratio of desirable to undesirable outcomes; expecting all outcomes in an experiment to be equally likely; holding a bias against certain numbers (e.g., believing that it is hardest to throw a 6 on a die); selecting on the basis of preference (e.g., color) rather than on probability (this is particularly true of young children); and failing to make inferences from data.

2. What can elementary school students learn about probability?
Cantor, Dunlap, and Rettie (1982); Jones (1974); Dunlap (1980); and McLeod (1971) used various types of probability devices with children and found that they learned. Jones found better learning if the devices show the unit (e.g., on spinners). Cantor, Dunlap, and Rettie (1982) recommend that instructional tasks and criterion tasks be kept the same (i.e., don't expect much transfer).

Hirst (1977), Schroeder (1983), and Armstrong (1972) support the use of experiments to teach probability. Armstrong cautions that units of two-and-one-half to three weeks in length are tiring. Doherty (1965) found that students learned more about probability *while taking* a test on the topic. Gipson (1971) taught some students in grades 3–5 and determined that they could learn the topics but had difficulty counting and listing all the possible outcomes of an experiment.

Shepler (1969, 1970) used mastery-learning techniques with sixth graders and found that they mastered ten of fourteen objectives related to probability, counting procedures, cumulative frequency graphs (showing experimental probabilities stabilizing), and the law of averages. The retention study reported by Romberg and Shepler (1973) demonstrated continued high-level performance four weeks later.

Thus the research indicates that students have some understanding of probability and related topics, that this understanding increases with age and instruction, and that probability can successfully be taught in the elementary school in carefully selected experiments. The curriculum groups and individuals recommending that probability be a topic for the elementary school are thus on target.

Bibliography

Armstrong, Prince Winston. "The Ability of Fifth and Sixth Graders to Learn Selected Topics in Probability." Ph. D. diss., University of Oklahoma, 1972.

Brainerd, Charles J. "Working Memory and the Developmental Analysis of Probability Judgment." *Psychological Review* 88 (November 1981): 463–502.

Cantor, Gordon N., Linda L. Dunlap, and Candice S. Rettie. "Effects of Reception and Discovery Instruction on Kindergartners' Performance on Probability Tasks." *American Educational Research Journal* 19 (Fall 1982): 453–63.

Doherty, Joan. "Level of Four Concepts of Probability Possessed by Children of Fourth, Fifth, and Sixth Grade before Formal Instruction." Ph.D. diss., University of Missouri, 1965.

Dunlap, Linda Louise. "First-Grade Children's Understanding of Probability." Ph.D. diss., University of Iowa, 1980.

Falk, Ruma, Raphael Falk, and Iris Levin. "A Potential for Learning Probability in Young Children." *Educational Studies in Mathematics* 11 (1980): 181–204.

Gipson, Joella Hardeman. "Teaching Probability in the Elementary School: An Exploratory Study." Ph.D. diss., University of Illinois, 1971.

Green, David. "From Thumbtacks to Inference." *School Science and Mathematics* 83 (November 1983a): 541–51.

———. "Shaking a Six." *Mathematics in School* 12 (November 1983b): 29–32.

Hirst, Helen. "Probability." *Mathematics Teaching* 80 (September 1977): 6–8.

Jones, Graham. "A Case for Probability." *Arithmetic Teacher* 26 (February 1979): 37, 57.

———. "The Performance of First, Second, and Third Grade Children on Five Concepts of Probability and the Effects of Grade, I. Q. and Embodiments on Their Performances." Ph.D. diss., Indiana University, 1974.

King, Cynthia Clarke. "Development and Evaluation of an Activity-based Probability Unit for Prospective Elementary Teachers Incorporating the Teaching of Mini-Lessons to Elementary School Children." Ph.D. diss., Florida State University, 1975.

McLeod, Gordon Keith. "An Experiment in the Teaching of Selected Concepts of Probability to Elementary School Children." Ph.D. diss., Stanford University, 1971.

Mullenex, James Lee. "A Study of the Understanding of Probability Concepts by Selected Elementary School Children." Ph.D. diss., University of Virginia, 1968.

Perner, Josef. "Discrepant Results in Experimental Studies of Young Children's Understanding of Probability." *Child Development* 50 (December 1979): 1121–27.

Piaget, Jean, and Bärbel Inhelder. *The Origin of the Idea of Chance in Children.* New York: W. W. Norton & Co., 1975.

Romberg, Thomas A., and Jack Shepler. "Retention of Probability Concepts: A Pilot Study into the Effects of Mastery Learning with Sixth-Grade Students." *Journal for Research in Mathematics Education* 4 (January 1973): 26–32.

Schroeder, Thomas Leonard. "An Assessment of Elementary School Students' Development and Application of Probability Concepts While Playing and Discussing Two Strategy Games on a Microcomputer." Ph. D. diss., Indiana University, 1983.

Shepler, Jack Lee. "A Study of Parts of the Development of a Unit in Probability and Statistics for the Elementary School." Ph. D. diss., University of Wisconsin, 1969.

———. "Parts of a System Approach to the Development of a Unit in Probability and Statistics for the Elementary School." *Journal for Research in Mathematics Education* 1 (November 1970): 197–205.

PART 3
Tools for Representing Mathematical Ideas

Section 10

Concrete Materials as Tools

Concrete Materials and Teaching for Mathematical Understanding

Patrick W. Thompson

> Learning without thought is labor lost.
> —*Confucius*

> An experience is not a true experience until it is reflective.
> —*John Dewey*

Today we find common agreement that effective mathematics instruction in the elementary grades incorporates liberal use of concrete materials. Articles in the *Arithmetic Teacher* no longer exhort us to use concrete materials, nor does the *Professional Standards for Teaching Mathematics* (NCTM 1991) include a standard on the use of concrete materials. The use of concrete materials seems to be assumed unquestioningly.

My aim in this article is to reflect on the role of concrete materials in teaching for mathematical understanding—not to argue against their use but instead to argue for using them more judiciously and reflectively. Our primary question should always be "What, in principle, do I want my students to *understand?*" Too often it is "What shall I have my students learn to *do?*" If we can answer only the second question, then we have not given sufficient thought to what we hope to achieve by a particular segment of instruction or use of concrete materials.

Preparation of this paper was supported by National Science Foundation (NSF) grants no. MDR 89-50311 and 90-96275 and by a grant of equipment from Apple Computer, Inc., Office of External Research. Any conclusions or recommendations stated here are those of the author and do not necessarily reflect official positions of NSF or Apple Computer.

Research on the Use of Concrete Materials

The use of concrete materials has always been intuitively appealing. The editors of a turn-of-the-century methods textbook stated, "Examples in the concrete are better for the student at this stage of his development, as he can more readily comprehend these" (Beecher and Faxon [1918, p. 47], as quoted in McKillip et al. [1978]). Their appearance accelerated in the 1960s, at least in the United States, with the publication of theoretical justifications for their use by Zoltan Dienes (1960) and by Jerome Bruner (1961).

A number of studies on the effectiveness of using concrete materials have been conducted since Dienes's and Bruner's publications, and the results are mixed. Fennema (1972) argued for their use with beginning learners while maintaining that older learners would not necessarily benefit from them. However, Suydam and Higgins (1977) reported a pattern of beneficial results for all learners. Middle and upper primary students observed by Labinowicz (1985) experienced considerable difficulties making sense of base-ten blocks, although Fuson and Briars (1990) reported astounding success in the use of the same materials in teaching addition and subtraction algorithms. Thompson (1992) and Resnick and Omanson (1987) reported that using base-ten blocks had little effect on upper-primary students' understanding or use of their already memorized whole-number addition and subtraction algorithms, whereas Wearne and Hiebert (1988) reported consistent success in the use of concrete materials to aid students' understanding of decimal fractions and decimal numeration (see also Hiebert, Wearne, and Taber [1991]).

These apparent contradictions are probably due to aspects of instruction and students' engagement to which studies did not attend. Evidently, just using concrete materials is not enough to guarantee success. We must look at the total instructional environment to understand effective use of concrete materials—especially teachers' images of what they intend to teach and students' images of the activities in which they are asked to engage.

Seeing Mathematical Ideas in Concrete Materials

It is often thought, for example, that an actual wooden base-ten cube is more concrete to students than is a picture of a wooden base-ten cube. When one is considering the physical characteristics of objects, this statement certainly seems true. But to students who are still constructing concepts of numeration, the "thousandness" of a wooden base-ten cube often is no more concrete than the "thousandness" of a pictured cube (Labinowicz 1985). To understand the cube—either actual or pictured—as representing a numeral value of 1000, students need to create an image of a cube that entails its relations to its potential parts (e.g., that it can be made of 10 blocks each having a value of 100, 100 blocks each having a value of 10, or 1000 blocks each having a value of 1). If their image of a cube is simply as a big block named *thousand*, then no substantive difference accrues to students between a picture of a cube or an actual cube—the issue of concreteness would be immaterial to their understanding of base-ten numeration. This comment is not to say that, to students, a substantive difference never occurs between pictures and actual objects. Rather, it says only that concrete materials do not automatically carry mathematical meaning for students. A substantive difference can exist between how students experience actual materials and how they experience depictions of materials, but the difference resides in how they are used. This point will be reexamined later in this article.

Seeing mathematical ideas in concrete materials can be challenging. The material may be concrete, but the idea that students are intended to see is not in the material. The idea is in the way the teacher understands the material and understands his or her actions with it. Perhaps two examples will illustrate this point.

A common approach to teaching fractions is to have students consider a collection of objects, some of which are distinct from the rest, as depicted in **figure 1**. If a student had a collection like that depicted in **figure 1**, it would certainly be concrete. But what might the collection illustrate to students? Three circles out of five? If so, they see a part and a whole but not a fraction. Three-fifths of one? Perhaps. But depending on how they think of the circles and collections, they could also see three-fifths of five, five-thirds of one, or five-thirds of three (see **fig. 2**).

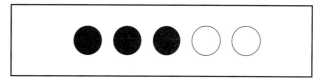

Fig. 1. What does this collection illustrate?

They could also see **figure 1** as illustrating that $1 \div 3/5 = 1\ 2/3$—that within 1 whole is one three-fifths and two-thirds of another three-fifths, or that $5 \div 3 = 1\ 2/3$—that within 5 is one three and two-thirds of another three. Finally, they could see **figure 1** as illustrating $5/3 \times 3/5 = 1$—that five-thirds of (three-fifths of 1) is 1. It is an error to think that a particular material or illustration, by itself, presents an idea unequivocally. Mathematics, like beauty, is in the eye of the beholder—and the eye sees what the mind conceives.

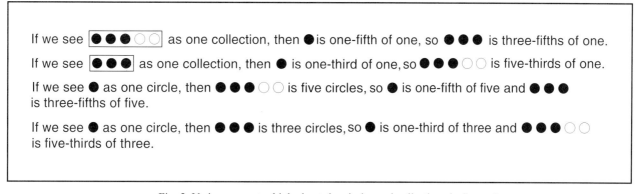

Fig. 2. Various ways to think about the circles and collections in **figure 1**

Teachers sometimes understand the discussion of **figure 1** and **figure 2** as saying that we need to take care that students form *correct* interpretations of materials—namely, the one we intend they have. In actuality the opposite is meant. It should be our instructional goal that students can make, in principle, *all* interpretations of **figure 1**.

A teacher needs to be aware of multiple interpretations of materials so as to hear hints of those that students actually make. Without this awareness it is easy to presume that students see what we intend they see, and communication between teacher and student can break down when students see something other than what we presume.

Also, it is important that students can create multiple interpretations of materials. They are empowered when they recognize the multiplicity of viewpoints from which valid interpretations can be made, for they are then alert to choose among them for the most appropriate actions relative to a current situation. However, it is a teacher's responsibility to cultivate this view. It probably will not happen if the teacher is unaware of multiple interpretations or thinks that the ideas are "there" in the materials.

Figure 1 is customarily offered by textbooks and by teachers to illustrate 3/5. Period. In fact we rarely find textbooks or teachers discussing the difference between thinking of 3/5 as "three out of five" and thinking of it as "three one-fifths." How a student understands **figure 1** in relation to the fraction 3/5 can have tremendous consequences. When students think of fractions as "so many out of so many," they are justified in being puzzled about fractions like 6/5. How do you take six things out of five?

The second example continues the discussion of fractions. It illustrates that how we think of the materials in a situation can have implications for how we may think about our actions with them.

Suppose, in **figure 3**, that the top collection is an example of 3/5 and the bottom collection is an example of 3/4. Next combine the two collections. Does the combined collection produce an example of 3/5 + 3/4? Yes and no. If we were thinking of 3/5 and 3/4 as ratios (so many out of every so many), then 3/5 + 3/4 = 6/9 (three out of every five combined with three out of every four gives six out of every nine, as when computing batting averages). If we understood 3/5 and 3/4 in **figure 3** as fractions, then it doesn't make sense to talk about combining them. It would be like asking, "If we combine three-fifths of a large pizza and three-fourths of a small pizza, then how much of a pizza do we have?" How much of *which kind* of pizza? It only makes sense to combine amounts measured as fractions when both are measured in a common unit. Both answers (6/9 and "the question doesn't make sense") are correct—each in regard to a particular way of understanding the concrete material at the outset.

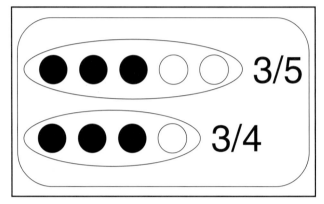

Fig. 3. Three-fifths and three-fourths combined into one collection

Using Concrete Materials in Teaching

It is not easy to use concrete materials well, and it is easy to misuse them. Several studies suggest that concrete materials are likely to be misused when a teacher has in mind that students will learn to perform some prescribed activity with them (Boyd 1992; Resnick and Omanson 1987; Thompson and Thompson 1994). This situation happens most often when teachers use concrete materials to "model" a symbolic procedure. For example, many teachers use base-ten blocks to teach addition and subtraction of whole numbers. Students often want to begin working with the largest blocks, such as adding or subtracting the thousands in two numbers. Teachers often say, "No, start with the smallest blocks, the ones. You have to go from right to left." Would it be incorrect to start with the largest blocks? No, it would be unconventional but not incorrect.

Since our primary question should be "What do I want my students to *understand?*" instead of "What do I want my students to *do?*" it is problematic when teachers have a prescribed activity in mind—a standard algorithm—and unthinkingly reject creative problem solving when it doesn't conform to convention or prescription. In situations in which students' unconventional, but legitimate, actions are rejected by a teacher, students learn once more that "to understand" means to memorize a prescribed activity.

Concrete materials are used appropriately for two purposes. First they enable the teacher and the students to have grounded conversations about something "concrete." The nature of the talk should be (*a*) how to think about the materials and (*b*) the meanings of various actions with them. Such conversations are part of what Wearne and Hiebert (1988) call the "connecting" phase of mathematical learning—constructing strong connections among ways of thinking about concrete situations and conventional mathematical language and notation. Asking the questions "How *do* we think about this? How *can* we think about this? How *shall* we think about this?" is also the heart of what Thompson et al. (1994) call "conceptually oriented" instruction.

Second, concrete materials furnish something on which students can act. The pedagogical goal is that they reflect on their actions in relation to the ideas the teacher has worked to establish and in relation to the constraints of the task as they have conceived it. The discussion of **figure 3** highlighted occasions where the teacher and the students might reflect on the implications of various understandings of fractional amounts for what it means to combine two collections.

Concrete materials can be an effective aid to students' thinking and to successful teaching. But effectiveness is contingent on what one is trying to achieve. To draw maximum benefit from students' use of concrete materials, the teacher must continually situate her or his actions with the question "What do I want my students to *understand?*"

References

Beecher, W. J., and G. B. Faxon, eds. *Practical Methods and Devices for Teachers.* Dansville, N.Y.: F. W. Owens Publishing Co., 1918.

Boyd, Barbara A. "The Relationship between Mathematics Subject Matter Knowledge and Instruction: A Case Study." Master's thesis, San Diego State University, 1992.

Bruner, Jerome. *The Process of Education.* New York: Vintage Books. 1961.

Dienes, Zoltan. *Building Up Mathematics.* London: Hutchinson Educational, 1960.

Fennema, Elizabeth H. "Models and Mathematics." *Arithmetic Teacher* 18 (December 1972): 635–40.

Fuson, Karen C., and Diane J. Briars. "Using a Base-Ten Blocks Learning/Teaching Approach for First- and Second-Grade Place-Value and Multidigit Addition and Subtraction." *Journal for Research in Mathematics Education* 21 (May 1990): 180–206.

Hiebert, James, Diana Wearne, and Susan Taber. "Fourth Graders' Gradual Construction of Decimal Fractions during Instruction Using Different Physical Representations." *Elementary School Journal* 91 (March 1991): 321–41.

Labinowicz, Edward. *Learning from Children: New Beginnings for Teaching Numerical Thinking.* Menlo Park, Calif.: Addison-Wesley Publishing Co., 1985.

McKillip, William D., Thomas J. Cooney, Edward J. Davis, and James W. Wilson. *Mathematics Instruction in the Elementary Grades.* Morristown, N.J.: Silver Burdett Co., 1978.

National Council of Teachers of Mathematics (NCTM). *Professional Standards for Teaching Mathematics.* Reston, Va.: NCTM, 1991.

Resnick. Lauren, and Susan Omanson. "Learning to Understand Arithmetic." In *Advances in Instructional Psychology,* vol. 3, edited by Robert Glaser, pp. 41–95. Hillsdale, N.J.: Lawrence Erlbaum Associates, 1987.

Suydam, Marilyn M., and John L. Higgins. *Activity-Based Learning in Elementary School Mathematics: Recommendations from the Research.* Columbus, Ohio: ERIC/SMEE, 1977.

Thompson, Alba G., Randolph A. Philipp, Patrick W. Thompson, and Barbara A. Boyd. "Calculational and Conceptual Orientations in Teaching Mathematics." In *Professional Development for Teachers of Mathematics,* 1994 Yearbook of the National Council of Teachers of Mathematics (NCTM), edited by Douglas B. Aichele, pp. 79–92. Reston, Va.: NCTM, 1994.

Thompson, Patrick W. "Notations. Conventions, and Constraints: Contributions to Effective Uses of Concrete Materials in Elementary Mathematics." *Journal for Research in Mathematics Education* 23 (March 1992): 123–47.

Thompson, Patrick W., and Alba G. Thompson. "Talking about Rates Conceptually, Part 1: A Teacher's Struggle. *Journal for Research in Mathematics Education* 25 (May 1994): 279–303.

Wearne, Diana, and James Hiebert. "A Cognitive Approach to Meaningful Mathematics Instruction: Testing a Local Theory Using Decimal Numbers." *Journal for Research in Mathematics Education* 19 (November 1988): 371–84.

Manipulatives Don't Come with Guarantees

Arthur J. Baroody

Recent research indicates that to promote meaningful learning, mathematical instruction should begin with experiences that are real to pupils and then proceed to the symbolic level (e.g., Baroody 1987; Davis 1984; Ginsburg 1982). Clearly, manipulatives can be an important tool in efforts to make mathematics more real to pupils. However, to discourage their uncritical use, perhaps manipulatives should carry the following warning label: The Secretary of Education [or other appropriate authority] has not determined that using manipulatives is either a sufficient or a necessary condition for meaningful learning.

Not sufficient?

Simply using manipulatives does not guarantee meaningful learning. Like any tool, manipulatives must be used judiciously and carefully for good results. If used inappropriately or without skill, they may not get the job done or, worse, may make a mess of things. Thoughtful use of manipulatives entails asking such questions as these: Can pupils use this manipulative in such a way that it connects with their existing knowledge and, hence, is meaningful to *them*? Is the manipulative used in such a way that it requires reflection or thought on the part of students?

One of my students, Genie Siedler, recently observed a first-grade class using colored rods to determine the sums of single-digit combinations. The pupils were supposed to use a number-to-rod chart to find the appropriate rods to represent each addend, place the two rods side-by-side, find a third rod with the same length, and then use the chart to decipher the sum. Most pupils followed this prescribed procedure. Others found their own way to use the manipulatives. Using a white cube (unit), they marked off units on the other rods (e.g., the brown rod contained 8 units). They then modeled the addends and found sums without recourse to the chart.

This incident raises the issue of whether colored rods are the most effective—or even an appropriate—tool for introducing the arithmetic operations in the primary grades (cf. Kidron [1985]). For pupils in these grades, arithmetic and counting are closely linked. Counting is their natural way of figuring out sums. The use of colored rods does not build on this informal foundation for arithmetic. The inventive pupils described in the foregoing redesigned the rods to fit their thinking. That is, these pupils apparently transformed a relatively unfamiliar model in which length and color represent number into a personally meaningful model in which a number is composed of so many units.

Certainly, colored rods are a good tool for some things, such as introducing concepts and operations involving fractions. However, other manipulatives, such as number sticks (Baroody 1989), may be more appropriate for introducing arithmetic operations. Number sticks can be constructed from interlocking blocks and then numbered. Thus this manipulative more closely embodies pupils existing unitary conception of numbers (e.g., 4 is four things or units).

The aforementioned incident raises another question: Did the pupils in the scenario who followed the prescribed procedure reflect on what they were doing and relate it to their understanding of addition? Just as with symbols, pupils can learn to use manipulatives mechanically to obtain answers. For example, research by Ross (e.g., [1986]) indicates that using Dienes blocks does not guarantee meaningful learning of place-value concepts. Research emphasizes the importance of relating manipulative-based models explicitly to pupils informal understandings (e.g., Fuson [1988]). Furthermore, to ensure that pupils understand what they are doing, a teacher might, for example, have them use their manipulative-based procedure to solve word problems and have them justify their solutions (cf. Lampert [1986]).

Not necessary?

Lesh, Landau, and Hamilton (1980) described five modes of presentation: real-world situations, manipulative models, pictures, spoken symbols, and written symbols. The first two modes are viewed by many as crucial steps for meaningful learning. For example,

Campbell (1988) argued that the microcomputer cannot and should not replace the use of manipulatives. Some (e.g., Jencks and Peck [1987]) suggest that pictorial representations (e.g., sketches of manipulative models to foster the development of mental imagery) are important to bridge the gap between concrete and abstract representations. Unfortunately, insufficient evidence is as yet available to determine what modes of presentations are crucial and what sequence of representations should be used before symbolic representations are introduced (e.g., Beishuizen [1985]; Post [1988]; Resnick and Ford [1981]).

One is tempted to conclude that pupils in the elementary grades require manipulatives because they are in the stage of concrete operational thought. Assuming that such a stage of mental development actually exists, it does not follow from Piagetian theory that children must actively manipulate something *concrete* and reflect on *physical* actions to construct meaning. It does suggest that they should actively manipulate *something familiar* and reflect on these physical *or mental* actions. The particular medium (objects, pictures of objects, or video displays of pictured objects) may be less important than the fact that the experience is meaningful to pupils and that they are actively engaged in thinking about it.

In conclusion, because we are still learning about what manipulatives should be used, how to use them effectively, and when they need to be used, we must be aware of the importance of keeping an open mind about using manipulatives.

References

Baroody, Arthur J. *Children's Mathematical Thinking: A Developmental Framework for Preschool, Primary, and Special Education Teachers.* New York: Teachers College Press, 1987.

———. *A Guide to Teaching Mathematics in the Primary Grades.* Boston: Allyn & Bacon, 1989.

Beishuizen, M. "Evaluation of the Use of Structured Materials in the Teaching of Primary Mathematics." In *Aspects of Educational Technology: New Directions in Education and Training Technology,* Aspects of Education, vol. 18, edited by B. S. Alloway, G. M. Mills, and A. J. Trott, pp. 246–58. New York: Nichols Publishing Co., 1985.

Campbell, Patricia F. "Microcomputers in the Primary Mathematics Classroom." *Arithmetic Teacher* 35 (February 1988): 22–30.

Davis, Robert B. *Learning Mathematics: The Cognitive Science Approach to Mathematics Education.* Norwood, N.J.: Ablex Publishing Corporation, 1984.

Fuson, Karen C. *Children's Counting and Concepts of Number.* New York: Springer-Verlag, 1988.

Ginsburg, Herbert P. *Children's Arithmetic: How They Learn It and How You Teach It.* Austin, Tex.: Pro Ed, 1982.

Jencks, Stanley M., and Donald F. Peck. *Beneath Rules.* Menlo Park, Calif.: Benjamin Cummings Publishing Co., 1987.

Kidron, Rivka. *Arithmetic Disabilities: Characterization, Diagnosis and Treatment.* Tel Aviv: Otzar Hamoreh, 1985.

Lampert, Magdalena. "Knowing, Doing and Teaching Multiplication." *Cognition and Instruction* 3 (fall 1986):305–42.

Lesh, Richard, Marsha Landau, and E. Hamilton. "Rational Number Ideas and the Role of Representational Systems." In *Proceedings of the Fourth International Conference for the Psychology of Mathematics Education,* edited by R. Karplus, pp. 50–59. Berkeley, Calif.: Lawrence Hall of Science, 1980.

Post, Thomas R. "Some Notes on the Nature of Mathematics Learning." In *Teaching Mathematics in Grades K–8: Research-based Methods,* edited by Thomas R. Post, pp. 1–19. Boston: Allyn & Bacon, 1988.

Resnick, Lauren B., and Wendy W. Ford. *The Psycholoogy of Mathematics for Instruction.* Hillsdale, N.J.: Lawrence Erlbaum Associates, 1981.

Ross, Sharon Hill, "The Development of Children's Place-Value Numeration Concepts in Grades Two through Five." Paper presented at the annual meeting of the American Educational Research Association, San Francisco, April 1986.

Rethinking "Concrete" Manipulatives

Douglas H. Clements and Sue McMillen

Close your eyes and picture students doing mathematics. Like many educators, the mental pictures may include manipulative objects, such as cubes, geoboards, or colored rods. Does the use of such concrete objects really help students learn mathematics? What is meant by "concrete"? Are computer displays concrete and can they play an important role in learning? By addressing these questions, the authors hope to change the mental picture of what manipulatives are and how they might be used effectively.

Are Manipulatives Helpful?

Helpful, yes . . .

Students who use manipulatives in their mathematics classes usually outperform those who do not (Driscoll 1983; Sowell 1989, Suydam 1986). This benefit holds across grade level, ability level, and topic, given that using a manipulative makes sense for the topic. Manipulative use also increases scores on retention and problem-solving tests. Finally, attitudes toward mathematics are improved when students are instructed with concrete materials by teachers knowledgeable about their use (Sowell 1989).

. . . but no guarantee.

Manipulatives, however, do not guarantee success (Baroody 1989). One study showed that classes not using manipulatives outperformed classes using manipulatives on a test of transfer (Fennema 1972). In this study, all teachers emphasized learning with understanding.

In contrast, students sometimes learn to use manipulatives only in a rote manner. They perform the correct steps but have learned little more. For example, a student working on place value with beans and bean sticks used the bean as 10 and the bean stick as 1 (Hiebert and Wearne 1992).

Similarly, students often fail to link their actions on base-ten blocks with the notation system used to describe the actions (Thompson and Thompson 1990). For example, when asked to select a block to stand for 1 then put blocks out to represent 3.41, one fourth grader put out three flats, four longs, and one single after reading the decimal as "three hundred forty-one."

Although research suggests that instruction begin concretely, it also warns that concrete manipulatives are not sufficient to guarantee meaningful learning. This conclusion leads to the next question.

What Is Concrete?

Manipulatives are supposed to be good for students because they are concrete. The first question to consider might be, What does *concrete* mean? Does it mean something that students can grasp with their hands? Does this sensory character itself make manipulatives helpful? This view presents several problems.

First, it cannot be assumed that when children mentally close their eyes and picture manipulative-based concepts, they "see" the same picture that the teacher sees. Holt (1982, pp.138–39) said that he and his fellow teacher "were excited about the rods because we could see strong connections between the world of rods and the world of numbers. We therefore

---

Time to prepare this material was funded in part by the National Science Foundation under grants no. MDR-9050210 and MDR 8954664. Any opinions, findings, and conclusions or recommendations expressed in this publication are those of the authors and do not necessarily reflect the views of the National Science Foundation.

The authors extend their appreciation to Arthur J. Baroody and several anonymous reviewers for their insightful comments and suggestions on earlier drafts of this article.

assumed that children, looking at rods and doing things with them, could *see* how the world of numbers and numerical operations worked. The trouble with this theory is that [my colleague] and I *already* knew how the numbers worked. We could say, 'Oh, the rods behaved just the way numbers do.' But if we *hadn't* known how numbers behaved, would looking at the rods enable us to find out? Maybe so, maybe not."

Second, physical actions with certain manipulatives may suggest mental actions different from those that teachers wish students to learn. For example, researchers found a mismatch when students used the number line to perform addition. When adding 5 and 4, the students located 5, counted "one, two, three, four," and read the answer. This procedure did not help them solve the problem mentally, for to do so they must count "six, seven, eight, nine" and at the same time count the counts—6 is 1, 7 is 2, and so on. These actions are quite different (Gravemeijer 1991, p. 59). These researchers also found that students' external actions on an abacus did not always match the mental activity intended by the teacher.

Although manipulatives have an important place in learning, they do not carry the meaning of the mathematical idea. They can even be used in a rote manner. Students may need concrete materials to build meaning initially, but they must reflect on their *actions* with manipulatives to do so. Later, they are expected to have a "concrete" understanding that goes beyond these physical manipulatives. For example, teachers like to see that numbers as mental objects—"I can think of 13 + 10 in my head"—are "concrete" for sixth graders. It appears that "concrete" can be defined in different ways.

Types of concrete knowledge

Students demonstrate *sensory-concrete* knowledge when they use sensory material to make sense of an idea. For example, at early stages, children cannot count, add, or subtract meaningfully unless they have actual objects to touch.

Integrated-concrete knowledge is built through learning. It is knowledge that is connected in special ways. This concept is the root of the word *concrete*—"to grow together." Sidewalk concrete derives its strength from the combination of separate particles in an interconnected mass. Integrated-concrete thinking derives its strength from the combination of many separate ideas in an interconnected structure of knowledge. While still in primary school, Jacob read a problem on a restaurant place mat asking for the answer to 3/4 + 3/4. He solved the problem by thinking about the fractions in terms of money: 75¢ plus 75¢ is $1.50, so 3/4 + 3/4 is 1 1/2. When children have this type of interconnected knowledge, the physical objects, the actions they perform on the objects, and the abstractions they make are all interrelated in a strong mental structure. Ideas such as "four," "3/4," and "rectangle" become as real and tangible as a concrete sidewalk. Each idea is as concrete to a student as a ratchet wrench is to a plumber—an accessible and useful tool. Jacob's knowledge of money was such a tool.

An idea, therefore, is not simply concrete or not concrete. Depending on what kind of relationship the student has with it (Wilensky 1991), an idea might be sensory-concrete, abstract, or integrated-concrete. The catch, however, is that mathematics cannot be packaged into sensory-concrete materials, no matter how clever our attempts are, because ideas such as number are not "out there." As Piaget has shown, they are constructions—reinventions—of each human mind. "Fourness" is no more "in" four blocks than it is "in" a picture of four blocks. The child creates "four" by building a representation of number and connecting it with either real or pictured blocks (Clements 1989; Clements and Battista 1990; Kamii 1973, 1985, 1986).

Mathematical ideas are ultimately made integrated-concrete not by their physical or real-world characteristics but rather by how "meaningful"—connected to other ideas and situations—they are. Holt (1982, 219) found that children who already understood numbers could perform the tasks with or without the blocks. "But children who could not do these problems without the blocks didn't have a clue about how to do them with the blocks.... They found the blocks... as abstract, as disconnected from reality, mysterious, arbitrary, and capricious as the numbers that these blocks were supposed to bring to life."

Are Computer Manipulatives Concrete?

The reader's earlier mental picture of students using manipulatives probably did not feature computer technology. But as has been shown, "concrete" cannot be equated simply with physical manipulatives. Computers might supply representations that are just as personally meaningful to students as are real objects; that is, they might help develop integrated-concrete knowledge. These representations may also be more manageable, "clean," flexible, and extensible. For example, one group of young students learned number concepts with a computer-felt-board environment. They constructed "bean stick pictures" by selecting and arranging beans, sticks, and number symbols. Compared with a real bean-stick environment, this computer environment offered equal, and sometimes greater, control and flexibility to students (Char 1989). The computer manipulatives were just as meaningful and were easier to use for learning. Both computer and physical bean sticks were worthwhile, but work with one did not need to precede work with the other.

The important point is that "concrete" is, quite literally, in the mind of the beholder. Ironically, Piaget's period of concrete operations is often used, incorrectly, as a rationalization for objects-for-objects'-sake activities in elementary school. Good concrete activity is good *mental* activity (Clements 1989; Kamii 1989).

This idea can be made more concrete. Several computer programs allow children to manipulate on-screen base-ten blocks. These blocks are not physically concrete. However, no base-ten blocks "contain" place-value ideas (Kamii 1986). Students must build these ideas from working with the blocks and thinking about their actions.

Actual base-ten blocks can be so clumsy and the manipulations so disconnected one from the other that students may see only the trees—manipulations of many pieces—and miss the forest—place-value ideas. The computer blocks can be more manageable and "clean."

In addition, students can break computer base-ten blocks into ones or glue ones together to form tens. Such actions are more in line with the mental actions that students are expected to learn. The computer also links the blocks to the symbols. For example, the number represented by the base-ten blocks is usually dynamically linked to the students' actions on the blocks; when the student changes the blocks, the number displayed is automatically changed, as well. This process can help students make sense of their activity and the numbers. Computers encourage students to make their knowledge explicit, which helps them build integrated-concrete knowledge. A summary of specific advantages follows.

Computers offer a manageable, clean manipulative. They avoid distractions often present when students use physical manipulatives. They can also mirror the desired mental actions more closely.

Computers afford flexibility. Some computer manipulatives offer more flexibility than do their noncomputer counterparts. For example, Elastic Lines (Harvey, McHugh, and McGlathery 1989) allows the student to change instantly both the size, that is, the number of pegs per row, and the shape of a computer-generated geoboard (**fig. 1**). The ease of accessing these computer geoboards allows the software user many more experiences on a wider variety of geoboards. Eventually, these quantitative differences become qualitative differences.

Computer manipulatives allow for changing the arrangement or representation. Another aspect of the flexibility afforded by many computer manipulatives is the ability to change an arrangement of the data. Most spreadsheet and database software will sort and reorder the data in numerous different ways. Primary Graphing and Probability Workshop (Clements, Crown, and Kantowski 1991) allows the user to convert a picture graph to a bar graph with a single keystroke (**fig. 2**).

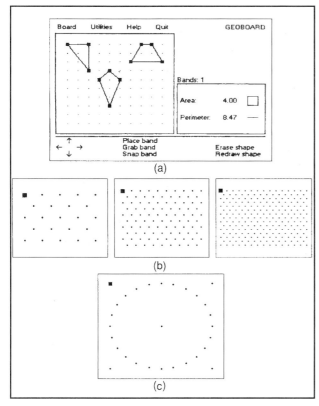

Fig.1. Elastic Lines (Harvey, McHugh, and McGlathery 1989) allow a variety of arrangements of "nails" on its electronic geoboard; size can also be altered.

Computers store and later retrieve configurations. Students and teachers can save and later retrieve any arrangement of computer manipulatives. Students who had partially solved a problem can pick up immediately where they left off. They can save a spreadsheet or database created for one project and use it for other projects.

Computers record and replay students' actions. Computers allow the storage of more than static configurations. Once a series of actions is finished, it is often difficult to reflect on it. But computers have the power to record and replay *sequences* of actions on manipulatives. The computer-programming commands can be recorded and later replayed, changed, and viewed. This ability encourages real mathematical exploration. Computer games such as Tetris allow students to replay the same game. In one version, Tumbling Tetrominoes, which is included in Clements, Russell et al. (1995), students try to cover a region with a random sequence of tetrominoes (**fig. 3**). If students believe that they can improve their strategy, they can elect to receive the same tetrominoes in the same order and try a new approach.

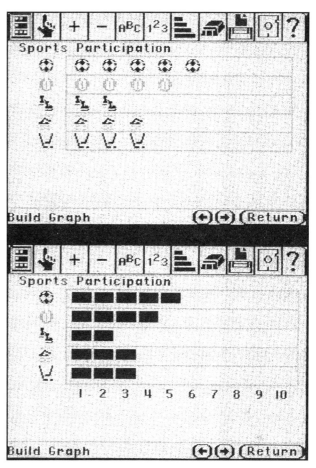

Fig. 2. Primary Graphic and Probability Workshop (Clements, Crown, and Kantowski 1991) converts a picture graph to a bar graph with a single keystroke.

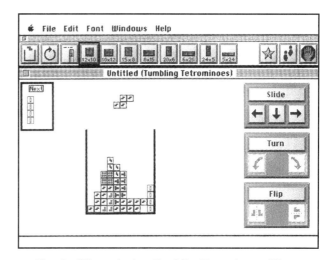

Fig. 3. When playing Tumblin Tetrominoes (Clements, Russell, et al. 1995), students attempt to tile tetrominoes— shapes that are like dominoes except that four squares are connected with full sides toucing. Research indicates that playing such games involves conceptual and spatial reasoning (Bright, Usnick, and Williams 1992). Students can elect to replay a game to improve their strategy.

Computer manipulatives link the concrete and the symbolic by means of feedback. Other benefits go beyond convenience. For example, a major advantage of the computer is the ability to associate active experience with manipulatives to symbolic representations. The computer connects manipulatives that students make, move, and change with numbers and words. Many students fail to relate their actions on manipulatives with the notation system used to describe these actions. The computer links the two.

For example, students can draw rectangles by hand but never go further to think about them in a mathematical way. In Logo, however, students must analyze the figure to construct a sequence of commands, or a procedure, to draw a rectangle (see **fig. 4**). They have to apply numbers to the measures of the sides and angles, or turns. This process helps them become explicitly aware of such characteristics as "opposite sides equal in length." If instead of fd 75 they enter fd 90, the figure will not be a rectangle. The link between the symbols and the figure is direct and immediate. Studies confirm that students' ideas about shapes are more mathematical and precise after using Logo (Clements and Battista 1989; Clements and Battista 1992).

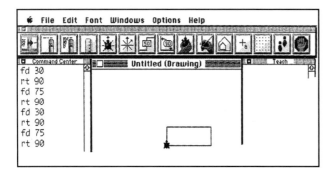

Fig. 4. Students use a new version of Logo, Turtle Math (Clements and Meredith 1994), to construct a rectangle. The commands are listed in the command center on the left (Clements, Battista, et al. 1995; Clements and Meredith 1994).

Some students understand certain ideas, such as angle measure, for the first time only after they have used Logo. They have to make sense of what is being controlled by the numbers they give to right- and left-turn commands. The turtle immediately links the symbolic command to a sensory-concrete turning action. Receiving feedback from their explorations over several tasks, they develop an awareness of these quantities and the geometric ideas of angle and rotation (Kieran and Hillel 1990).

Fortunately, students are not surprised that the computer does not understand natural language and that they must formalize their ideas to communicate them. Students formalize about fives times more often using computers than they do using paper (Hoyles, Healy, and Sutherland 1991). For example, students struggled to express the number pattern that had they explored on spreadsheets. They used such phrases as "this cell equals the next one plus 2; and then that result plus this cell plus 3 equals this." Their use of the structure of the spreadsheet's rows and columns, and their incorporation of formulas in the cells of the spreadsheet, helped them more formally express the generalized pattern they had invented.

But is it too restrictive or too hard to operate on symbols rather than to operate directly on the manipulatives? Ironically, less "freedom" might be more helpful. In a study of place value, one group of students worked with a computer base-ten manipulative. The students could not move the computer blocks directly. Instead, they had to operate on symbols—digits—as shown in **figure 5** (Thompson 1992; Thompson and Thompson 1990). Another group of students used physical base-ten blocks. Although teachers frequently guided students to see the connection between what they did with the blocks and what they wrote on paper, the physical-blocks group did not feel constrained to write something that represented what they did with blocks. Instead, they appeared to look at the two as separate activities. In comparison, the computer group used symbols more meaningfully, tending to connect them to the base-ten blocks.

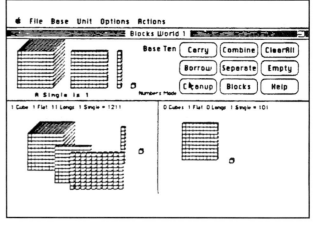

Fig. 5. A screen display of the base-ten-blocks computer microworld (Thompson 1992)

In computer environments, such as computer base-ten blocks or computer programming, students cannot overlook the consequences of their actions, which is possible to do with physical manipulatives. Computer manipulatives, therefore, can help students build on their physical experiences, tying them tightly to symbolic representations. In this way, computers help students link sensory-concrete and abstract knowledge so they can build integrated-concrete knowledge.

Computer manipulatives dynamically link multiple representations. Such computer links can help students connect many types of representations, such as pictures, tables, graphs, and equations. For example, many programs allow students to see immediately the changes in a graph as they change data in a table.

These links can also be dynamic. Students might stretch a computer geoboard's rectangle and see the measures of the sides, perimeter, and area change with their actions.

Computers change the very nature of the manipulative. Students can do things that they cannot do with physical manipulatives. Instead of trading one hundred-block for ten ten-blocks, students can break the hundred-block pictured on the screen into ten ten-blocks, a procedure that mirrors the mathematical action closely. Students can expand computer geoboards to any size or shape. They can command the computer to draw automatically a figure symmetrical to any they create on the geoboard.

Advantages of Computer Manipulatives for Teaching and Learning

In addition to the aforementioned advantages, computers and computer manipulatives possess other characteristics that enhance teaching and learning mathematics. Descriptions of these features follow.

Computer manipulatives link the specific to the general. Certain computer manipulatives help students view a mathematical object not just as one instance but as a representative of an entire class of objects. For example, in Geometric Supposer (Schwartz and Yerushalmy 1986) or Logo, students are more likely to see a given rectangle as one of many that could be made rather than as just one rectangle.

This effect even extends to problem-solving strategies. In a series of studies, fourth-grade through high school students who used Logo learned problem-solving strategies better than those who were taught the same strategies with noncomputer manipulatives (Swan and Black 1989). Logo provided malleable representations of the strategies that students could inspect, manipulate, and test through practice. For example, in Logo, students broke a problem into parts by disembedding, planning, and programming each piece of a complex picture separately. They then generalized this strategy to other mathematics problems.

Computer manipulatives encourage problem posing and conjecturing. This ability to link the specific to the general also encourages students to make their own conjectures. "The essence of mathematical creativity lies in the making and exploring of mathematical conjectures" (Schwartz 1989). Computer manipulatives can furnish tools that allow students to explore their own conjectures while also decreasing the psychological cost of making incorrect conjectures.

Because students may themselves test their ideas on the computer, they can more easily move from naive to empirical to logical thinking as they make and test conjectures. In addition, the environments appear conducive not only to posing problems but to wondering and to playing with ideas. In early phases of problem solving, the environments help students explore possibilities, not become "stuck" when no solution path presents itself. Overall, research suggests that computer manipulatives can enable "teaching children to be mathematicians vs. teaching about mathematics" (Papert 1980, 177).

For example, consider the following dialogue, in which a teacher was overheard discussing students' Logo procedures for drawing equilateral triangles.

Teacher: Great. We got the turtle to draw bigger and smaller equilateral triangles. Who can summarize how we did it?

Monica: We changed all the forward numbers with a different number—but all the same. But the turns had to stay 120, 'cause they're all the same in equilateral triangles. (See **fig. 6**.)

Chris: We didn't make the biggest triangle.

Teacher: What do you mean?

Chris: What's the biggest one you could make?

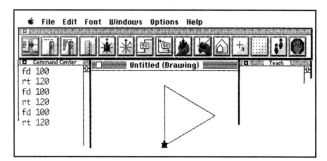

Fig. 6. Students use Turtle Math (Clements and Meredith 1994) to construct an equilateral triangle.

Teacher: What do people think?

Rashad: Let's try 300.

The class did (see **fig. 7**).

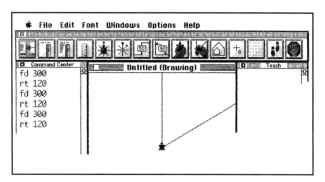

Fig. 7. When commands are changed, the figure is automatically changed—here, to an equilateral triangle with sides of length 300 turtle steps.

Monica: It didn't fit on the screen. All we see is an angle.

Teacher: Where's the rest?

Beth: Off here [gesturing]. The turtle doesn't wrap around the screen.

Tanisha: Let's try 900!

The student typing made a mistake and changed the command to fd 3900.

Teacher: Whoa! Keep it! Before you try it, class, tell me what it will look like!

Ryan: It'll be bigger. Totally off the screen! You won't see it at all!

Jacob: No, two lines will still be there, but they'll be way far apart.

The children were surprised when it turned out the same! (See **fig. 8**.)

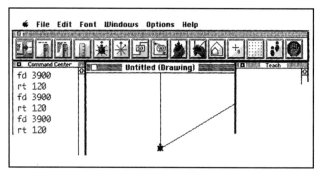

Fig. 8. Why did the figure not change when the side lengths were changed to 3900 turtle steps?

Teacher: Is that what you predicted?

Rashad: No! We made a mistake

Monica: Oh, I get it. It's right. It's just farther off the screen. See, it goes way off, there, like about past the ceiling.

The teacher challenged them to explore this and other problems they could think of.

Jacob: I'm going to find the smallest equilateral triangle.

Rashad: We're going to try to get all the sizes inside one another.

Computer manipulatives build scaffolding for problem solving. Computer environments may be unique in furnishing problem-solving scaffolding that allows students to build on their initial intuitive visual approaches and construct more analytic approaches. In this way, early concepts and strategies may be precursors of more sophisticated mathematics. In the realm of turtle geometry, research supports Papert's (1980) contention that ideas of turtle geometry are based on personal, intuitive knowledge (Clements and Battista 1991; Kynigos 1992). One boy, for example, wrote a procedure to draw a rectangle. He created a different variable for the length of each of the four sides. He gradually saw that he needed only two variables, since the lengths of the opposite sides are equal. In this way, he recognized that the variables could represent *values* rather than specific sides of the rectangle. No teacher intervened; Logo supplied the scaffolding by requiring a symbolic representation and by allowing the boy to link the symbols to the figure.

Computer manipulatives may also build scaffolding by assisting students in getting started on a solution. For example, in a spreadsheet environment, typing headings or entering fixed numbers might help students organize their ideas.

Computer manipulatives focus attention and increase motivation. One group of researchers studied pairs of students as they worked on computers and found that the computer "somehow draws the attention of the pupils and becomes a focus for discussion," thus resulting in very little off-task talk (Hoyles, Healy, and Sutherland 1991). Although most children seem to enjoy working on the computer, such activity can be especially motivating for some students who have been unsuccessful with mathematics. For example, two such third graders were observed as they eagerly worked in a Logo environment. They had gone forward twenty-three turtle steps but then figured out that they needed to go forward sixty turtle steps in all. They were so involved that both of them wanted to do the necessary subtraction. One grabbed the paper from the other so he could compute the difference.

Computer manipulatives encourage and facilitate complete, precise explanations. Compared with students using paper and pencil, students using computers work with more precision and exactness (Butler and Close 1989; Clements and Battista 1991; Gallou-Dumiel 1989). For example, students can use physical manipulatives to perform such motions as slides, flips, and turns. However, they make intuitive movements and corrections without being aware of these geometric motions. Even young children can move puzzle pieces into place without a conscious awareness of the geometric motions that can describe these physical movements. In one study, researchers attempted to help a group of students using noncomputer manipulatives become aware of these motions. However, descriptions of the motions were generated from, and interpreted by, physical motions of students who understood the task. In contrast, students using the computer specified motions to the computer, which does not "already understand." The specification had to be thorough and detailed. The results of these commands were observed, reflected on, and corrected. This effort led to more discussion of the motions themselves, not just the shapes (Butler and Close 1989).

Firming Up Ideas about the Concrete

Manipulatives can play a role in students' construction of meaningful ideas. They should be used before formal instruction, such as teaching algorithms. However, teachers and students should avoid using manipulatives as an end—without careful thought—rather than as a means to that end.

The appropriate use of representations is important to mathematics learning. In certain topics, such as early number concepts, geometry, measurement, and fractions, the proper use of manipulatives is especially crucial. However, manipulatives alone are not sufficient—they must be used to actively engage children's thinking with teacher guidance—and definitions of what constitute a "manipulative" may need to be expanded. Research offers specific guidelines for selecting and using manipulatives.

How Should Manipulatives Be Selected?

The following guidelines are offered to assist teachers in selecting appropriate and effective manipulatives.

Select manipulatives primarily for children's use. Teacher demonstrations with manipulatives can be valuable; however, children themselves should use the manipulatives to solve a variety of problems.

Select manipulatives that allow children to use their informal methods. Manipulatives should not prescribe or unnecessarily limit students' solutions or ways of making sense of mathematical ideas. Students should be in control.

Use caution in selecting "prestructured" manipulatives in which the mathematics is built in by the manufacturer, such as base-ten blocks as opposed to interlocking cubes. They can become what the colored rods were for Holt's students—"another kind of numeral, symbols made of colored wood rather than marks on paper" (1982, p. 170). Sometimes the simpler, the better. For example, educators from the Netherlands found that students did not learn well using base-ten blocks and other structured base-ten materials. A mismatch may have occurred between trading one base-ten block for another and the actions of mentally separating a ten into ten ones or thinking of the same quantity simultaneously as "one ten" and "ten ones." The Netherlands students were more successful after hearing a story of a sultan who often wants to count his gold. The setting of the story gave students a reason for counting and grouping: the gold had to be counted, packed, and sometimes unwrapped—and an inventory had to be constantly maintained (Gravemeijer 1991). Students, therefore, might best start by using manipulatives with which they create and break up groups of tens into ones, such as interlocking cubes, instead of base-ten blocks (Baroody 1990). Settings that give reasons for grouping are ideal.

Select manipulatives that can serve many purposes. Some manipulatives, such as interlocking cubes, can be used for counting, place value, arithmetic, patterning, and many other topics. This versatility allows students to find many different uses. However, a few single-purpose devices, such as mirrors or Miras, make a significant contribution.

Choose particular representations of mathematical ideas with care. Perhaps the most important criteria are that the experience be meaningful to students and that they become actively engaged in thinking about it.

To introduce a topic, use a single manipulative instead of many different manipulatives. One theory held that students had to see an idea presented by several different manipulatives to abstract the essence of this idea. However, in some circumstances, using the same material consistently is advantageous. "Using the tool metaphor for representations, perhaps a representation becomes useful for students as they handle it and work with it repeatedly" (Hiebert and Wearne 1992, p. 114). If the tool is to become useful, perhaps an advantage accrues in using the same tool in different situations rather than in using different tools in the same situation. Students gain expertise through using a tool over and over on different projects.

Should only one manipulative be used, then? No, different children may find different models meaningful (Baroody 1990). Further, reflecting on and discussing different models may indeed help students abstract the mathematical idea. Brief and trivial use, however, will not help; each manipulative should become a tool for thinking. Different manipulatives allow, and even encourage, students to choose their own representations. New material can also be used to assess whether students understand the idea or just have learned to use the previous material in a rote manner.

Select computer manipulatives when appropriate. Certain computer manipulatives may be more beneficial than any physical manipulative. Some are just the sort of tools that can lead to mathematical expertise. The following recommendations and special considerations pertain to computer manipulatives. Select programs that—

- have uncomplicated changing, repeating, and undoing actions;
- allow students to save configurations and sequences of actions;
- dynamically link different representations and maintain a tight connection between pictured objects and symbols;
- allow students and teachers to pose and solve their own problems; and
- allow students to develop increasing control of a flexible, extensible, mathematical tool. Such programs also serve many purposes and help form connections among mathematical ideas.

Select computer manipulatives that—

- encourage easy alterations of scale and arrangement,
- go beyond what can be done with physical manipulatives, and
- demand increasingly complete and precise specifications.

How Should Manipulatives Be Used?

The following suggestions are offered to assist teachers in effectively using manipulatives in their classrooms.

Increase the students' use of manipulatives. Most students do not use manipulatives as often as needed. Thoughtful use can enhance almost every topic. Also, short sessions do not significantly enhance learning. Students must learn to use manipulatives as tools for thinking about mathematics.

Recognize that students may differ in their need for manipulatives. Teachers should be cautious about requiring all students to use the same manipulative. Many students might be better off if allowed to choose their manipulatives or to use just paper and pencil. Some students in the Netherlands were more successful when they drew pictures of the sultan's gold pieces than when they used any physical manipulative. Others may need manipulatives for different lengths of time (Suydam 1986).

Encourage students to use manipulatives to solve a variety of problems and then to reflect on and justify their solutions. Such varied experience and justification help students build and maintain understanding. Ask students to explain what each step in their solution means and to analyze any errors that occurred as they use manipulatives—some of which may have resulted from using the manipulative.

Become experienced with manipulatives. Attitudes toward mathematics, as well as concepts, are improved when students have instruction with manipulatives, but only if their teachers are knowledgeable about their use (Sowell 1989).

Some recommendations are specific to computer manipulatives.

- Use computer manipulatives for assessment as mirrors of students' thinking.
- Guide students to alter and reflect on their actions, always predicting and explaining.
- Create tasks that cause students to see conflicts or gaps in their thinking.
- Have students work cooperatively in pairs.
- If possible, use one computer and a large-screen display to focus and extend follow-up discussions with the class.
- Recognize that much information may have to be introduced before moving to work on computers, including the purpose of the software, ways to operate the hardware and software, mathematics content and problem-solving strategies, and so on.
- Use extensible programs for long periods across topics when possible.

Final Words

With both physical and computer manipulatives, teachers should choose meaningful representations, then guide students to make connections between these representations. No one yet knows what modes of presentations are crucial and what sequence of representations should be used before symbols are introduced (Baroody 1989; Clements 1989). Teachers should be careful about adhering blindly to an unproved, concrete-pictorial-abstract sequence, especially when more than one way of thinking about "concrete" is possible. It is known that students' knowledge is strongest when they connect real-world situations, manipulatives, pictures, and spoken and written symbols (Lesh 1990). They should relate manipulative models to their intuitive, informal understanding of concepts and translate between representations at all points of their learning. This process builds integrated-concrete ideas.

Now when teachers close their eyes and picture children doing mathematics, manipulatives should still be in the picture, but the mental image should include a new perspective on how to use them.

References

Baroody, Arthur J. "One Point of View: Manipulatives Don't Come with Guarantees." *Arithmetic Teacher* 37 (October 1989): 4–5.

———. "How and When Should Place-Value Concepts and Skills Be Taught?" *Journal for Research in Mathematics Education* 21 (July 1990): 281–86.

Bright, George, Virginia E. Usnick, and Susan Williams. *Orientation of Shapes in a Video Game.* Hilton Head, S.C.: Eastern Educational Research Association, 1992.

Butler, Deirdre, and Sean Close. "Assessing the Benefits of a Logo Problem-Solving Course." *Irish Educational Studies* 8 (1989): 168–90.

Char, Cynthia A. *Computer Graphics Feltboards: New Software Approaches for Young Children's Mathematical Exploration.* San Francisco: American Educational Research Association, 1989.

Clements, Douglas H. *Computers in Elementary Mathematics Education.* Englewood Cliffs. N.J.: Prentice Hall, 1989.

Clements, Douglas H., and Michael T. Battista. "Learning of Geometric Concepts in a Logo Environment." *Journal for Research in Mathematics Education* 20 (November 1989):450–67.

———. "Research into Practice: Constructivist Learning and Teaching." *Arithmetic Teacher* 38 (September 1990): 34–35.

———. "The Development of a Logo-Based Elementary School Geometry Curriculum." Final report for NSF grant no. MDR-8651668. Buffalo, N. Y.: State University of New York at Buffalo, and Kent, Ohio: Kent State University, 1991.

———. "Geometry and Spatial Reasoning." In *Handbook of Research on Mathematics Teaching and Learning,* edited by Douglas A. Grouws, pp. 420–64. New York: Macmillan Publishing Co., 1992.

Clements, Douglas H., Michael T. Battista, Joan Akers, Virginia Woolley, Julie Sarama Meredith, and Sue McMillen. *Turtle Paths,* Cambridge, Mass.: Dale Seymour Publications, 1995. Includes software.

Clements, Douglas H., Warren D. Crown, and Mary Grace Kantowski. *Primary Graphing and Probability Workshop.* Glenview, Ill.: Scott, Foresman & Co., 1991, Software.

Clements, Douglas H., and Julie Sarama Meredith. *Turtle Math.* Montreal: Logo Computer Systems (LCSI). 1994, Software.

Clements, Douglas H., Susan Jo Russell, Cornelia Tierney, Michael T. Battista, and Julie Sarama Meredith. *Flips, Turns, and Area,* Cambridge, Mass.: Dale Seymour Publications, 1995. Includes software.

Driscoll, Mark J. *Research within Reach: Elementary School Mathematics.* St. Louis: CEMREL, 1983.

Fennema, Elizabeth. "The Relative Effectiveness of a Symbolic and a Concrete Model in Learning a Selected Mathematical Principle." *Journal for Research in Mathematics Education* 3 (November 1972): 233–38.

Gallou-Dumiel, Elisabeth. "Reflections, Point Symmetry and Logo." In *Proceedings of the Eleventh Annual Meeting. North American Chapter of the International Group for the Psychology of Mathematics Education,* edited by C.A. Maher, G. A. Goldin, and R. B. Davis, pp. 140–57. New Brunswick, N.J.: Rutgers University Press, 1989.

Gravemeijer, K. P. E. "An Instruction-Theoretical Reflection on the Use of Manipulatives." In *Realistic Mathematics Education in Primary School,* edited by L. Streefland, pp. 57–76. Utrecht, Netherlands: Freudenthal Institute, Utrecht University, 1991.

Harvey, Wayne, Robert McHugh, and Douglas McGlathery. Elastic Lines. Pleasantville. N.Y.: Sunburst Communications, 1989. Software.

Hiebert, James, and Diana Wearne. "Links between Teaching and Learning Place Value with Understanding in First Grade." *Journal for Research in Mathematics Education* 23 (March 1992): 98–122.

Holt. John. *How Children Fail.* New York: Dell Publishing Co., 1982.

Hoyles, Celia, Lulu Healy, and Rosamund Sutherland. "Patterns of Discussion between Pupil Pairs in Computer and Non-Computer Environments." *Journal of Computer Assisted Learning* 7 (1991): 210–28.

Kamii, Constance. "Pedagogical Principles Derived from Piaget's Theory: Relevance for Educational Practice." In *Piaget in the Classroom,* edited by M. Schwebel and J. Raph, pp. 199–215. New York: Basic Books, 1973.

———. *Young Children Reinvent Arithmetic: Implications of Piaget's Theory.* New York: Teachers College Press, 1985.

———. "Place Value: An Explanation of Its Difficulty and Educational Implications for the Primary Grades. *Journal of Research in Childhood Education* 1 (August 1986): 75–86.

———. *Young Children Continue to Reinvent Arithmetic: 2nd Grade. Implications of Piaget's Theory.* New York: Teachers College Press, 1989.

Kieran, Carolyn, and Joel Hillel. " 'It's Tough When You Have to Make the Triangles Angles': Insights from a Computer-Based Geometry Environment." *Journal of Mathematical Behavior* 9 (October 1990): 99–127.

Kynigos, Chronis. "The Turtle Metaphor as a Tool for Children's Geometry." In *Learning Mathematics and Logo,* edited by C. Hoyles and R. Noss, pp. 97–126. Cambridge, Mass.: MIT Press, 1992.

Lesh, Richard. "Computer-Based Assessment of Higher Order Understandings and Processes in Elementary Mathematics." In *Assessing Higher Order Thinking in Mathematics,* edited by G. Kulm, pp. 81–110. Washington D.C.: American Association for the Advancement of Science, 1990.

Papert, Seymour, *Mindstorms: Children, Computers, and Powerful Ideas.* New York: Basic Books, 1980.

Schwartz, Judah L. "Intellectual Mirrors: A Step in the Direction of Making Schools Knowledge-Making Places." *Harvard Educational Review* 59 (February 1989): 51–61.

Schwartz, Judah L., and Michal Yerushalmy. The Geometric Supposer Series. Pleasantville, N. Y.: Sunburst Communications, 1986. Software.

Sowell, Evelyn J. "Effects of Manipulative Materials in Mathematics Instruction." *Journal for Research in Mathematics Education* 20 (November 1989): 498–505.

Suydam, Marilyn N. "Research Report: Manipulative Materials and Achievement." *Arithmetic Teacher* 33 (February 1986): 10, 32.

Swan, Karen, and John B. Black. "Logo Programming Problem Solving, and Knowledge-Based Instruction." University of Albany, Albany, N.Y., 1989. Manuscript.

Thompson, Patrick W. "Notations, Conventions, and Constraints: Contributions to Effective Uses of Concrete Materials in Elementary Mathematics." *Journal for Research in Mathematics Education* 23 (March 1992): 123–47.

Thompson, Patrick W., and Alba G. Thompson. *Salient Aspects of Experience with Concrete Manipulatives.* Mexico City: International Group for the Psychology of Mathematics Education, 1990.

Wilensky, Uri. "Abstract Mediations on the Concrete and Concrete Implications for Mathematics Education." In *Constructionism,* edited by I. Harel and S. Papert, pp. 193–99. Norwood, N.J.: Ablex Publishing Corporation, 1991.

Section 11

Electronic Tools

Research on Calculators in Mathematics Education

Ray Hembree and Donald J. Dessart

In the spring of 1984, we began to study all the research that could be found on the effects of calculator use in precollege mathematics. Our effort was prompted by what seemed to be conflicting circumstances: large potential benefits from the devices on the one hand; little actual impact on the curriculum on the other.

What had caused this lack of attention to calculators in the schools? As early as 1974, the National Council of Teachers of Mathematics (NCTM) had issued a statement that urged calculator use (p. 3), and ten years later, with shrinking costs, the device seemed ready to permeate American society. Negative forces appeared, however, that countered the positive attitudes. Mainly, these forces seemed to evolve from concerns that the hand-held computing machine would displace students' skills in mental arithmetic and paper-and-pencil algorithms. The resulting debate about calculator effects induced a flurry of research, one of the largest efforts for any topic in mathematics education (Suydam 1982). Do calculators threaten basic skills? The answer consistenly seemed to be no, provided those basics have been developed first in a calculator-free environment (Suydam 1979). Nonetheless, the question persisted. Some studies had yielded ambiguous findings. Other studies, although recording no harmful effects from calculators, failed to show improvement in either student achievement or attitude. Moreover, these findings were scattered and disparate in the literature, not centralized and integrated to tell a common story. What seemed to be needed was a rigorous, formal study of the overall body of research literature on calculator effects. Perhaps such an analysis would help resolve the controversy.

We found seventy-nine reports of experiments and relational investigations. In each of those studies, one group of students had been permitted to use calculators within a "treatment" period (in the usual case, about thirty school days), using the devices for computation or to help develop concepts and problem-solving strategies. In each study, during the same period, a comparison group received instruction on the same mathematical topics but had no in-class access to calculators. At the end of the treatment, both groups were examined; their average scores could then be compared.

The studies used a broad range of scales and instruments to measure outcomes. To synthesize the various findings, we transformed results to a common numerical base called *effect size,* with positive effect size indicating a study favoring the calculator treatment. In all, 524 "effects" were measured in the seventy-nine studies. These effects were then partitioned into subsets by grouping studies that focused on common aspects of performance or attitude. Within each subset, studies were clustered according to school grade and student ability levels. Average effect sizes were determined for each subset, and these effect-size averages were tested for statistical significance. The resulting average values were *the effects* of calculators in precollege classrooms. (See Hembree and Dessart [1986] for a formal report.)

Table 1 summarizes the findings. Performance had been measured in three aspects of mathematics: computation, concepts, and problem solving (as gathered, for example, from subtests of the Iowa Test of Basic Skills). In most of the studies, neither group was permitted to use calculators on tests. Thus, differences in their average scores revealed the effects of calculator use during instruction on students' skills with paper and pencil. In some of the studies, the experimental group was also allowed to use calculators during tests, whereas comparison students worked the same tests with only paper and pencil. The difference in average scores in those studies showed the advantage of using calculators on tests that followed instruction with calculators.

The seventy-nine studies showed no effects of calculator use on tests of conceptual knowledge. Effects for computation and problem solving were specific to student ability and school grade levels. For tests *with*

Table 1

Calculator Effects on Student Achievement (Original Study)

Test Condition	Student Ability	Skills Area	
		Computation	Problem Solving
With calculators*	Low	Moderate (+)	Moderate (+)
	Average	Large (+)	Small (+)
	High	No data	Moderate (+)
Without calculators	Low	Not significant	Not significant
	Average		
	Grade 4	Small (−)	Small (+)
	Other grades	Small (+)	Small (+)
	High	Not significant	Not significant

*Calculator groups only. Noncalculator groups used paper and pencil on tests.

Note: (+) and (−) denote respectively higher and lower scores of calculator groups compared to noncalculator groups.

calculators, scores were improved for low- and average-ability students, with effects that seemed moderate to large (3 to 8 points on a 100-point scale, converted from effect-size notation). None of the studies provided data on high-ability students in computation, but scores of these students in problem solving displayed a moderate improvement as a result of calculator use. Regarding tests *without* calculators:

1. For average students, small but significant effects (1 or 2 points on a 100-point scale) were observed at all grade levels. Three of the four effects were positive; the use of calculators during instruction advanced the students' skills with paper-and-pencil algorithms. The sole negative effect size (based on a total of seven studies) regarded computation in grade 4. In that situation, calculators appeared to detract from the growth of students' computational skills.

2. No apparent effects were observed for low- or high-ability students in either computation or problem solving. The prior use of calculators had neither damaged nor improved the students' performance.

We interpreted these findings as encouragement toward calculator use in the classroom with a modest caution in some areas. On the positive side, the calculator could apparently advance the average student's computational skills while doing no harm to the computational skills of low- and high-ability students. Moreover, at all ability levels, calculators could provide a clear advantage when used during tests. The single negative finding serves to remind us that calculators, though generally beneficial, may not be appropriate for use at all times, in all places, and for all subject matters. Discretion in using calculators was advised.

Along with effects on performance, the body of studies had also yielded effects of calculator use on students' attitudes. Again, comparisons were drawn at the end of a treatment period after one student group had used calculators, whereas a second group had no such access. Results appear in **table 2**. Those students using calculators displayed a better attitude toward mathematics and an especially better self-concept in mathematics than students who had no formal contact with the devices.

Extending the Study

Subsequent to our meta-analysis, we have found nine additional studies that probed effects of calculator use in precollege mathematics. We used the results of these studies to extend our previous findings. **Table 3** updates the outcomes related to student achievement. In every instance, new data either supported or

Table 2

Calculator Effects on Student Attitudes (Original Study)

Dimension of Attitude	Size of Effect
Attitude toward mathematics	Small (+)
Self-concept in mathematics	Moderate (+)
Anxiety toward mathematics	Not significant

Note: (+) denotes a better attitude or feeling of calculator groups compared to noncalculator groups.

Table 3

Updated Calculator Effects on Student Achievement

		Skills area	
Test Condition	Student Ability	Computation	Problem Solving
With calculators*	Low	[Moderate (+)]**	[Moderate (+)]***
	Average	[Large (+)]**	[Small (+)]***
	High	No new data	[Moderate (+)]***
Without calculators	Low	[Not significant]**	[Not significant]**
	Average		
	Grade 4	No new data	No new data
	Other grades	[Small (+)]**	[Small (+)]**
	High	[Not significant]**	[Not significant]**

*Calculator groups only. Noncalculator groups used paper and pencil on tests.

Notes: 1. Previous findings are shown in brackets. (+) and (−) denote respectively higher and lower scores of calculator groups compared to noncalculator groups.
2. **denotes support of previous findings.
***denotes extension of previous findings in the same direction.

enhanced the previous findings. For tests *with* calculators, the new data showed—

1. continued advantage from calculators in computation;
2. better advantage from the devices in problem solving.

For tests *without* calculators, the data suggested that using the calculator during instruction may improve paper-and-pencil skills for low- and high-ability students in addition to those of average ability. However, these findings were not strong enough to revise the prior interpretations.

With regard to student attitudes, the data supported previous findings that calculators help promote a better attitude toward mathematics and an especially better self-concept in mathematics.

Student Attitudes toward Calculators

Attitudes toward calculators seem to have become more positive at the start of the 1990s. **Table 4** shows results of a recent survey of nearly 500 middle school students (Bitter and Hatfield 1991). In large proportions, the students favored the presence of calculators during most mathematics activities. However, they

Table 4
Percent of Students Responding "Agree" or "Strongly Agree" to Attitude Items (from Bitter and Hatfield 1991)

Attitude Item	Percent
Calculators make mathematics fun.	79.1
Mathematics is easier if a calculator is used to solve problems.	86.3
It is important that everyone learn how to use a calculator.	85.6
I would do better in math if I could use a calculator.	72.7
I prefer working word problems with a calculator.	69.6
I would try harder in math if I had a calculator to use.	49.3
Students should not be allowed to use a calculator while taking math tests.	28.3
The calculator will hinder students' understanding of the basic computation skills.	36.6
Since I will have a calculator, I do not need to learn to do computation on paper.	12.7

seemed to display more reserve regarding calculators on tests and regarding questions on whether or not the calculator will hinder the understanding of basic computational skills. Most students believed that calculators ought to be used in restricted conditions, to check arithmetic computations, and to learn particular topics.

Conclusions

We conclude this article with the following observations:

1. The preponderance of research evidence supports the fact that calculator use for instruction and testing enhances learning and the performance of arithmetical concepts and skills, problem solving, and attitudes of students. Further research should dwell on the best ways to implement and integrate the calculator into the mathematics curriculum.

2. In some of the synthesized research, calculators were used for instruction and then their use was denied for testing. To deny the use of calculators for both instruction and testing is an antiquated policy that should be changed.

3. The use of calculators in the early grades is frequently for familiarization, for checking work, and for problem solving. The senior high school seems to emphasize using calculators as tools for calculation and reference. There have been no empirical studies on how to integrate the calculator directly into the learning process.

4. The attitudes of students using calculators are favorable, but apparently some students still feel that using a calculator is tantamount to "cheating." The latter is a most unfortunate attitude for students to develop.

5. Calculators are gaining wide acceptance by teachers. The findings of research probably have accelerated that acceptance.

References

Bitter, Gary G., and Mary M. Hatfield. "The Calculator Project: Assessing School-wide Impact of Calculator Integration on Mathematics Achievement and Attitude." Unpublished manuscript.

National Council of Teachers of Mathematics (NCTM). "NCTM Board Approves Policy Statement on the Use of Minicalculators in the Mathematics Classroom." *NCTM Newsletter* 11 (December 1974): 3.

Suydam, Marilyn N. *The Use of Calculators in Precollege Education: A State-of-the-Art Review.* Columbus, Ohio: Calculator Information Center, 1979. (ERIC Document Reproduction Service No. ED 171 573)

———. *The Use of Calculators in Precollege Education: A State-of-the-Art Review.* Columbus, Ohio: Calculator Information Center, 1982. (ERIC Document Reproduction Service No. ED 220 273)

Calculators and Constructivism

Grayson H. Wheatley and Douglas H. Clements

Many conflicting views have emerged about the place of calculators in elementary school mathematics. Some teachers and many parents believe that the use of calculators will undermine mastery of the "basics" and thus should not be used, at least until students "know their facts" and are proficient with paper-and-pencil computations. Others suggest that in today's society, facility with calculators is essential. The National Council of Teachers of Mathematics has, for many years now, held the position that calculators should be used at all grade levels.

From a constructivist perspective (von Glasersfeld 1990), a calculator can aid mathematics learning when it—

1. permits meaning to be the focus of attention;
2. creates problematic situations;
3. facilitates problem solving;
4. allows the learner to consider more complex tasks; and
5. lends motivation and boosts confidence.

The use of a calculator can present potential mathematics learning opportunities by creating problematic situations. Consider the following two-person activity. (Use nonscientific, algebraic-logic calculators; check to make certain this activity will work.)

Guess my number

Pat chooses 59 as a mystery number and enters [59] [÷] [59] [=]. The calculator display shows "1." She hands the calculator to Wanda and asks, "Can you find my number? You may enter your guess and push the equals key. When you get 'one' you have found my mystery number."

Wanda replied, "I'll guess forty."

[40] [=] [0.6779661]

At this point, Wanda has little idea what all these digits mean. Her "problem" is giving meaning to the display of digits to achieve her goal. She next tries 90.

[90] [=] [1.5254237]

To Wanda, this looks like a bigger number. She follows this trial with the following entries.

[70] [=] [1.1864406]
[60] [=] [1.0169491]
[50] [=] [0.8474576]
[56] [=] [0.9491525]
[59] [=] [1]

Through this game, Wanda was able to give meaning to strings of digits showing on the calculator's display. The two numbers on each side of the "dot" took on special significance while the other digits were ignored. Successive experiences playing this game would deepen Wanda's concept of decimals. (This activity works because the calculator is storing the operation ÷ 59 and performs that operation on each of Wanda's entries.)

Jenny, a third-grade student, responded to the task shown in **figure 1** by adding a number to 85 and 63 until she found one that gave 244 as a sum. For her this task was not one for which the subtraction key on a calculator could be used. She entered the addends in the order they appeared on the page from left to right, 85, her number, and 63. In just a few minutes she was able to determine the number that gave the required sum. The calculator was useful in freeing Jenny from

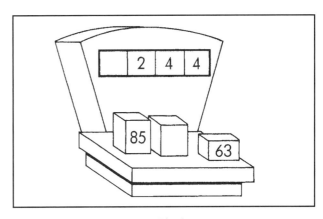

Fig. 1

computational rules and allowed her to focus on numerical relationships. In fact, Jenny did not have a reliable method for determining differences with three-digit numerals.

Patrick, another student in the same class, first tried to solve this problem without a calculator but soon asked, "Is there any way the calculator can do 148 plus how much will give 244?" In asking this question, he was taking a first step in relating a how-many-more question to a subtraction task. When he learned how to use the subtraction key to achieve his goal, Patrick was enthusiastic about this newfound capability and proceeded to try other subtractions. For each of these students, the calculator created a problematic situation and led to meaningful mathematical activity.

Considerable time is required for students to construct action schemas that include calculator use (Wheatley and Wheatley 1982). Students have been observed doing long division by paper and pencil while a calculator was lying near their hand. This situation is analogous to a carpenter's putting a power saw in the toolbox but continuing to use a familiar hand saw. With calculator availability and opportunities to use a calculator, students will slowly build patterns of action that include reaching for a calculator. Activities such as the "range game" facilitate this transition.

The "range game"

The range game (Reys et al. 1979) is a versatile calculator activity that can be used at any grade level. It has proved to be a powerful experience in building number sense. The task shown in **figure 2** is to find a number that when multiplied by 75 will give a result in the range from 400 to 625. Students are instructed to enter 75 in their calculator and choose a number they think will put them in the range. Ricardo chose 10 and found it was too large. He then tried 6, arriving at $75 \times 6 = 450$. But other numbers besides 6 will work. Students are encouraged to find all numbers (thinking whole numbers) that will satisfy the condition. The question "What is the smallest number that works?" can generate an interesting discussion. As students begin trying decimals such as 5.4, getting 405, and then 5.35, getting 401.25, they have an opportunity to construct meaning for decimals and develop number sense in the way the term is used in the NCTM's *Curriculum and Evaluation Standards for School Mathematics (Standards)* (1989).

Because calculators display quotients in decimal form, students may be productively perturbed when using a calculator. What does 3.456138 mean? In solving the problem "One bus can carry 34 students. How many buses will be required to carry 489 students?" Julie divided 489 by 34 on a calculator, looked at the display, saw a string of numbers with a dot, and exclaimed, "Goodness!" She quickly put the calculator down and did the division using the standard paper-and-pencil procedure. Being unfamiliar with a calculator, she expected the display to show her the type of information she would have attained if paper-and-pencil methods were used. She was unprepared for a decimal form and did not give it numerical meaning. A calculator fosters rich opportunities to reason mathematically, in this example to reflect on quotients expressed in decimal notation. The calculator activities "guess my number" and "range game" described in the foregoing offer just such opportunities.

In the 1990s, the view of mathematics as a fixed set of procedures to be mastered by practice is giving way to a different conceptualization, that is, mathematics as the activity of constructing patterns and relationships. Calculators can play an important role in students' construction of mathematical relationships. However, we must think of their use in a quite different instructional setting and with activities markedly different from explain-and-practice lessons or even so-called discovery teaching where the teach-

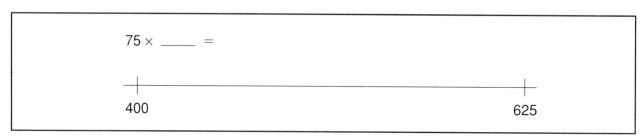

Fig. 2

er directs students to a principle to be learned. Students must be free to construct meaning in ways that make sense to them. Shifting attention from carrying out tedious paper-and-pencil computations to solving problems and building relationships results in conceptually powerful learning. When used in a learning environment compatible with constructivism (Cobb et al. 1988) as illustrated in this article, calculators can facilitate students' construction of number patterns that form the basis of mathematical reasoning by allowing many complex computations to be performed quickly and accurately while attention remains focused on meaning. The activities described herein suggest how calculators can be used to achieve these goals.

References

Cobb, Paul, Erna Yackel, Terry Wood, and Grayson Wheatley. "Research into Practice: Creating a Problem-Solving Atmosphere." *Arithmetic Teacher* 36 (September 1988): 46–47.

National Council of Teachers of Mathematics (NCTM), *Curriculum and Evaluation Standards for School Mathematics.* Reston, Va.: NCTM, 1989.

Reys, Robert, Barbara Bestgen, Terrence Coburn, Harold Schoen, Richard Shumway, Charlotte Wheatley, Grayson Wheatley, and Arthur White. *Keystrokes: Multiplication and Division.* Palo Alto, Calif.: Creative Publications, 1979.

von Glasersfeld, Ernst. "An Exposition of Constructivism: Why Some Like It Radical." In *Constructivist Views on the Teaching and Learning of Mathematics. JRME* Monograph Number 4, pp.19–29 Reston, Va.: National Council of Teachers of Mathematics, 1990.

Wheatley, Grayson, and Charlotte Wheatley. *Calculator Use and Problem Solving Strategies of Grade Six Pupils.* Final Report. Washington, D.C.: National Science Foundation, 1982.

Calculators and Computers: Tools for Mathematical Exploration and Empowerment

Michael T. Battista

Over the last decade, major advances in technology have brought new and exciting possibilities to mathematics education. The universal availability of calculators—both as stand-alones and as built-ins for such devices as cash registers—is having a profound impact on what should be taught in mathematics curricula. Computers are furnishing increasingly more powerful learning environments for students. Both devices require and support enhanced development of students' mathematical reasoning. This article discusses a number of ways that calculators and computers can be used as tools for exploration and empowerment in school mathematics.

Calculators

Although many people originally feared that ready access to calculators would harm students' mathematical development, especially their ability to compute, much research suggests that calculator use in mathematics instruction enhances rather than inhibits students' learning (e.g., see Hembree and Dessart [1992]). In fact, such fear seems inappropriate now that the widespread availability of calculators has made traditional skill with paper-and-pencil computational algorithms, and therefore much of the current school mathematics curriculum, obsolete. Instead of worrying about the possible ill effects of calculator use on students' computation, we should be investigating how a modern curriculum can best employ calculators to enhance students' overall mathematical thinking. The following example illustrates how such thinking might develop if calculators are fully embraced in mathematics teaching.

What pairs of numbers have a product of 9?

Jonathan is a fifth grader who has explored mathematics meaningfully and used calculators throughout his school years. He overheard the interviewer ask a third grader to find pairs of numbers whose product is 9. After hearing the products 9×1, 1×9, and 3×3. Jonathan posed his own question. "Are there more numbers? You can use decimals or fractions." When no one replied, he offered the answer that seemed to prompt his query: $4\ 1/2 \times 2$.

Interviewer (I): Are there others?

Jonathan: Yeah. 6×1.5, $8 \times 1\ 1/8$, $4 \times 2\ 1/4$, $7 \times 1\ 2/7$ (working mentally, with obvious pauses between answers)

I: Great!

Jonathan: Are there more?

I: Lots more.

Jonathan: [Pause] $30 \times .3$.

I: How did you get that?

Jonathan: One tenth (of 30), times 3 is 9.

Jonathan then worked by himself with paper, pencil, and a calculator that did not "do" fractions. After about thirty minutes, he had completed a chart listing the numbers from 1 to 21 on one side, and beside each number was listed the other number in the pair whose product is 9. Some of these numbers were in decimal form, some were in fractional form, and some were expressed both ways. See **figure 1**.

When the interviewer asked Jonathan about specific products, it was clear that he had calculated most of the products mentally, using the calculator only when he was uncertain of his computations. For example, when he was asked how he had gotten $10 \times .9$, he said that he just knew it. He derived $20 \times .45$ from $10 \times .9$ by taking half of .9 and doubling 10. He

---

Support for part of the work reported in this paper was provided by grants MDR 8954664 and 9050210 from the National Science Foundation. The options expressed, however, do not necessarily reflect the views of that foundation.

x	y
1	9
2	4.5
3	3
4	2.25
5	1.80
6	1.5
7	1 2/7
8	1 1/18 or 1.125
9	1
10	.9
11	9/11
12	3/4 or .75
13	9/13
14	9/14
15	3/5 or .6
16	9/16
17	9/17
18	1/2 or .5 or 3/6
19	9/19
20	.45
21	3/7 or 9/21

Fig. 1. Jonathan's number pairs for 9

also switched between fractions and decimals quite easily. For instance, even though he wrote 4.5, he referred to this number as "four and a half."

I: How many pairs do you think there are?

Jonathan: Billions (short pause). There's one for every number. Like a googolplex. It would be a googolplex times 9 "googolplex-ths [sic]." (To Jonathan, a googolplex was an unimaginably large number; he had encountered the term in his reading.)

I: You mean 9 over a googolplex?

Jonathan: Yes.

I: Could you find pairs of numbers that multiply together to give 11?

Jonathan: Sure. 1×11, $2 \times 5\ 1/2$, $3 \times 3\ 2/3$.

I: How are you getting these? How did you get that last one?

Jonathan: Well, 3×3 is too small (less than 11), and 3×4 is too much. So it's 3. Then 3×3 is 9, so you need *two*-thirds.

In this episode, Jonathan demonstrated excellent mental-mathematics skills—skills built on a firm conceptual understanding of fraction and decimal multiplication and of the relationship between fractions and decimals. He also generalized his ideas to discover that all nonzero whole numbers could be multiplied by some number to get 9 and that the same principle applied to other whole numbers—important steps in understanding the system of rational numbers.

Furthermore, the powerful conceptual knowledge that Jonathan developed about rational numbers served as a basis for moving him toward developing a symbolic multiplication algorithm for decimals. Indeed, when the interviewer later asked him if he could figure out how to multiply decimals with paper and pencil, his first attempt was $75 \times .5 = 35.25$. He tried to separate whole number and decimal parts of the product with a decimal point. Such use of the decimal point to separate whole number and fractional parts of a number is consistent with his textbook series' treatment of money, in which the decimal point separates dollars and cents. So the interviewer attempted to activate his conceptual knowledge about the situation by asking, "What is .5 times 75? What's half of 75?" He replied, "37.5" and recognized the error in his original answer. He was then asked to try $1.5 \times .5$, to which be replied, ".75." Finally, he was given $2.5 \times .7$. He multiplied the whole numbers 25 and 7 using the traditional whole-number algorithm, then said, "One decimal number here and one here (pointing to 2.5 and .7), so two (decimal places)—(the answer is) 1.75." So Jonathan was well on his way to discovering the traditional symbolic algorithm for multiplying decimals.

Jonathan's work suggests several important points

about the interrelationships among computation, numerical reasoning, and calculator use. First, Jonathan accomplished his powerful reasoning about products of rational numbers *without* knowing computational algorithms for multiplying either decimals or fractions. He computed most products mentally, using a variety of numerical relationships. Thus, this episode demonstrates that students who have developed fluent conceptual knowledge of mathematics do not need to know computational algorithms to enable them to think powerfully about operations on numbers. Conceptual knowledge and reasoning, not knowledge of computational algorithms, empowers students working on real mathematical problems.

Second, this example suggests several positive effects of Jonathan's frequent use of a calculator. Because Jonathan has spent much time working on problems using a calculator, he feels comfortable using a calculator for all types of mathematical problem solving. The calculator serves as a tool that supports his mathematical reasoning and concept development. For instance, when he was unsure of his mental computations involving factors of 9, having a calculator available and knowing how to use it for operating on rational numbers allowed him to check and reflect on his ideas, supporting his exploration. He has, in fact, discovered many numerical relationships by exploring ideas with a calculator. Also, Jonathan has learned when mental computation and calculator use are most appropriate. For instance, after he learned how to find the decimal equivalent of a fraction on the calculator, he spent time exploring decimal equivalents of common fractions, making lists of those that interested him. So at this point, when he needs to find the answer to a product or quotient that involves a fraction, if he cannot do the computation in his head he converts it to a decimal and uses a calculator. Moreover, Jonathan's use of a calculator has permitted his knowledge of fractions and decimals to become far more integrated than typically happens in a traditional curriculum.

In conclusion, it is obvious that Jonathan is not typical of fifth graders educated in the traditional, computationally focused elementary mathematics curriculum. Even in an improved curriculum that focuses on understanding and inquiry, we would not expect all students to make the discoveries that Jonathan did. But in an improved curriculum, we can and should expect that students develop the conceptual understanding and type of mental-mathematics skills that Jonathan exhibited. Thus, the discussion of Jonathan's work illustrates the type of mathematical reasoning that a modern mathematics curriculum should strive to develop in students and the role that calculators can play in supporting that development. It should also dispel the myth that becoming skilled with computational algorithms is a prerequisite for more advanced concept development and problem solving. Fluency with conceptual ideas can develop into powerful mental techniques that often obviate the need for computational devices or algorithms.

Several suggestions for classroom instruction with calculators are given here. However, the most important suggestion for using calculators in the classroom is to view the calculator not as a danger or as a mere computational device but rather as a tool that can support students' constant involvement in mathematical reasoning and exploration.

Suggestions for the classroom

1. Numerous suggestions for calculator use are given in the NCTM's 1992 Yearbook (Fey 1992) and in articles published in the *Arithmetic Teacher*. For example, the range game has been found to promote an understanding of numerical relationships, estimation skills, decimals, and the idea of limits (Wheatley and Shumway 1992). In the multiplication version of this game, students might be asked to enter 34 into a calculator then determine which numbers, when multiplied by 34, will result in a product between 400 and 500. After students have found all the whole-number solutions to the problem, if none of them have discovered decimal numbers, the teacher might introduce one: "I tried 12.5 and it worked. Might other numbers like this also work? Why?" Note that students do not have to wait until they "cover" decimals in the textbook before they become involved in this game. The game can be used to introduce students to decimals.

2. Several worthwhile classroom activities are suggested by Jonathan's discussion. *(a)* Students can be introduced to mathematical concepts by having them explore real-world problems with a calculator. For instance, with appropriate guidance, they can be asked to make sense of computing with money or calculating baseball batting averages before they have been taught decimals. *(b)* Having

students investigate how to work with fractions using calculators that don't "do" fractions offers a rich source of opportunities for exploring the connection between fractions and decimals. The teacher can ask students, "How can I put the fraction 1/2 into the calculator? How can I multiply 10 by 1/2 on the calculator?" *(c)* The problem pursued by Jonathan is also rich: How many different pairs of numbers can you find that will multiply to make 9? A similar problem might be posed for addition. Each activity affords students an opportunity to explore and make sense of sophisticated mathematical ideas. We should not, however, expect such activities to produce full understanding of these mathematical concepts. Rather, they present exploratory opportunities that contribute to that understanding.

Computers

Computer microworlds consist of objects and actions for manipulating those objects. Such microworlds are used to enhance students' development of mathematical concepts and reasoning. They do so by supporting mathematical exploration and by furnishing environments in which students can more easily reflect on manipulations of mathematical objects they have created. Two examples are given here for geometry.

Logo

In Logo (LCSI 1986; Terrapin Software 1993), students create geometric figures by typing commands that direct the movement of a "turtle" on the computer screen. As the turtle responds to such commands as FORWARD 50 and RIGHT 90, it leaves a trace of the path it has followed. Research has shown that properly designed Logo activities can encourage students to think about geometric objects in terms of their components and geometric properties rather than simply as visual wholes, encouraging them to move to higher levels of geometric thought (Clements and Battista 1990; Clements and Battista 1992). And because Logo is based on continuous measurement, work with it gives students a chance to recreate in a different context numerical ideas that they have previously used only in discrete situations. This experience broadens students' understanding of number and gives them a powerful tool for geometric thinking.

Much Logo work reported in research has been conducted with specially designed microworlds. For instance, *Logo Geometry* (Battista and Clements 1991a) is a K–6 curriculum that has special microworlds for different grade levels and for investigating such topics as paths, turns and angles, shapes, motion geometry, and geometric constructions. These microworlds serve as the basis for sequences of activities that focus on geometry, not computer programming. Some activities from this curriculum can be done in any version of Logo. Several of these are described next.

Classroom activities with Logo

1. Have students use Logo to draw squares, nonsquare rectangles, and other polygons. This pursuit encourages them to think about the parts of these shapes and how they are related, as well as to use numbers to describe and analyze geometric shapes.

2. When asked to draw squares and rectangles in Logo, almost all students will use 90-degree turns to make horizontal and vertical sides. But many will not explicitly understand that the right angles formed are a characteristic of all squares and rectangles. Asking them to draw tilted squares and rectangles (e.g., by showing them a square rotated 45 degrees from the horizontal) encourages them to reflect on this idea more carefully (Battista and Clements 1991b). Such tasks help students progress beyond simple procedural knowledge of making shapes in Logo, encouraging them to think of Logo commands as expressions of properties of shapes.

3. Ask students how many rectangles with perimeter 200 can be drawn in Logo. This problem encourages reflection on properties of rectangles (e.g., that opposite sides are equal). It also promotes a deeper understanding of the connection between manipulations of numbers and the resulting changes to the geometric objects the numbers describe.

Promoting dynamic imagery that supports geometric reasoning

In one important type of computer microworld, the objects and actions are designed to be compatible with the mental operations students use in meaningful mathematical activity. Researchers have conjectured that performing or contemplating performing the computer's actions will encourage students to create and apply the corresponding mental operations (Steffe 1993).

The Shape Makers program (Battista 1998) is one example of such a microworld. In this microworld, each class of common geometric shapes has a "shape maker," a construction by The Geometer's Sketchpad (Klotz 1992) that can be dynamically transformed in various ways, but only into different shapes in the same class. For instance, by using the computer's mouse to grab and move small circles, which we call "handles," located at its vertices, the rectangle maker can be manipulated to make any desired rectangle, no matter what its shape, size, or orientation—but only rectangles. See **figure 2**. Shape makers can be duplicated and moved around on the computer screen. Quadrilateral shape makers are available for squares, rectangles, parallelograms, kites, rhombuses, trapezoids, and quadrilaterals. Triangle shape makers are available for general, isosceles, equilateral, and right triangles.

The Shape Maker microworld is designed to promote the development of dynamic mental models for thinking about geometric shapes. The power of such models is illustrated by a second grader who had been contemplating the notion that squares are special types of rectangles. She made sense of the idea not by referring to verbally stated properties of these shapes but by thinking about how some "stretchy square bathroom things" could be stretched into rectangles. Thus, she reasoned by performing a simulation of changing a square into a rectangle using a special type of visual transformation, one that incorporated some geometric constraints—preserving 90 degree angles—on her mental objects and actions (Battista 1994). Such reasoning can be extremely important in mathematical thinking (Steffe 1990; Battista 1994).

Observations of students suggest how the shape makers can encourage the use of such imagery and become powerful tools with which students can reason about shapes. Three fifth graders were considering whether the parallelogram maker could be used to make trapezoid I in **figure 3**. They did not know enough about the properties of parallelograms to predict that this construction was impossible. However, as they manipulated the parallelogram maker—which for students can become a concrete representation of the *class* of parallelograms—they discovered something special about parallelograms that enabled them to solve the problem. Pointing to

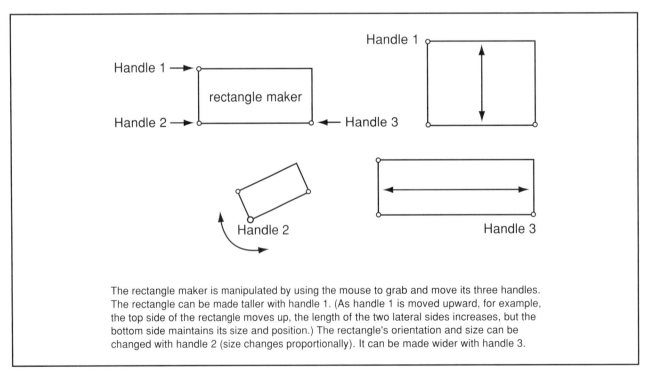

Fig. 2. Rectangle maker

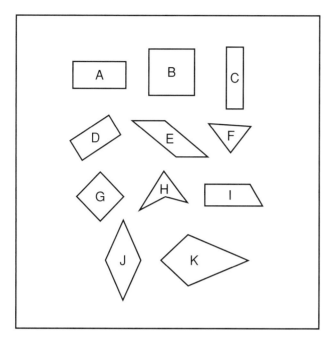

Fig. 3. Find all quadrilateral shape makers that can be used to make each shape above.

one pair of opposite sides, one student said, "No, it won't work. See, this one and this one stay the same, you know, together. If you push this one [one side] out, this one [the opposite side] goes out.... This side moves along with this side." Significantly, this student inferred a solution from her discovery and imagery rather than from a failure to make the trapezoid with the parallelogram maker.

Activities with Shape Makers

1. Ask students to make a picture of a robot or building with the rectangle maker. They should use at least eight rectangles. This activity helps students develop a more diversified conceptual image for rectangles. That is, the ease with which the shape makers can be used to generate numerous and varied examples of shapes can help encourage students to associate more varied images with classes of shapes, an ability that has been found to be important in the use of imagery in geometric thinking (Clements and Battista 1992).

2. Ask students to make five different rectangles with their rectangle maker. Ask: How are these five shapes the same and different? What stayed the same and what changed as you manipulated the rectangle maker? What's the same about all shapes that can be made with the rectangle maker? Repeat this task with other shapes. This activity encourages students to think about the properties of shapes, an important notion in moving to more sophisticated geometric thinking (Clements and Battista 1992). And it does so in a way that encourages the integration of this property-based knowledge with imagery.

3. Ask students which of the shapes in **figure 3** can be made with each shape maker. They should make predictions and then check with the shape makers. If a figure cannot be made by a shape maker, they should explain why not. This activity encourages students to think about the properties of shapes. For example, for shape E, students might conclude, "You can't do it with the rectangle maker because it doesn't have 'square corners' (90-degree vertex angles)." The activity can also encourage students to think about classifying shapes, promoting an even higher level of geometric thinking. For example, some students will predict that shape B cannot be made by the rectangle maker "because a square is not a rectangle."

4. The following questions further encourage students' thinking about the classification of shapes.

   a) How are the square and rectangle makers related? Can every shape made by the square maker be made by the rectangle maker? Why? Can every shape made by the rectangle maker be made by the square maker? Why?

   b) How are the parallelogram and rectangle makers related? Can every shape made by the parallelogram maker be made by the rectangle maker? Why? Can every shape made by the rectangle maker be made by the parallelogram maker? Why?

Conclusion

Calculators and computers are important computational devices for children and adults alike. Mathematics instruction should help students learn to take advantage of the computational power that such devices offer. However, for mathematics teachers, calculators and computers should be much more. They are invaluable tools for supporting the type of exploration and problem solving that is requisite for developing powerful mathematical reasoning. We must continue to explore ways in which these devices can best be used to empower students mathematically.

References

Battista, Michael T. "On Greeno's Environmental/Model View of Conceptual Domains: A Spatial/Geometric Perspective." *Journal for Research in Mathematics Education* 25 (January 1994):86–94.

———. Shape Makers: Developing Geometric Reasoning with The Geometer's Sketchpad. Berkeley, Calif.: Key Curriculum Press, 1998. Software

Battista, Michael T., and Douglas H. Clements, *Logo Geometry*. Morristown, N.J.: Silver Burdett & Ginn, 1991a.

———. "Using Spatial Imagery in Geometric Reasoning." *Arithmetic Teacher* 39 (1991b):18–21.

Clements, Douglas H., and Michael T. Battista: "The Effects of Logo on Children's Conceptualizations of Angle and Polygons." *Journal for Research in Mathematics Education* 21 (1990):356–71.

———. "Geometry and Spatial Reasoning." In *Handbook of Research on Mathematics Teaching,* edited by Douglas A. Grouws, pp. 420–64. Reston, Va.: National Council of Teachers of Mathematics and Macmillan, 1992.

Fey, James T., ed. *Calculators in Mathematics Education.* 1992 Yearbook of the National Council of Teachers of Mathematics (NCTM). Reston, Va.: NCTM, 1992.

Hembree, Ray, and Donald J. Dessart. "Research on Calculators in Mathematics Education." In *Calculators in Mathematics Education,* 1992 Yearbook of the National Council of Teachers of Mathmatics (NCTM), edited by James T. Fey, pp. 23–32. Reston, Va.: NCTM, 1992.

Klotz, Eugene. The Geometer's Sketchpad. Berkeley, Calif.: Key Curriculum Press, 1992. Software.

LCSI. Logowriter. Montreal: Logo Computer Systems, 1986. Software.

Steffe, Leslie P. "Inconsistencies and Cognitive Conflict: A Constructivist's View." *Focus on Learning Problems in Mathematics* 12 (1990):99–109.

———. "Interaction and Children's Mathematics." Paper presented at the annual meeting of the American Educational Research Association, Atlanta, April 1993.

Terrapin Software. Logo. Portland, Maine: Terrapin Software, 1993. Software.

Wheatley, Grayson H., and Richard Shumway. "The Potential for Calculators to Transform Elementary School Mathematics." In *Calculators in Mathematics Education,* 1992 Yearbook of the National Council of Teachers of Mathematics (NCTM), edited by James T. Fey, pp. 1–8. Reston, Va.: NCTM, 1992.

Constructing Geometric Concepts in Logo

Michael T. Battista and Douglas H. Clements

Why is it that students often do not learn what they are taught? On what do they base their thinking? What can we as teachers do to help them construct accurate and robust understandings?

Our research has convinced us that furnishing Logo tools for manipulating embodiments of geometric objects helps students construct more abstract and coherent concepts. It also consistently supports several related constructivist themes. First, contrary to intuition, students frequently do not use verbal definitions and rules when they think (Clements and Battista 1990). They use conceptual structures that they have constructed out of pieces of in- and out-of-school experiences. Even when students' verbal definitions or descriptions are correct, these conceptual structures—often faulty—actually rule their thinking. Second, students need personally to manipulate ideas and reflect on them to construct concepts, even after "clear" explanations and demonstrations have been given by the teacher. We shall illustrate our findings with several examples of students' learning.

Constructing Concepts versus Learning Verbal Definitions

In one study we conducted, one group of students had worked with Logo during a free period of the day: the other had not (Clements and Battista 1989). The two groups received identical classroom instruction in geometry. Even *after* being taught the standard textbook definition, most of the non-Logo students' definitions of angles reflected mainly the influence of common language uses. Typical responses included the following: "It's like a line"; "Like a line that goes on a slant or a corner"; and "Like if you were in the classroom here (to the side) and the chalkboard were here, then that would be a bad angle to see." Students who had worked with Logo displayed different conceptual structures. Their responses emphasized more dynamic and abstract ideas, for example, "An angle is a turn"; "Sort of a line from a turn"; and "It's when two lines meet each other and they come from two different ways."

Students exhibited other clear signs that they were not tied to verbal definitions—even their own. For instance, we asked students to identify figures that were angles. Several Logo students had defined an angle only as a turn, but they nevertheless correctly identified all the angles. Other Logo students correctly defined angles but also included diagonal line segments. Note that drawing such segments in Logo would require a turn if the turtle started from its "home" heading.

More than twice as many Logo students responded correctly to the question "How many angles in this figure (an 'X')?" Again, however, alternative conceptions emerged in both groups. One Logo student responded "zero." She explained, "When you turn after drawing a straight line, *that's* an angle." For her, the figure had no angles but would have had two "if you had turned." A non-Logo student had a different notion: "If it were two halves of two triangles, it would be two angles. If it was an 'X', it wouldn't be any angles." Thus, the *way* a figure was made or viewed was important to these students in deciding whether or not it was an "angle." Neither their textbook nor their teachers had ever mentioned such an idea.

Tasks for Integrating Turns and Angles

In our present work with Logo, we encourage students to reflect on their physical and mental actions. Logo then serves as a transitional device between physical experiences, such as moving and turning, and abstract mathematics. It allows students actively to explore and invent mathematical models and ideas.

For example, you might ask your students the following questions: "What is a turn? Can you turn a whole turn—all the way around? Can you turn a half turn? Turn to the right or do a 'right face'; how much of a whole turn is that? Which way would you be facing if you turned four (or three or two) of these turns? So what portion of a whole turn is it?"

For some students it is beneficial to introduce turns

that are smaller than quarter turns with a "floor clock." Focus on the amount of turn from one "clock number" to another. Say, "Point at two o'clock. Turn right two units. Where are you pointing now? How many turns of this type must you make before you are facing in the same direction you started?"

Explain that the Logo turtle uses much smaller units. Show a transparency with 360 radii superimposed on the clock transparency. Explain that 360 turns of this size (degrees) are required to turn all the way around. Ask students how many degrees are needed to turn halfway around; a quarter way around; and one to six clock units.

Have students play simple Logo games, such as "Hit the Spot," that emphasize estimation of the amount of turn. One partner places a finger or small sticker anywhere on the screen. The other types in as few Logo commands as possible to place the turtle directly beneath the other's marker. A related activity has one partner create a list of commands off the computer. The other partner puts a sticker on the screen at the place he or she believes the turtle will stop. The prediction is checked by typing in the commands. In these and all following work with Logo, encourage students to reflect on their use of reference measures (e.g., "That looks like ninety degrees and about half of ninety degrees more. I'll try one hundred thirty-five degrees"), thereby laying a foundation for the notion of angle classification.

Furnish activities to help students integrate their understanding of turn with the concept of angle. For example, give students sets of commands (in the pattern FD 50 RT __ FD 50) and ask them to predict the path the turtle will draw for various entries of RT. They check their predictions on the computer. Explain that these special paths are called *angles*. Ask the students to describe angles in path language. Older students can use protractors to measure the angles formed. By examining a table with the headings "amount of turn along path" and "measure of angle formed," they can discover the relationship between these two quantities.

Group Discussions and Personal Constructions

A fifth-grade teacher is sitting at a computer that is connected to a screen that all students can see. She is typing commands given by the students. They are attempting to write a procedure to cause the turtle to maneuver through a maze to each of several points on the screen, returning home after each. The path home is to be the same as the path *to* the destination. The students have directed the turtle to the first destination, a restaurant.

Teacher: How are we going to get the turtle back?

Jonathan: Everything that was a forward is a backward, and everything that was a right is a left, and everything that was a left is a right, probably.

Teacher: So you're telling me what?

Sally: Reverse the stuff.

Jonathan: You reverse, go from the bottom (of the procedure) to the top.

Teacher: What comes first?

The students now correctly start giving the commands to "undo" the original path.

Jonathan: We've got a problem here because here's the restaurant; it goes out of the restaurant BK 10; then right would be up here, so it's going to get screwed up.

Robin: Now it's going *backward*.

As the students discuss Jonathan's misgiving about the proposed solution, Andy figures out his mistake by standing up and acting out the commands. A consensus is reached about the commands to enter. As the commands are then entered into the computer, the students agree that their proposed solution is correct.

In this episode, the students are intensely involved with solving a problem. Consistent with the NCTM's curriculum standards (1989), the students are discussing mathematical ideas and making and evaluating conjectures and arguments. Although the teacher is entering the students' commands into the computer, she does not judge the correctness of the students' ideas. The students judge their ideas by examining the results on the computer screen.

At this point, students are sent to the computers in pairs to continue the activity. Emily and Ryan excitedly attempt to get the turtle to the second destination. As they try the commands for the path they have planned, they express no surprise or disappointment when it does not work. They anticipated that it would not be perfect and that they would have to alter it. Because the computer permits Emily and Ryan to evaluate their conjectures, it enables them to strike out on their own in solving the problem—promoting the development of autonomous thinking in mathematics.

Immediately on seeing that their solution is correct, Emily says excitedly, "Let's get him back," Initially, she thinks that the first command in getting the turtle back to its starting point is the reverse of the first command in the procedure for getting the turtle to the destination. But Ryan says that it must be the reverse of the last command. They examine the instructions for returning the turtle from its first destination (written as a class). When asked, "Why does it have to be this [the last] command?" Emily and Ryan simultaneously offer several explanations, all of which have correct elements but do not answer the question. Finally, Emily explains the decision: "We started from here [turtle's starting point] and ended here [draws in the *air* a curved path that is not a copy of the screen path], so we want to start here this time [indicates destination], so we have to start with the last one [points to the last command in the original procedure]."

Even though the idea of undoing a path had been discussed the previous day in class, Emily and Ryan needed to apply and reflect on it themselves to make the idea personally meaningful. Thus, this episode illustrates the power of individual work on computers and the relationship between this work and class discussions. The individual work seems essential to students' constructions of ideas. It is as if the class discussions offer blueprints for the students' constructions. The actual construction does not take place until the students themselves manipulate the ideas.

Logo allows them to do so—personally to manipulate the ideas and decide on their validity and exact form.

Maze Tasks

Maze tasks like those we have used in our project can be created in various ways. The easiest might be to draw mazes on overhead transparencies and tape them to the computer screens. We used Logo and "paint" programs to create and save screen mazes.

Be sure to direct students not only to get the turtle to various locations in the screen mazes but to return it along the same path. This task lets them encounter the fundamental idea of "undoing" a sequence of operations. It is the same idea encountered in such problems as "I'm thinking of a number. When I multiply it by three, then subtract five, I get sixteen. What is my number?"

An extension of the original maze problem is possible in versions of Logo that have a "window" command that allows the turtle to go off the screen. Give students the directions for some destination that is off the screen. Ask them to return the turtle to its starting point. You might also give students directions from one off-screen destination to another. Have the students get the turtle to one destination after the other, then return the turtle to its starting point.

References

Clements, Douglas H., and Michael T. Battista. "Learning of Geometric Concepts in a Logo Environment." *Journal for Research in Mathematics Education* 20 (1989): 450–67.

———. "Research into Practice: Constructivist Learning and Teaching." *Arithmetic Teacher* 38 (September 1990): 34–35.

National Council of Teachers of Mathematics (NCTM). *Curriculum and Evaluation Standards for School Mathematics.* Reston, Va.: NCTM, 1989.

Section 12

Other Representational Tools

Giving Pupils Tools for Thinking

Robert B. Davis

Research ...

Pupils often invent mathematical ideas and methods that go beyond what has been taught in class. This inventiveness is frequently noted in research studies based on careful observation of pupils talking about mathematical problems. See, for example, the videotaped lesson *Double-Column Addition: A Teacher Uses Piaget's Theory* by Constance Kamii (1989). See also Cochran, Barson, and Davis (1970) and Whitney (1985). Clearly we want to build on these pupils' ideas, not discourage them (Corwin 1989). They do, however, pose an interesting problem for teachers: When pupils invent pieces of mathematics on their own, they usually do not have a command of the standard words and they often are not entirely clear about the concepts. How are we going to accept the pupil's original ideas and show our approval of them yet help the pupil move on to more sophisticated and more powerful concepts? Research suggests an answer.

I want to look at an example of a pupil who is beginning to invent negative numbers but has not quite managed to get the concept right. Help is needed, and I want to suggest a possible strategy that teachers can use to give this help.

First, let's look at some pupils. On the Kamii videotape a second-grade class, taught by teacher Linda Joseph, is given this problem:

$$\begin{array}{r} 26 \\ -17 \\ \hline \end{array}$$

(Despite the title of the videotape, the lesson includes some problems in subtraction). Gary, a second grader, solves the problem by saying that 20 minus 10 is 10; then he subtracts 7, getting 3; finally, he adds 6, for an answer of 9. This method is his own invention, not one he had been taught. Gary's work probably does not call for any immediate intervention by the teacher, and the teacher gives none.

Elizabeth is the next second grader to describe a method of solving this problem. This time the pedagogical situation is different. Elizabeth says that 10 from 20 leaves 10. Then she says, in effect, that 6 minus 7 is 0. In fact, Elizabeth has a correct idea, which becomes clear when she continues her solution: She now subtracts 1 from the 10, getting the final answer, 9. (Presumably Elizabeth had in mind the idea that if I have 6 and I owe you 7, I'll have none left and I'll owe you 1, to be made good at some time in the future.) Elizabeth is beginning to invent negative numbers, as children do quite often, sometimes calling them "below zero" numbers (presumably referring to thermometers) or "numbers that I owe you."

Here we see a teacher's dilemma: Elizabeth is thinking about the problem in a creative and essentially correct way, which we surely do not want to discourage. But her language is likely to get her into trouble sooner or later. What should a teacher do? Fortunately, research does suggest an answer. The teacher can help Elizabeth, and the other students, by suggesting a metaphor or mental representation that they can use in thinking about such problems.

... Into Practice

We all think of new matters by relating them to something that we already know (Davis 1984, 1990; Lakoff and Johnson 1980; Johnson 1987), and so do pupils when they speak of "below zero" numbers in terms of the already familiar notion of a thermometer. This reference sounds commonplace enough—but in classrooms one rarely sees it used to maximum effect. The implication for teachers is that one should help a pupil create a very clear mental picture (called an *assimilation paradigm* [Davis 1984]) on the basis of very familiar objects whose possibilities are well understood by the child. In this example, Elizabeth has started to invent—and make use of—negative numbers, but she has not clearly identified some familiar idea to which she can relate her new thoughts. If the teacher does not help her to develop an appropriate assimilation paradigm, Elizabeth will probably develop some not-quite-correct ways of thinking about negative numbers, which, unfortunately, is also a very common occurrence, as research shows.

An assimilation paradigm

Think about the different kinds of numbers that pupils encounter. Most familiar are the "how many" numbers, which answer questions like "How many brothers do you have?" These are clearly the numbers 0, 1, 2, 3, ... Other numbers arise in measurement: things don't always turn out to be exactly two cups, or exactly seven inches, and we have to deal with more than two cups but less than three. These "measurement" numbers are clearly whole numbers plus nonnegative fractions or decimals. How, then, will a pupil encounter negative numbers? Not from "how many" questions and not from the usual kinds of measurement. But all is not lost. The "signed numbers" (positive, negative, and zero) arise in reference-point situations, in which the starting point is arbitrary, as in the example of temperature in which no convenient starting point—zero—can be made to represent "the temperature on the coldest day I've ever seen," or something of that sort. Instead, we establish a clear and definite reference point, perhaps based on the temperature of water and ice in equilibrium.

So if we are going to give pupils a way to think about signed numbers, we need to give them a convenient example of a reference-point situation, something that can very easily be visualized. One that has been used effectively was introduced in Accra, Ghana, and uses a bag and a pile of pebbles (Davis 1967, chap. 4; 1984, p. 312). At the start of the activity, some of the pebbles are in the bag and some are outside of the bag, in a pile on the table. We mark a reference point in time by having some child—Mary, say—clap her hands. Suppose we then put five pebbles into the bag. To keep track, we write "5" on the chalkboard. We next ask, "How many pebbles are in the bag?" Of course we don't want to empty out the bag and count—that would never lead us to negative numbers. Instead, we refer to our reference point (which, of course, is why it is called a *reference point*). We ask, "Are more or fewer pebbles in the bag now than when Mary clapped her hands?" Since we have put five pebbles into the bag, the answer is that more are in the bag. We ask, "How many more?" Again, no child doubts the answer: Five more are in the bag. We write this result as "+5" to indicate that we have five *more* (+5 is read as "positive five").

Notice that we are not saying that the bag contains five pebbles; it contains many more than that because it held a large but unknown number of pebbles to start with. What we are saying is that five *more* pebbles are in the bag than when Mary clapped her hands.

Then we get to the exciting part. Suppose we next remove six pebbles from the bag. To summarize: Mary clapped her hands; we put five pebbles into the bag; we took six pebbles out of the bag. We can write

$$5 - 6.$$

Are more or fewer pebbles in the bag than when Mary clapped her hands? Answer: Fewer, since we put five in and took six out. How many fewer? Answer: One less than when Mary clapped her hands. We describe this state of affairs by writing

$$5 - 6 = {}^{-}1$$

(we read the symbol $^{-}1$ as "negative one"). In trials with thousands of pupils, this metaphor, or special case (or, to give it its correct name, this "assimilation paradigm"), has worked reliably with pupils, provided that it is presented carefully. A pupil can now think about, say, the $6 - 7$ problem that had concerned Elizabeth by imagining that Tony claps his hands, we put six pebbles into the bag, we take seven out, and we ask, "Are there more or less?" Answer: Less. "How many less?" Answer: One less. "How do we write that?" Answer: $6 - 7 = {}^{-}1$.

Why does this assimilation paradigm, or tool with which to think, allow pupils to think about such problems for themselves? *Because they already know about bags, and putting things into bags, and taking things out of bags.* They do not have equal experience with stock prices, which is why we use pebbles and a bag instead of the closing price of IBM common stock on the New York Stock Exchange, even though those examples work exactly the same way that the pebbles do.

Of course, after we show Elizabeth the "pebbles in the bag" idea, she must still do a lot more thinking before she has a full command of the use of negative numbers in solving arithmetic problems. We have not shown her how to solve all these problems, but we have given her some intellectual tools that she can use to think about such problems because she already knows about bags and pebbles and about putting things in and taking things out.

Returning to the subject of pupils' procedures for two-digit subtraction, we see the powerful thinking of those who had been given such tools. Among the algorithms that have been created by these pupils, three applied to the problem $64 - 28$ (Cochran, Barson, and Davis 1970). First, a method invented by a third-grade boy, Kye Hedlund: "Four minus eight is negative four. Sixty minus twenty is forty. Forty and negative four is thirty-six." Second, a method that many pupils invent (using the idea that "64 − 28" means "how much more is 64 than 28?" or "if I bought something that cost twenty-eight dollars and gave the salesperson a check for sixty-four dollars, how much change should I get?"): "Twenty-eight and two is thirty [write '2'], and thirty is sixty [write '30'], and four is sixty-four [write '4']. Now add: Two plus thirty plus four equals thirty-six [write '2 + 30 + 4 = 36']." (This is the method often used by cashiers in actually making change in stores, so this familiar example may be what the pupils have in mind when they use it.) Finally, a method invented by a sixth grader in Syracuse, New York, and reported by Leah Horwitz: "I can't take eight from four, so I take ten from the sixty, making it fifty. But I don't add the ten to the four [as the usual method does]. I subtract the eight from the ten, getting two. Next I add the four to the two, getting six. The tens digit is fifty minus twenty, therefore thirty. So the answer is thirty-six. This way, I never have to subtract from any number except ten, so I am less likely to make a mistake." These pupils have developed paradigms of subtracting that assimilate their experiences with numbers and computation.

References

Cochran, Beryl S., Alan Barson, and Robert B. Davis. "Child-Created Mathematics." *Arithmetic Teacher* 17 (March 1970):211–215.

Corwin, Rebecca Brown. "Multiplication as Original Sin." *Journal of Mathematical Behavior* 8 (August 1989):223–25.

Davis, Robert B. *Explorations in Mathematics.* Palo Alto, Calif.: Addison-Wesley Publishing Co., 1967.

———. "The Knowledge of Cats: Epistemological Foundations of Mathematics Education." In *Proceedings of the Conference of the International Group for the Psychology of Mathematics Education.* Conference held 15–20 July 1990, Oaxtepec, Mexico.

———. *Learning Mathematics: The Cognitive Science Approach to Mathematics Education.* Norwood, N.J.: Ablex Publishing Corporation, 1984.

Johnson, Mark. *The Body in the Mind.* Chicago: University of Chicago Press, 1987.

Kamii, Constance. *Double-Column Addition: A Teacher Uses Piaget's Theory.* Videotape. 1989. Distributed by NCTM and Teachers College Press.

Lakoff, George, and Mark Johnson. *Metaphors We Live By.* Chicago: University of Chicago Press, 1980.

Whitney, Hassler. "Taking Responsibility in School Mathematics Education." *Journal of Mathematical Behavior* 4 (December 1985): 219–35.

Students' Use of Symbols

Deborah A. Carey

Learning to communicate mathematically is one of the goals of the mathematics curriculum (NCTM 1989). One way to communicate mathematical ideas is through the use of symbols. In the primary grades, number sentences are among students' first introduction to symbolic representation. Number sentences include two kinds of symbols: (1) numerals, which represent quantity, and (2) signs for operations, such as + or −, or signs for relationships, such as =, that describe associations between or among quantities. Young students use these written symbols to record what is already known about an event or a situation (Hiebert 1989). For young students, symbols initially do not have meaning in and of themselves. Rather, meaning for symbols is developed through the process of interpreting and recording what is known about a problem and the solution. Symbols also supply students with a framework for investigating and understanding number, such as the part-whole relationship of quantity.

Children's Problem-Solving Abilities

Research on young students' problem solving furnishes a detailed analysis of addition and subtraction word problems and the strategies students use to solve problems (Carpenter and Moser 1984; Riley, Greeno, and Heller 1983). This information has been useful to classroom teachers as they plan effective instruction (Peterson, Fennema, and Carpenter 1991). Research on addition and subtraction also has implications for interpreting students' developing understanding of symbols across the primary grades. The number sentences that students write can reflect their thinking about word problems in much the same way that their solution strategies reflect their thinking. Both are affected by the structure of word problems.

Linking Word Problems and Symbols

The word problems in **table 1** are subtraction problems and can be represented by the number sentence $12 - 7 = \underline{}$. However, the semantic structure, or meaning, of each problem is different. This difference is made evident through observation of students who successfully solve these problems by directly model-

Table 1
Examples of Three Subtraction Problems

	Representation	
Problem Type	Arithmetic Solution	Semantic Structure
Separate, result unknown		
Anat had 12 balloons. She gave 7 balloons to Tom. How many balloons does Anat have left?	$12 - 7 = \underline{}$	$12 - 7 = \underline{}$
Join, change unknown		
Anat had 7 balloons. How many more balloons does she need to have 12 balloons altogether?	$12 - 7 = \underline{}$	$7 + \underline{} = 12$
Join, start unknown		
Anat had some balloons. Tom gave her 7 more balloons. Now she has 12 balloons. How many balloons did Anat have to begin with?	$12 - 7 = \underline{}$	$\underline{} + 7 = 12$

ing the situation (Carpenter and Moser 1984). For example, students may solve the "separate, result-unknown" problem by first making a set of twelve objects, then removing seven objects and counting the remaining objects for the solution. Students may solve the "join, change-unknown" problem by making a set of seven objects, then adding on a second set until they have twelve objects and counting the second set for the solution. Students often solve the "join, start-unknown problem" with a trial-and-error strategy. In this example, a set of objects is constructed (the number of objects may be the student's "best guess" or estimation), then seven are added and the total is counted to see if twelve objects result altogether. If the total is not twelve, then the student may adjust the beginning quantity and repeat the process. Research indicates that students use these strategies to model the structure of word problems directly with concrete objects, and these strategies reflect students' careful analysis of each problem.

The number sentences that support these solution strategies reflect the semantic structure of each problem and are appropriate alternative representations. Open-number sentences, such as $7 + \_\_ = 12$ for the "join, change-unknown" problem and $\_\_ + 7 = 12$ for the "join, start-unknown" problem, are direct translations between the students' actions on manipulatives and the symbols. If students are familiar with these alternative number sentences, they initially tend to write them for these types of word problems (Carey 1991).

Recognizing that all the problems in **table 1** are descriptions of subtraction events requires a sophisticated understanding of the part-whole relationship of number. Recall that young students use number sentences to record their solution, a solution reached by applying a strategy that reflects the problem's structure. As young students begin to use symbols, it is difficult for them to express with the same number sentence the different strategies they use for word problems that have different structures. When young students are restricted to using number sentences that reflect only the arithmetic solution to problems (such as $12 - 7 = \_\_$ for the problems in **table 1**), they have limited opportunity to develop meaning for symbols and to relate them to real-world events (Carpenter, Hiebert, and Moser 1983). Although representing many situations with the same number sentence is efficient, young students need to spend time developing meaningful number sentences that direct-

ly record their actions on objects. It is necessary for them to relate their actions on objects and symbols (Heibert 1988).

Some addition and subtraction word problems contain no action to model. For example, the problem "Rosa has 12 balloons. Seven are red and the rest are blue. How many blue balloons does Rosa have?" describes a collection of balloons. For problems with no implied action, the order of the numbers given in the problem can influence students' solution strategies (De Corte and Verschaffel 1987; Riley and Greeno 1988) and the number sentences they write for the problems (Carey 1991). Each of the number sentences $12 - 7 = \_\_$, $7 + \_\_ = 12$, and $\_\_ + 7 = 12$ expresses the part-whole relationship between the seven red balloons and the total collection of twelve balloons. But for this problem, none of these number sentences records a sequence of events. These representations allow students to discuss the part-whole relationship and to build on their knowledge of number. In this word problem, students are working with the number triple 12, 7, and 5. Students can generate different ways to express a number in terms of its parts, then make these relationships explicit by representing them with symbols.

The Importance of Context for Understanding Symbols

A recent study, conducted with sixty-four first-grade students found that the context of word problems could support students' work with symbols (Carey 1989). During regular classroom instruction, these students solved a variety of addition and subtraction word problems, discussed their varying solution strategies, and were free to write any appropriate representation for the problems. Both traditional and open-number sentences were acceptable. In a paper-and-pencil assessment task, the students were first asked to solve open-number sentences. They were then read word problems and were asked to write number sentences for the word problems and to solve the word problems. The word problems that were selected had the potential to stimulate the number sentences included in the paper-and-pencil task. Comparisons were made between students' performance on the number sentences with and without context. For example, the open-number sentence

___ − 9 = 3 was completed correctly by 31 percent of the students. The associated word problem in **table 2** was solved correctly by 72 percent of the students. The number sentence written by the majority of the students (59%) was in the form ___ − 9 = 3. Only one-third of the students had been able to solve this same representation when it was presented to them as an open-number-sentence task.

Students were more successful in solving the word problem and writing an appropriate representation for the word problem than they were in solving a number sentence that was not related to any context. Success with symbols depended on the presence of the word-problem context. Only 9 percent of the students wrote 9 + 3 = ___ for the word problem in **table 2**. This number sentence represents the arithmetic operation used to solve the problem; however, this number sentence requires an explicit understanding of the part-whole relationship of the numbers in the problem (Riley, Greeno, and Heller 1983).

Familiarity with open-number sentences for problems with sums less than 18 can help students represent similar problems that involve larger numbers. For example, the "join, change-unknown" problem "Sam has 36 stickers. How many more stickers does she need to have 75 stickers altogether?" tends to be solved with an adding-on or counting-up strategy and can be represented as 36 + ___ = 75 (Carey 1991). Students can relate their work with smaller numbers to this new problem. They may know that in a similar problem, 6 + ___ = 15 was solved by computing 15 − 6 = 9. Classroom discussion can help the students make connections between their approach to solving a problem with small numbers and this similar situation, in which 36 + ___ = 75 may also be represented and solved as 75 − 36 = ___.

Opportunities and Support for Developing Meaning for Symbols

Young students can solve problems without using symbols by relying on objects, counting strategies, or related facts ("6 + 7 = 13 because 6 + 6 = 12 and 1 more is 13"). Symbolic representation becomes useful when the operations become more complex and when the ideas expressed by symbols are less closely related to physical objects (Hiebert 1988). The use of number sentences in the primary grades essentially helps students extend their mathematical ideas to a system of conventional symbols and assists them in developing meaning for symbols.

When students are asked to represent word problems, they need to be familiar with a variety of open-number sentences so they can accurately represent their thinking. Although students have a difficult time dealing with open-number sentences in isolation (Kamii 1987), these representations can be natural extensions of word problems. Elements of a number

Table 2

Percent of Correct and Incorrect Responses and the Number Sentences Written for a "Separate, Start-unknown" Problem

Word Problem	Performance Scores	
	Correct	Incorrect
Pam had some kites. She gave 9 to Bjorn. Then she had 3 left. How many kites did Pam have to begin with?	72%	28%
Number Sentence Written	Percent Written	
9 + 3 = ___	9%	
9 − 3 = ___	11%	
___ − 9 = 3	59%	
Other	21%	

sentence, such as the numerals, signs, and a square (□) indicating the unknown quantity in a problem, should be available for students as they construct different representations. One technique is for the symbols to be written on two-inch-by-three-inch cards that the students can arrange on their desks (Carpenter, Carey, and Kouba 1990).

During a lesson it is very important to have students solve a word problem before asking them to decide whether the problem represents addition or subtraction. This decision can be made after the students have discussed their solution strategies and have written their own representations for the problem. Initially, it is not productive to think about problems only in terms of $a + b =$ _____ or $a - b =$ _____, because many different strategies and many ways of thinking about a given problem are valid. It is important that students' number sentences coincide with their solution strategies. In time, students may write more than one number sentence for a problem. Then they can discuss the similarities and differences in the representations.

Classroom discussions about how individual students represent problems can help students develop meaning for symbols. These discussions also can inform teachers about their students' mathematical thinking as students extend their knowledge to symbols.

References

Carey, Deborah. *Number Sentences: Linking Addition and Subtraction Word Problems and Symbols.* Doctoral dissertation, University of Wisconsin—Madison, 1989. *Dissertation Abstracts International,* 1989.

Carey, Deborah A. "Number Sentences: Linking Addition and Subtraction Word Problems and Symbols." *Journal for Research in Mathematics Education* 22 (July 1991):266–80.

Carpenter, Thomas P., Deborah A. Carey, and Vicky L. Kouba. "A Problem-Solving Approach to the Operations." In *Mathematics for the Young Child,* edited by Joseph N. Payne, pp. 111–31. Reston, Va.: National Council of Teachers of Mathematics, 1990.

Carpenter, Thomas P., James Hiebert, and James M. Moser. "The Effect of Instruction on Children's Solutions of Addition and Subtraction Word Problems." *Educational Studies in Mathematics* 14 (1983): 55–72.

Carpenter, Thomas P., and James M. Moser. "The Acquisition of Addition and Subtraction Concepts in Grades One through Three." *Journal for Research in Mathematics Education* 15 (May 1984):179–202.

De Corte, Erik, and Lieven Verschaffel. "The Effect of Semantic Structure on First Graders' Strategies for Solving Addition and Subtraction Word Problems." *Journal for Research in Mathematics Education* 18 (November 1987):363–81.

Hiebert, James. "A Theory of Developing Competence with Written Mathematical Symbols." *Educational Studies in Mathematics* 19 (1988):333–55.

———."The Struggle to Link Form and Understanding: An Update." *Arithmetic Teacher* 36 (March 1989):38–44.

Kamii, Constance. "One Point of View: Arithmetic: Children's Thinking or Their Writing of Correct Answers?" *Arithmetic Teacher* 35 (November 1987):2.

National Council of Teachers of Mathematics (NCTM). *Curriculum and Evaluation Standards for School Mathematics.* Reston, Va.: NCTM, 1989.

Peterson, Penelope L., Elizabeth Fennema, and Thomas P. Carpenter. "Teachers' Knowledge of Students' Mathematics Problem-Solving Knowledge." In *Advances in Research on Teaching: Teachers' Knowledge of Subject Matter As It Relates to Their Teaching Practice,* vol. 2. edited by Jere Brophy, pp. 49–86. Greenwich, Conn.: JAI Press, 1991.

Riley, Mary S., and James G. Greeno. "Developmental Analysis of Understanding Language about Quantities and of Solving Problems." *Cognition and Instruction* 5 (1988):49–101.

Riley, Mary S., James G. Greeno, and Joan G. Heller. "Development of Children's Problem-Solving Ability in Arithmetic." In *The Development of Mathematical Thinking,* edited by Herbert Ginsburg, pp. 153–96. New York: Academic Press, 1983.

PART 4
Mathematics for All

Section 13

Equity

Making Equity a Reality in Classrooms

Patricia F. Campbell and Cynthia Langrall

The NCTM's *Curriculum and Evaluation Standards for School Mathematics* (1989) speaks of the necessity of providing effective mathematics education for all students. Noting that "the social injustices of past schooling practices can no longer be tolerated" (p. 4). the standards document calls for a mathematics content that is "what we believe all students will need if they are to be productive citizens in the twenty-first century. If all students do not have the opportunity to learn this mathematics, we face the danger of creating an intellectual elite and polarized society" (p. 9). Similarly, the National Research Council's Mathematical Sciences Education Board noted that two themes underlie current analysis of American education: "equity in opportunity and . . . excellence in results" (1989, pp. 28–29). Although the NCTM's standards and other reform documents have been critiqued as addressing the issue of equity in terms of "enlightened self-interest" as opposed to seeking justice (Secada 1989), these documents have called attention to educational disparity. The issue today is how to make the goal of equity a reality in classrooms. To do otherwise would be to assign "mathematics for all" to the status of a slogan, a catchy phrase but having no meaning in practice.

What Is Equity?

Although historically and legally the definition of equity has revolved around equal opportunity, today one finds more widespread acceptance of a stronger criteria: equal treatment and equal outcomes (Fennema and Meyer 1989). The standard is that educational experiences in the classroom support equal achievement and future participation in mathematics for all students regardless of their gender, ethnicity and race, or socioeconomic background. Within this perspective, equity refers to justice (Secada 1989). The criteria for opportunity moves beyond availability of instruction to that of taking steps "to ensure that students have a real chance to become engaged in and to learn from the academic core that they encounter" (Secada 1989, p. 40).

Research offers no formulas or checklists to guarantee equity in mathematics classrooms. However, research does offer an indication about what approaches seem to be ineffective and what approaches may hold promise. Consider the evidence of ineffective practice.

An examination of national entry and achievement data indicates that those students who are academically disadvantaged at the beginning of the school year are likely to show "progressive retardation" as they continue in school (Walberg 1988). An examination of educational practice in these classrooms often reveals different expectations. For example, students deemed as being less capable are taught less mathematics or are presented with skill-oriented, direct instruction and practice as opposed to conceptually focused instruction promoting problem solving and understanding. The premise is that the students are deficient and therefore need remediation directed at the root of their problem, their lack of basic or prerequisite skills. This deficit model of policy has been critiqued for many years (e.g., Bronfenbrenner [1979]), although it remains the basis of many state and federal programs.

What other approach can be used? The alternative to the deficit model is to change the environment of the students' instruction to focus on the character of the instruction as opposed to the perceived deficiencies of the student. This approach requires a nonthreatening classroom environment, an environment marked by an atmosphere of trust and respect, respect between the teacher and each student as well as respect between each of the students (Baptiste 1992).

The preparation of this material was supported in part by a grant from the National Science Foundation under grant No. MDR-8954652. Any opinions, findings, and conclusions expressed in this material are thse of the authors and do not necessarily reflect the views of the National Science Foundation.

Applying the principles suggested in research on students' learning of mathematics as well as the vision offered in both the *Curriculum and Evaluation Standards* (NCTM 1989) and the *Professional Teaching Standards* (NCTM 1991), the intent is to organize mathematics instruction around the notion of each student's "making sense" of mathematics. Recently, an NSF-funded project has been attempting to implement this approach to mathematics equity in three predominantly minority urban public schools outside Washington, D.C. These schools reflect diversity, both in terms of minority racial-ethnic groups and in terms of language.

Supporting Each Child's Engagement in the Classroom

Build on prior knowledge

Project IMPACT (Increasing the Mathematical Power of All Children and Teachers) is a school-based project wherein all kindergarten through third-grade teachers work together as grade-level teams with a mathematics specialist to establish a mathematical culture in each classroom. These classrooms are becoming places where ideas are accepted, where suggestions are investigated, and where meaningful problems are solved. Teachers in Project IMPACT have come to recognize that every student, no matter what economic or social conditions he or she endures outside of school, has mathematical knowledge. This knowledge may not be the formal mathematical knowledge of school curricula. In fact, it may be understood by students only in terms of a particular setting or application. Nevertheless, it is knowledge. The emphasis in IMPACT is to teach students mathematics by building on their existing knowledge. Teachers focus on students' thinking and direct attention to that thinking, as opposed to focusing on the correctness of their answer. These principles are firmly grounded in research on mathematical learning.

Use meaningful, shared contexts

When students are presented mathematical ideas, or even realistic problems, in contexts or phrasings that are not meaningful or relevant to them, it may limit some students' potential to connect or use their informal knowledge with the mathematics or problem under discussion. This lack of connection in turn may discourage a student from attempting a solution or suggesting an approach. However, if the mathematical idea is set in a meaningful context, either because it is a common experience in the community outside of school or because it is a problem that is set in the shared culture of the school or classroom, all students are more likely to offer information or suggest approaches. In IMPACT classrooms, teachers attempt to propose mathematical investigations in context. Then they will frequently say or ask, "What do you see?... Tell us something else about this problem.... What do you know?... What else do you know?... Tell us your thinking.... So what is the problem?... What could we try?... Someone else tell the class what you are thinking about.... Does anyone else notice something else about this problem?... Does anyone have another idea?" These questions serve to legitimize students' thoughts and to encourage students to verbalize or show their ideas. They also give teachers a sense of the meanings that students are attributing to the problem or situation presented.

Celebrate each student's thinking

Just as important, all responses are valued; questioning does not stop when an expected response is offered; no responses are judged. The intent is to celebrate each student's thinking, whether that thinking is a solution process, a recognition of a relationship, or simply a statement of fact about the problem. The expectation in IMPACT is that all students can learn to communicate mathematical ideas and to participate actively in mathematical inquiry. Teachers continue to solicit responses as long as different ideas continue to be offered by the students.

In many of IMPACT's urban classrooms are students who did not initially respond to questions. Either their prior perception was that mathematics instruction meant waiting for the teacher to tell you what to do, or they did not yet trust either their teacher or their classmates sufficiently. We must recognize that students take a risk when they expose their thinking or approaches. However, IMPACT teachers resist the impulse to excuse any student from participation. Instead they might note almost parenthetically to the class that the student is certainly thinking and that the class will wait quietly to hear what he or she is thinking about. The difficult step for IMPACT teachers and students is actually to wait at this point. But waiting has a reward. Even the reluctant student eventually responds in some form, and when that response is accepted and valued by the teacher, that student learns to persevere and participate in the class.

Permit sufficient time for investigation

In IMPACT classrooms, following the discussion of a problem or a topic, time is permitted for the students to solve mathematics problems with the expectation that the students will express and explain their thinking. This thinking might occur in cooperative pairs or groups with the group charged with the responsibility of figuring out a solution for a problem. At other times, students work on problems individually. But eventually, either during the same class period or the next day, students express and justify their approach, either in writing or in classroom discussion.

Value explanation without judgment

IMPACT teachers attempt to refrain from evaluating an approach as being correct or incorrect; however, this nonjudgmental pose can be very difficult. At first, the tendency of many IMPACT teachers was to praise correct responses and to accept incorrect responses by either offering no comment or by tendering the refrain, "Good thinking." Their intent was to encourage problem-solving attempts, but instead they were implicitly communicating a judgment without explanation and simultaneously generating confusion. Eventually the IMPACT teachers decided that they still did not want to be the authority passing judgment but that any student should be expected to explain or justify his or her approach. Thus the classroom routine now promotes questioning of *all* responses within classroom discourse: "I'm not sure I understand. Explain that again.... Did anyone think of that in a different way?... So that is Darrell's idea. What do you think? Does anyone want to ask Darrell a question?... Does everyone agree that you could think about this idea that way? Or does someone disagree or have a question?... Wait a minute. I'm not sure I follow this. How did you know?..." In this way, students are encouraged to reflect on their thinking as they verbalize or demonstrate their strategy. They learn to disagree without insult. They also learn to ask questions without losing esteem. Sometimes, as a result of discussion, a student may realize that his or her initial thinking was erroneous or that he or she prefers another student's approach. In IMPACT classrooms, students are told that it is okay at any time to change one's mind, that even teachers sometimes change their mind. The venture, "I changed my mind," permits a student to modify or reverse an opinion without losing respect. The positive classroom atmosphere generated in this manner fosters motivation and responsibility.

Encourage listening and participation

Teachers use two approaches to encourage listening and participation. The teacher might ask if anyone else had solved a problem in the same way and then ask clarifying questions of *any* of those students. Or the teacher might ask if anyone had solved a problem in a different way. Following explanation, students might be asked to compare those two approaches. Or, if a student offered a previously described approach, that student would be challenged to explain how it was similar to, or different from, the prior strategy. Thus, a classroom norm is established wherein knowing what has already been discussed is significant if one wants to share one's ideas.

Integrate mathematical ideas

When asked how their mathematics teaching has changed, IMPACT teachers are unanimous in their reports that they have extended the time spent on mathematics. They not only have extended the length of time for mathematics lessons, they have related mathematics to other subject areas. Graphing and data collection are standard approaches for investigating science and social studies, even in kindergarten. Through a calendar center with full-class sharing, some topics are continually examined (e.g., revisiting place value in second grade by bundling tongue depressors, one tongue depressor for each day of school all year long) and other topics are anticipated (e.g., expressing completed and future days of school in a week or until a vacation or celebration as a fraction in third grade). The result is that students' access to mathematics is maximized and revisited.

Raise expectations

Experience in IMPACT classrooms demonstrates that generally preexisting assumptions regarding expectations for urban students and mathematics are low expectations. The phenomenon of "But not *my* students! *My* students can't do that" is all too alive and well in the primary classrooms of predominantly urban public schools. The intent must be to offer challenging, but not frustrating, mathematics, often using meaningful problem contexts. The determination of what is "challenging, but not frustrating" comes from aiming a little higher than one would expect and then honestly reflecting on the strengths of, and ideas offered by, the students.

Implications from Research

Researchers hypothesize that this approach to teaching has two benefits. First, because the perspective of each student is valued and the involvement of each student is expected in the problem-solving investigation, the classroom environment may encourage students to question and construct mathematical relationships. This atmosphere, in turn, promotes mathematical understanding. Second, because students express their ideas, teachers are able to reflect on their students' understandings, to ask questions to foster further thinking, and to recognize that instruction needs to be modified in some way to support a student's learning. This reflection assists teachers' decision making. The result is an equitable classroom (Carey et al. 1995).

Research indicates that ability grouping in elementary school mathematics can exaggerate differences between students and lead to differentiated outcomes (Oakes 1985, 1990). Goodlad (1984, p. 141) noted that after first grade, ability groupings tend to become established and foster limited opportunities for learning:

> One of the reasons for this stability in group membership is that the work of upper and lower groups becomes more sharply differentiated with each passing day. Since those comprising each group are taught as a group most of the time, it is difficult for any one child to move ahead and catch up with children in a more advanced group, especially in mathematics. It is not uncommon for a child in the most advanced group to have progressed five times as fast as a child in the least advanced group over the course of the year.

IMPACT classrooms do not implement homogeneous mathematics groups. Frequently, whole-class instruction is used, but sometimes students are grouped, either to permit more teacher observation and interaction or to vary the rigor of the mathematics problems being solved. However, even when grouping is used to vary problems, the mathematical topic being addressed remains constant across the groups. Further, the whole-class sharing still occurs.

Concern is voiced that current efforts toward reform in mathematics education will widen the differential achievement gap because the potential exists that those students "who are situated to take advantage of educational innovations receive a disproportionate amount of their benefits" (Secada 1989, p. 40). But this situation need not happen. The challenge is for teachers to "emancipate, empower and transform both themselves and their students" (Ladson-Billings 1992, p. 109), recognizing that teaching cannot be relegated to "prescriptive steps and techniques to be learned and demonstrated, . . . that the way social interaction takes place in the classroom is important to student success, . . . [and that] knowledge is continuously recreated, recycled and shared" (pp. 113–14). Teachers in IMPACT have taken up this challenge. We urge the reader to do the same.

References

Baptiste, H. Prentice Jr. "Conceptual and Theoretical Issues." In *Students at Risk in At-Risk Schools: Improving Environments for Learning,* edited by Hersholt C. Waxman, Judith W. de Felix, James E. Anderson, and H. Prentice Baptiste Jr., pp. 11–16. Newbury Park, Calif.: Corwin Press, 1992.

Bronfenbrenner, Uri. "Beyond the Deficit Model in Child and Family Policy." *Teachers College Press* 81 (fall 1979): 95–104.

Carey, Deborah A., Elizabeth Fennema, Thomas P. Carpenter, and Megan L. Franke. "Cognitively Guided Instruction: Towards Equitable Classrooms." In *New Directions in Equity for Mathematics Education*, edited by W. Secada, E. Fenneman, and L. Byrd, pp. 93–125. Cambridge: Cambridge University Press, 1995.

Fennema, Elizabeth, and Margaret R. Meyer. "Gender, Equity, and Mathematics." In *Equity in Education,* edited by Walter G. Secada, pp. 146–57. Bristol, Pa.: Falmer Press, 1989.

Goodlad, John J. *A Place Called School: Prospects for the Future.* New York: McGraw-Hill. 1984.

Ladson-Billings, Gloria. "Culturally Relevant Teaching: The Key to Making Multicultural Education Work." In *Research and Multicultural Education: From the*

Margins to the Mainstream, edited by Carl A. Grant, pp. 106–21. Bristol, Pa.: Falmer Press, 1992.

National Council of Teachers of Mathematics (NCTM). *Curriculum and Evaluation Standards for School Mathematics.* Reston, Va.: NCTM, 1989.

———. *Professional Standards for Teaching Mathematics.* Reston, Va.: NCTM, 1991.

National Research Council, Mathematical Sciences Education Board. *Everybody Counts: A Report to the Nation on the Future of Mathematics Education.* Washington, D.C.: National Academy Press, 1989.

Oakes, Jeannie. *Keeping Track: How Schools Structure Inequality.* New Haven: Yale University Press, 1985.

———. *Multiplying Inequalities: The Effects of Race, Social Class, and Tracking on Opportunities to Learn Mathematics and Science.* Santa Monica, Calif.: Rand Corporation, 1990.

Secada, Walter G. "Agenda Setting, Enlightened Self-Interest, and Equity in Mathematics Education." *Peabody Journal of Education* 66 (winter 1989):22–56.

Walberg, Herbert J. "Synthesis of Research on Time and Learning." *Educational Leadership* 45 (March 1988):76–85.

Gender and Race Equity in Primary and Middle School Mathematics Classrooms

Laurie Hart Reyes and George M. A. Stanic

A number of concerns have been expressed about the mathematics performance of all students in our schools (see, e.g., McKnight et al. [1987]), but the performance of certain groups is particularly troublesome. For example, on standardized tests of mathematics achievement, Hispanic students and black students consistently score below their white counterparts during the primary and middle school years. In addition, although female students perform at least as well as male students during the primary school years, some evidence suggests that gender differences in favor of boys begin to appear during the middle school years, particularly on problem-solving and applications tasks. Are female students, blacks, and Hispanics naturally less able in mathematics? Or do families, schools, and the rest of society offer experiences that create these differences? Individual differences in mathematics performance are normal, inevitable, and obviously related to natural ability; but no reason exists to believe that female students, blacks, and Hispanics, as *groups,* are by nature less able in mathematics. The problem is one of equity.

All members of society must share the blame for the inequitable treatment of girls and women, blacks, and Hispanics. Teachers cannot be expected to overcome all the unjust treatment received by these groups, but they can, within their own classrooms, deal with attitudes and actions that hamper and limit the mathematics performance of students. In this article, we shall describe the results of relevant research and offer concrete suggestions that can be drawn from this research.

Relevant Research

- On the basis of a review of studies published from 1975 to 1984, Lockheed et al. (1985) claim that on the types of mathematics skills taught in grades 4 to 8, few, if any, consistent gender differences exist in mathematics performance.

- The National Assessment of Educational Progress (NAEP) furnishes information about the mathematics achievement of representative nationwide samples of nine-, thirteen-, and seventeen-year-olds. Students are assessed on knowledge, skills, understanding, and applications. Mathematics performance was assessed in 1973, 1978, and 1982. In 1978, at ages nine and thirteen, girls scored slightly higher than boys on the knowledge and skills exercises, and boys had somewhat higher averages in the understanding and applications categories. But no really clear pattern of differences emerges at these ages. At age seventeen, even when the number of mathematics courses taken was considered, males scored higher than females at all four cognitive levels. Moreover, the differences were greater in favor of males as the cognitive level of the questions increased and as the number of courses taken increased (Fennema and Carpenter 1981). According to Meyer and Fennema (1987), the 1982 results are basically the same as those from 1978.

- Race differences in mathematics performance are larger than gender differences. The NAEP mathematics results have consistently shown that black and Hispanic students are achieving at levels well below the national average. However, these students have shown improvement; in fact, black and Hispanic students made greater gains in mathematics achievement than did white students between the second and third national assessments done in 1978 and 1982, respectively (Matthews et al. 1984).

- Lockheed et al. (1985) also reviewed sixteen studies of race differences in mathematics performance in grades 4 through 8. A clear pattern of performance among various student groups was found. Asian students usually outperformed white students; Asian students and white students performed better than Hispanic students; and all the other groups outperformed black students.

- Female students, blacks, and Hispanics have traditionally enrolled in fewer optional mathematics courses than white male students (Fennema 1984; Marrett 1981; Matthews 1984). Differences in enrollment, however, do not explain all differences in achievement. Other factors, more relevant for students in primary and middle schools, that may contribute to differences in mathematics achievement include teacher attitudes, student attitudes (e.g., confidence and beliefs about reasons for success and failure), student achievement-related behaviors (e.g.,persistence and independence), and classroom processes. A detailed discussion of these factors can be found in Reyes and Stanic (1987).

- Socioeconomic status (SES) is an important factor in academic achievement (e.g., White [1982]). Because blacks and Hispanics are disproportionately represented in lower SES groups, the difference in achievement that we attribute to race may be explained in large part by the disadvantages associated with lower SES levels.

Suggestions for Practice

Teachers can begin to deal with the issue of equity by studying the achievement patterns within their classrooms, schools, and school districts. If the achievement of certain groups of students is not what it should be, a variety of actions are possible:

1. Teachers should become more aware of their own attitudes, what they feel and what they believe about different groups of children. Before the treatment of students can become more equitable, teachers need to believe that a student's race and gender should not determine how well she or he will perform in mathematics. Perhaps the best way for a teacher to examine and change attitudes is to get together with other teachers in a setting that encourages open and honest discussion.

2. Teachers should become more aware of the attitudes of their students. Attitudes important to success in mathematics include confidence in learning mathematics, perceived usefulness of mathematics, beliefs about the appropriateness of mathematics as an area of study, attributions of success and failure in mathematics, attitudes toward other students, and attitudes toward teachers. A formal instrument is a valuable means for examining student attitudes; however, most of the available instruments (e.g., Fennema and Peterson [1984]; Fennema and Sherman [1976]; Fennema, Wolleat, and Pedro [1979]) focus on gender equity and were designed for middle and high school students. Teachers would need to revise these instruments for use in other settings. Other ways to assess attitudes include interviewing students and having students keep journals about their experiences in mathematics class. Day-to-day interactions in the classroom will certainly tell teachers something about the attitudes of their students, but paper-and-pencil instruments, interviews, and journals can reveal much that would not be apparent under normal classroom circumstances. Knowing what students believe and feel about mathematics and about themselves as learners of mathematics and trying to change inappropriate attitudes may not lead to immediate changes in mathematics performance; the exact relationship between attitudes and achievement is not known. Nonetheless, some relationship exists, and we believe that developing positive attitudes is a valuable goal in its own right.

3. Persistence and a certain level of independence are characteristics of good problem solvers. Important questions for teachers are "What level of independence do my students demonstrate when they do not know immediately how to proceed on a problem?" "Once a problem is begun, what level of peristence do my students demonstrate in working on it?" "How do my actions influence students' approaches to problem solving?" The best way to gather information about these and other achievement-related behaviors is carefully to observe students as they work on problems. Teachers can gather information about their influence on these behaviors by using the techniques described in the next section on classroom processes.

4. Finally, teachers need to analyze what goes on in their classrooms during mathematics lessons. An individual teacher needs to ask, "How often do I interact with each student in my class? Do I tend to talk about different things with different students? What intended and unintended messages about themselves and mathematics am I giving students? How do my students interact with each other? What do they talk about? What is the general climate in my classroom? Do some students seem to

perform better in that climate than others?" Information that can be used to answer such questions can come simply from teachers' thinking about the questions and trying to remember what happens during class, but additional ways of gathering information can be pursued. Teachers can keep daily journals; they can audiotape class sessions; they can have a student or another teacher videotape class sessions; they can ask fellow teachers to come in and observe their classes; and they can use student journals and interviews to gather information about classroom processes. All students do not have to be treated in exactly the same way by a teacher; in fact, teachers are told that they should adjust to individual differences. But students should be treated equitably; everything a teacher chooses to do with a student should help the student develop positive attitudes and perform at her or his highest possible level. In short, teachers need to ask whether they are treating students differently for clearly justifiable reasons that stand up to the tests of available research and moral and ethical standards.

References

Fennema, Elizabeth. "Girls, Women, and Mathematics." In *Women and Education: Equity or Equality,* edited by Elizabeth Fennema and M. Jane Ayer, pp. 137–64. Berkeley, Calif.: McCutchan Publishing Corporation, 1984.

Fennema, Elizabeth, and Thomas P. Carpenter. "Sex-Related Differences in Mathematics: Results from National Assessment." *Mathematics Teacher* 74 (October 1981):554–59.

Fennema, Elizabeth, and Penelope L. Peterson. *Classroom Processes, Sex Differences, and Autonomous Learning Behaviors in Mathematics.* Contract No. SED 8109077. Washington, D.C.: National Science Foundation, 1984.

Fennema, Elizabeth, and Julia A. Sherman. "Fennema-Sherman Mathematics Attitude Scales." *JSAS Catalog of Selected Documents in Psychology* 6 (1976):31. (Ms. No. 1225)

Fennema, Elizabeth, Patricia Wolleat, and Joan Daniels Pedro. "Mathematics Attribution Scale." *Journal Supplement Abstract Service* 9 (1979):88. (Ms. No. 1837)Lockheed, Marlaine E., Margaret Thorpe, J. Brooks-Gunn, Patricia Casserly, and Ann McAloon. *Sex and Ethnic Differences in Middle School Mathematics, Science and Computer Science: What Do We Know?* Princeton, N.J.: Educational Testing Service, 1985.

Marrett, Cora B. *Patterns of Enrollment in High School Mathematics and Science.* Final Report. Madison, Wis.: Wisconsin Center for Education Research, 1981.

Matthews, Westina. "Influences on the Learning and Participation of Minorities in Mathematics." *Journal for Research in Mathematics Education* 15 (March 1984):84–95.

Matthews, Westina, Thomas P. Carpenter, Mary Montgomery Lindquist, and Edward A. Silver. "The Third National Assessment: Minorities and Mathematics." *Journal for Research in Mathematics Education* 15 (March 1984):165–71.

McKnight, Curtis, F. Joe Crosswhite, John A. Dossey, Edward Kifer, Jane O. Swafford, Kenneth J. Travers, and Thomas J. Cooney. *The Underachieving Curriculum: Assessing U.S. School Mathematics from an International Perspective.* Champaign, Ill.: Stipes Publishing Co., 1987.

Meyer, Margaret R., and Elizabeth Fennema. "Girls, Boys, and Mathematics." In *Research-based Methods for Teachers of Elementary and Middle School Mathematics,* edited by Thomas R. Post. Newton, Mass.: Allyn & Bacon, 1987.

Reyes, Laurie Hart, and George M. A. Stanic. "Race, Sex, Socioeconomic Status, and Mathematics." *Journal for Research in Mathematics Education* 19 (January 1988):26–43.

White, Karl R. "The Relation between Socioeconomic Status and Academic Achievement." *Psychological Bulletin* 91 (May 1982):461–81.

Parental Involvement in a Time of Changing Demographics

Walter G. Secada

The involvement of parents in their children's education can take a number of forms. In their synthesis of the research literature, Tangri and Moles (1987) outlined three dimensions of parental involvement. First, it can refer to service in schools (e.g., participating in school governance activities, working in classrooms as paid aides or volunteers). Second, it can refer to home-school relationships (e.g., written and phone communications, home visits by teachers, parent-teacher conferences at school, parent education and training sponsored by the school). Finally, parental involvement can refer to support of learning activities at home (e.g., assisting with homework, tutoring, providing educational enrichment activities). It should be noted that in this article, "parent" refers to any adult caregiver in the home.

The Benefits and Importance of Parental Involvement

Among the benefits related to parental involvement are increased parental satisfaction with children's educations, better working relationships between home and school, enhanced ability of schools to offer educational and other services for students, and a better focusing of parental efforts to help their children with schoolwork. Benefits for students include decreased absenteeism, enhanced motivation to learn and do well in school, and improved achievement (Tangri and Moles 1987). These outcomes vary depending on the type of involvement, the length and intensity of involvement, the racial and ethnic backgrounds of the parents, and the grade level of the students. For example, giving middle school and high school students opportunities to learn by having books in the home and by generally monitoring school performance and after-school activities seems to be more effective "than specific actions like checking homework, signing it, or having a special place to study" (Tangri and Moles 1987, p. 530). Younger pupils may benefit from such "specific actions," but the research on this point is not clear. It is clear, however, that offering enrichment activities at home "can be effective even for busy and poorly educated parents and their elementary school children" (p. 530).

Research support for the importance and benefits of parental involvement comes from a variety of sources. The effective-schools research attests to that importance and also shows that effective parental involvement varies as a function of the backgrounds of students and parents (Brookover et al. 1979). For example, in predominantly white, middle-class effective schools, parents became more involved when their children's achievement was perceived to be unsatisfactory. In predominantly black schools that were labeled effective, parental involvement was not limited to cases of poor achievement by children: parents were more involved in the day-to-day functioning of these schools. The reasons for these differences are not clear. Possibly, schools are organized to match the backgrounds of white, middle-class students; hence, parents in predominantly black schools, through their involvement, help schools adapt to the cultural backgrounds of their children.

Evidence in favor of parental involvement also comes from research involving programs for at-risk students. Studies of Head Start and Follow Through have supported the importance of such programs for avoiding educational problems associated with children from non-middle-class backgrounds (Stallings and Stipek 1986, White and Buka 1987). After reviewing the outcomes of the Perry Preschool Program (probably the best-known Head Start program), the Parent Follow Through Education Program, and other programs, Stallings and Stipek (1986) argued that early intervention was not the only factor that led to such long-term benefits as lower rates of delinquency, fewer years in special education, higher rates of completing school, and higher rates of college attendance and employment. Rather, "over the 15 years of the study, the home [was] the more constant factor" (p. 741). Stallings and Stipek concluded that "family involvement activities help foster positive attitudes toward school and in turn support children to be successful in school and to be persistent enough to graduate" (p. 741).

Yet another reason to be concerned about parental involvement is suggested by studies that have shown that parents exert a strong influence on their children's attitudes toward mathematics and their decisions to take additional coursework. For example, cross-cultural studies have documented that children adopt beliefs about achievement that are similar to those of their parents (Holloway 1988). "American children placed greater emphasis on lack of ability than any other reason to explain low performance in mathematics, whereas Japanese children placed greater emphasis on lack of effort" (p. 330). Although "American mothers placed most importance on lack of effort . . . [they] placed only slightly less importance on lack of ability and inadequate training in school. In Japan, mothers focused almost exclusively on lack of effort" (p. 330). Another example of the influence of parents on their children comes from research on gender-role socialization in this country. Stage and others (1985) reviewed that research and concluded that "sex stereotypes held by parents are small and perhaps diminishing, but [they] favor boys when they are present. . . . Parents, teachers, and counsellors have all been found to provide boys more explicit rewards, encouragement, and reinforcements for learning math and for considering math-related careers" (p. 242).

Hence, parental involvement is not only important in general but also plays a critical role in ensuring that everyone participates in mathematics. Moreover, recent demographic projections indicate that in absolute numbers and proportionately, our schools are enrolling more of the kinds of students who are likely to be unsuccessful. More students will come from single-parent homes, be poor, and be from nonwhite or non-English-speaking backgrounds (Hodgkinson 1985, National Research Council 1989). Many observers (e.g., Slavin and Madden [1989]) are recommending early intervention programs to ensure that such children start school with an even chance and do not fall behind. The most successful of these programs include parental-involvement components, but establishing meaningful involvement for single, or poor, or nonwhite, or non-English-speaking parents is not a trivial task.

Practical Suggestions

Although most of the research on parental involvement tends to be general or to focus on reading, a number of suggestions based on the research are relevant to mathematics. Parental involvement should be broadly considered, however; not all parents can or wish to participate in the more narrowly defined ways. Parents should have options; schools and teachers should respect parental choice in such matters.

Service in schools

Parents could serve on mathematics curriculum-review and adoption committees. Parents from various racial and ethnic backgrounds who work in different fields (e.g., manufacturing, services, the trades) could be asked to participate in class activities involving the mathematics of those careers.

Home-school relationships

A recorded message giving the mathematics assignments for a day or a week could become part of a "homework hotline." For non-English-speaking parents, hotlines in their native languages might also be set up. Special classes on important topics in mathematics could be offered to parents. Videotapes such as *Multiplying Options and Subtracting Bias* (Fennema et al. 1980) and *The Polished Stones* (Secada 1989) could be used to alert parents to how their beliefs and attitudes influence their children's later-life opportunities in mathematics and mathematics-related fields and to inform them of how they can help their children achieve in mathematics.

Parental support of home-learning activities

This aspect of parental involvement is extremely important, and it should be understood as meaning more than just helping with homework. The research suggests that parents from various educational and socioeconomic levels can furnish enrichment activities to their children. These activities could be as sophisticated as those found in a mathematics textbook series or in supplementary materials. For example, parents of preschoolers might be encouraged to engage their children in activities found in *Math Their Way* (Baratta-Lorton 1976) or *Family Math* (Stenmark, Thompson, and Cossey 1986). Parents of older students could pose problems found in many of the professional materials available through NCTM.

Alternatively, parents might be helped and encouraged to identify mathematics in the everyday world of their children and to engage them in the mathematics arising from such situations. Young children will often spontaneously count things (e.g., as they walk or run up and down stairs, when they set the table for dinner). Somewhat older children might be asked to solve word problems, such as those studied by Carpenter and Moser (1983) or Ginsburg (1977). Video games typically involve the movement of objects in patterns across screens; though one might question the socially redeeming value of looking for geometric patterns and figures in a game like Super Mario Brothers, this game and others like it have potential for such activities.

Conclusion

Different kinds of parental involvement will lead to different results. On the one hand, if parents are asked to render services to the classroom or school, parental satisfaction may be enhanced and students may improve their attendance, but it is not clear that students' achievement will rise. On the other hand, the creation of a mathematically rich environment at home does seem like a promising tactic to increase students' engagement and achievement in mathematics; the hope is that just as children in homes with many books do better in reading (Tangri and Moles 1987), children in mathematically rich home environments will do better in mathematics.

In addition to being aware of the various results that arise from different kinds of parental involvement, teachers and schools must be prepared to deal with the barriers to parental involvement. For example, because of work schedules, some parents may not be able to attend parent-teacher conferences at regularly scheduled times. Language can also be a powerful barrier to involvement, blocking attempts of both parents and teachers to communicate with each other. Working with parents, schools must identify and try to overcome such barriers, because it is clear that parental involvement leads to desirable outcomes. Schools must make available a broad range of opportunities for parental involvement, present and explain those opportunities to parents, and allow parents to select forms of participation with which they are most comfortable and to suggest alternatives. Parents are a resource for helping all students participate and achieve in mathematics; especially in a time of increasing student diversity, they are a resource too precious to waste.

Bibliography

Baratta-Lorton, Mary. *Mathematics Their Way.* Palo Alto, Calif.: Addison-Wesley Publishing Co., 1976.

Brookover, Wilbur, Carol Beady, Patricia Flood, John Schweitzer, and Joseph Wisenbaker. *School Social Systems and Student Achievement: Schools Can Make a Difference.* New York: Praeger Publishers, 1979.

Carpenter, Thomas P., and James M. Moser. "The Acquisition of Addition and Subtraction Concepts." In *Acquisition of Mathematics Concepts and Processes,* edited by Richard M. Lesh and Marsha Landau, pp. 7–44. New York: Academic Press, 1983.

Fennema, Elizabeth, Ann D. Becker, Patricia L. Wolleat, and Joan Daniels Pedro. *Multiplying Options and Subtracting Bias.* Reston, Va.: National Council of Teachers of Mathematics, 1980. Videotape.

Flexer, Roberta J., and Carolyn L. Topping. "Mathematics on the Home Front." *Arithmetic Teacher* 36 (October 1988):12–19.

Ginsburg, Herbert P. *Children's Arithmetic: The Learning Process.* New York: Van Nostrand Reinhold, 1977.

Hodgkinson, Harold L. *All One System: Demographics of Education—Kindergarten through Graduate School.* Washington, D.C.: Institute for Educational Leadership, 1985.

Holloway, Susan D. "Concepts of Ability and Effort in Japan and the U.S." *Review of Educational Research* 58 (fall 1988):327–45.

Mathematical Sciences Education Board and National Research Council. *Everybody Counts: A Report to the Nation on the Future of Mathematics Education.* Washington, D.C.: National Academy Press, 1989.

Secada, Walter G. *The Polished Stones: K–5 Math Achievement in Japan and Taiwan.* Ann Arbor, Mich.: University of Michigan, 1989. Videotape.

Slavin, Robert, and Nancy Madden. "What Works for Students at Risk: A Research Synthesis." *Educational Leadership* 46 (February 1989):4–13.

Stage, Elizabeth K., Nancy Kreinberg, Jacqueline Eccles, and Joan Rossi Becker. "Increasing the Participation and Achievement of Girls and Women in Mathematics, Science, and Engineering." In *Handbook for Achieving Sex Equity through Education,* edited by Susan S. Klein, pp. 237–68. Baltimore: Johns Hopkins University, 1985.

Stallings, Jane A., and Deborah Stipek. "Research on Early Childhood Education and Elementary School Teaching

Programs." In *Handbook of Research on Teaching,* 3d ed., edited by Merlin C. Wittrock, pp. 727–53. New York: Macmillan, 1986.

Stenmark, Jean K., Virginia Thompson, and Ruth Cossey. *Family Math.* Berkeley, Calif.: University of California—Berkeley, 1986. (Note: A Spanish version is also available.)

Tangri, Sandra, and Oliver Moles. "Parents and the Community." In *Educators' Handbook: A Research Perspective,* edited by Virginia Richardson-Koehler, pp. 519–50. New York: Longman, 1987.

White, Sheldon H., and Stephen L. Buka. "Early Education: Programs, Traditions, and Policies." In *Review of Research in Education,* vol. 14, edited by Ernst Z. Rothkopf, pp. 43–91. Washington, D.C.: American Educational Research Association, 1987.

PART 5
Teaching and Research

Section 14

Using Research to Improve Teaching and Learning

Teachers Researching Their Mathematics Classrooms

Neil Pateman

Teachers typically are treated as consumers of what other people produce. For example, as a consumer, the teacher is supposed to work from ready-made materials, using accompanying teacher's guides, even though the entire curriculum has been prepared in a different place. Such materials are often characterized as being "teacher proof."

Just as curriculum development is often considered separate from the classroom practice of teaching, most research in mathematics education begins with the assumption that research and practice are separate activities that are carried out by people with different responsibilities. The writers of research articles sometimes try to bridge this research-practice gap by presenting classroom implications of their research and suggesting what teachers, still viewed as consumers, should do to change their daily practice.

Teachers and teaching can, however, be thought of in a different way. We might consider the teacher as not simply a consumer but also a producer of research. Such a classroom teacher is guided in the practice of teaching by the results of his or her own classroom research and does not rely exclusively on the research results of others. Research and practice might then be viewed as two sides of the same coin, and classroom teachers would be treated very differently by university faculty members who, like teacher-researchers, are also interested in studying the teaching and learning of mathematics.

One way to initiate this approach is by creating "collaboratives" between teachers and between teachers and other professionals interested in education. Two examples of how such collaboratives were developed in Australia are included here. The purpose of this article, then, is not to present specific research results and implications but to help us begin to think of researchers and teachers—of research and teaching—in another way.

A Small Parish Primary School in "Quietdale"

In 1985, the local Catholic Education Region office in a relatively remote area of rural Victoria (an Australian state similar in size and population to Georgia in the United States) applied to the Australian federal government for funding of professional development. The teachers in the region's parish primary school had expressed dissatisfaction with conventional forms of in-service training (basically one-day talks or brief presentations at staff meetings) and wanted to try something different. The focus the staff chose was mathematics in kindergarten through grade 3. The project was funded, with the crucial element of the funding being the provision of money to allow for substitute teachers to be brought into the school on a regular basis so that the classroom teachers involved could spend time on project activities.

Two faculty members from Deakin University were asked to work with the teachers. Our intent was that the locus of control of project activities should be with the teachers. However, the teachers did not fully accept this control from the start. The idea that they could effectively set their own agenda, in terms of deciding what the problems were and how the problems could be studied, took a while to take root. The role of the university staff was to facilitate this transition to teachers' taking control of the project.

During the planning stage, we established that the necessary condition for the project to get started was present: The teachers knew that change was needed in their mathematics classes, and they agreed to open their practice to the extent that they were willing to have teacher colleagues come into their classrooms and act as "critical friends." Perhaps the single most important feature of the project was giving teachers this opportunity to work together—to observe each other teach and then talk about what happened and about the problems they wanted to study further. Two classes at each of the levels, K–3, were involved, so teachers were teamed by grade levels. Because of the availability of substitute teachers, the project participants could observe each other teach and then spend time immediately afterward discussing what had just occurred and planning the next step.

The teachers studied a variety of problems related to classroom practice. For example, one teacher expressed concern about the balance between direction by the teacher and discovery by pupils when using structured materials. Two other teachers observed, in each other's classrooms, pupils' use of strategies of which they had been unaware; both said that as a result, they began to listen to pupils much more closely. Another teacher focused on the mathematical language of the pupils in her classroom by using a tape recorder. Yet another teacher asked her observer to listen to the types of questions she used when teaching mathematics.

Notice the different approaches to classroom research: the teacher studying mathematical language was focusing on the pupils, but in a way that allowed her to collect "real" data that she and her colleague could use to reflect on teaching practice. Other teachers were asking questions more directly about their own practice. As time went on, most of the teachers decided that they needed to focus on their own classroom practice rather than simply concentrate on the pupils. That is, they began to see themselves, rather than the pupils, as the source of problems to study.

The role of the university staff continued to be one of facilitating this research and interaction. We talked with the teachers about our experiences observing and studying classrooms and teaching mathematics. We knew of other work that had been done on, for example, pupils' mathematical language and teachers' questions and made the teachers aware of this work *as they asked for it in the context of studying their own classrooms*. The first year was judged to be very successful by the teachers, and they suggested that the school apply for funds for a second year. The successful application led to continued collaboration for the next year.

Key Groups: Creating and Networking Small Collaboratives

The style of collaboration in the Key Groups project was different from that of the Quietdale project, primarily because the size of the projects was so different. Instead of two university faculty members dealing with eight teachers in one school, the Key Groups project involved almost 200 people, including teachers from forty-eight schools distributed across the state of Victoria, teacher-consultants (i.e., classroom teachers appointed by the Victoria ministry of education to work with teachers in a region of the state for a period of two years), staff members from the Victoria ministry of education, and college and university faculty. A committee of 12 representing each of the groups involved helped organize and facilitate the project activities. A key group made up of 3 teachers and a consultant was established at each of the forty-eight schools. An individual consultant served on three or four key groups in a region of the state. One of the aims of the project was the sharing of information among the key groups through the development of networks, so collaboration took place within and among groups.

One of my roles in the project was the organization of a two-day conference for key groups from eighteen schools. The goal of this conference was to have each key group put together a plan for its activities, or what we called its "action research" plan. The only requirements for a plan were that it be directed toward improving some aspect of mathematics teaching in the school and that it outline explicitly the action steps each member of the key group would undertake to put the plan in motion. Workshops—conducted by teachers, consultants, or ministry staff—were held during the conference to stimulate ideas, but most of the time was given to discussions and planning within the eighteen key groups. Teaching about record keeping and data collection was a strong focus of the conference.

Just as was true in the Quietdale project, a variety of problems were studied when key groups returned to their schools. For example, one group looked at the model of "process writing" now predominant in many schools in Australia and attempted to have students approach mathematics in a similar way. That is, after some mathematical experience with materials, the students were expected to write about the experience, either by making a direct report or by writing an essentially fictional story involving the mathematics they had learned. The students then participated in a writing conference with the teacher and eventually "published" their work for other students to read.

Several groups worked on developing curricula based on extensive use of manipulative materials.

One school introduced a problem-solving approach in which all problems were presented in natural language familiar to the students with the expectation that the students answer in similar natural language. The problems were not presented after some mathematics was learned but were used instead as a source from which to extract mathematics. Manipulative materials were an essential part of this group's activities.

Six months later the eighteen previously mentioned groups held a follow-up conference in which group members shared their experiences, their data, and their conclusions. The point mentioned most often by the teachers was the importance of the provision of money for substitute teachers and for travel expenses; this funding allowed them to talk about their projects and to make visits to other schools to strengthen links between groups.

What Has Been Learned

Both projects were based in schools and on the assumption that teachers can develop professionally—that they can conduct important, useful, and rigorous research—in the context of their daily work. Both situations subscribed to the need for collaboration between teachers and between teachers and other professionals interested in education if lasting change is to be at all possible. A strong feature of both situations is that they allowed for a substantial time period so that teachers could set their own pace. Each of the situations took a bottom-up approach to innovation: no readymade curriculum changes were handed on to the teachers. Instead, the teachers acted as the primary sources of innovation.

The teachers involved talked about a spirit of collegiality that developed within their schools and more generally across the profession. They learned that an important aspect of taking a professional outlook on teaching is to get involved in curricular decision making. Research was viewed by the teachers in these projects as something familiar and useful rather than as the private province of a group of specialists whose main function is to translate the results of their research into other people's practice.

The most important lesson is that teachers are able to do this kind of research as part of their own practice provided they are given support and time to talk and work with colleagues. Teachers given the opportunity can become extended professionals and can further develop their own ideas about teaching and the nature of education.

Putting Research into Practice

Donald L. Chambers

After reading even one chapter of this book, you may be eager to make some changes in your instruction by putting some of those research ideas into practice. How could you begin?

Several articles contain explicit suggestions for putting into practice the research described in those articles. The suggestions are located near the end of the article in the "Action Research Ideas" section. For the articles that do not contain action research ideas, the following generic approach may help you develop your own action research program.

First, become familiar with the different patterns of thinking described and the ways in which children reveal their thinking. For example, in "Place Value and Addition and Subtraction," Wearne and Hiebert describe the differences in thinking of two students who both get correct answers to 347 + 48 in grade 2 but who get different answers to 3.5 + .62 in grade 4. The two approaches, one conceptual and the other procedural, are both typical of students.

Second, use the ideas from the chapter to develop an assessment that will enable you to determine what each of your students already knows about the topic under discussion and how he or she thinks about it. In the instance of addition of multidigit numbers, asking students to state their answers will not reveal the differences in their thinking. Students must be asked to describe what they did and why. Often the same example from the chapter may be used in your classroom. You can easily distinguish between those students who give conceptual explanations ("I put the 4s together because they are both tens and the 7 and the 8 together because they are both ones") and those who give procedural explanations ("First you line up the numbers and then you add"). Initially, your objective is to learn what the students think, not to correct errors in their thinking. Make written or mental notes about each child's thinking to help you see evidence of growth over time.

Third, use the ideas in the article to develop a plan of instruction on the topic discussed. Keep in mind that conceptual understanding emerges over an extended period of time, not as a result of one or two intensive lessons. During instruction, ask questions that elicit descriptions of student thinking. Reflect on what each student's work reveals about his or her understanding, continuing to probe for more information when the student's thinking is not clear to you.

Fourth, continue to assess each student's understanding at periodic intervals, in each instance relating the child's current understanding to his or her previous level of understanding.

Fifth, reflect on what learning experiences seem to be the most valuable to students and what periods of time are necessary for conceptual understanding to develop. In "Direct Modeling and Invented Procedures: Building on Students' Informal Strategies," Gretchen subtracts 23 from 70 and gets 53. Many teachers think that they can correct Gretchen's thinking in a few minutes, but in fact Gretchen requires about six months to develop a conceptual understanding of this subtraction problem.

Sixth, modify your instruction to capitalize on what you have learned through your action research. This phase will probably involve asking more questions, spending more time on student explanations, and allowing more time for conceptual understanding to develop.

The Role of *JRME* in Advancing Learning and Teaching Elementary School Mathematics

Michael T. Battista and Carol Novillis Larson

The *Journal for Research in Mathematics Education (JRME)* is the research journal of the National Council of Teachers of Mathematics. It is a "forum for disciplined inquiry into the learning and teaching of mathematics at all levels—from preschool through adult" (*JRME* Editorial Board 1993, p. 3). *JRME*'s twenty-fifth anniversary year is 1994. As part of the celebration of this event, members of *JRME*'s editorial panel have written articles for the Council's other journals that describe the role that *JRME* and research have played in promoting teaching and learning mathematics. This article will focus on *JRME*'s contribution to the view of learning and teaching elementary school mathematics embodied in current curricular recommendations for school mathematics.

Research on Children's Learning

Research in *JRME* illustrates how prevailing theories about how students learn mathematics have moved away from a behaviorist tradition, which focuses on what students do, to a more constructivist view, which focuses on how students think. According to a constructivist view, students learn mathematics meaningfully as they personally construct mental structures and operations that enable them to deal with problematic situations, organize their ideas about the world, and make sense of their interactions with others. The constructive process occurs as students reflect on and abstract the mental and physical actions they perform on their current representations of the world (see, e.g., Clement [1991]; Cobb et al. [1991]; Cobb et al. [1992]).

In a constructivist approach to teaching, the goal is to guide students in building mathematical structures that are more complex, abstract, and powerful than those they already possess. Such guidance can occur only if it is based on a knowledge of what conceptual structures the students might make and how the students might change during the course of instruction (Cobb et al. 1991). Along this line, *JRME* has reported much research that has investigated the cognitive structures and operations that students bring to bear on problems in elementary school mathematics. In this article, several important themes that recur in the research will be described, an example of an essential conceptual mathematics whose in-depth research analysis has important implications for curriculum redesign will be given, and some specific classroom applications of the research will be listed.

Recurrent themes

- Students often apply informal, common-sense knowledge rather than formal, school-taught procedures when solving mathematical problems. Sometimes this informal knowledge is powerful enough to solve the problems; sometimes it is not. For instance, students in grades 1–3 often solve addition and subtraction word problems by employing informally learned counting strategies to model the actions described in the problems (Carpenter and Moser 1984; De Corte and Verschaffel 1987). Similarly, for a pictorially presented problem, Lamon (1993) asked sixth graders if 19 food pellets would be enough for 9 aliens, given that 3 aliens need 5 pellets. She found that rather than comparing ratios, most students modeled the problem by matching 3 aliens with 5 pellets three times, correctly determining that 4 pellets were left over. However, both Lamon (1993) and Mack (1990) found that many of the students' informal procedures were inadequate for solving more complex fraction and ratio problems. Moreover, Clements and Battista (1990) found that fourth graders have many informal, everyday conceptualizations of geometric ideas that actually compete with formal concepts. For instance, students often think of angles as tilted lines, *straight*

as meaning perpendicular, and a rectangle as having two long sides and two short sides, which means that for them a square is not a rectangle.
- Students' informal, intuitive ideas are, for the most part, unconnected to the formal concepts, procedures, and symbols they learn in school. For example, when asked which is bigger, 1/6 or 1/8, four of five students who had just correctly solved an analogous verbal problem involving pizzas chose 1/8 (Mack 1990). Explaining why similar types of errors, such as using the wrong operation, occurred in primary-age children's solution of verbal addition and subtraction problems, Carpenter, Hiebert, and Moser (1981) blamed traditional instruction in operations and word problems: "Because the operations are initially learned outside the context of verbal problems and children are simply told that addition and subtraction can be used to solve these problems, they have no basis for using their natural intuition to relate the problem structure to the operations they have learned" (p. 37).
- Learning the procedures taught in school often interferes with students' efforts to make sense of mathematics (Mack 1990; Wearne and Hiebert 1988). For instance, for problems involving fractions, Mack found that students' use of procedures either preempted their attempts to apply informal knowledge or yielded incorrect answers that they trusted more than informally reasoned, correct answers. "The evidence suggests that learners who possess well-practiced, automatic rules for manipulating symbols are reluctant to connect the rules with other representations that might give them meaning" (Hiebert and Carpenter 1992, p. 78). In fact, the traditional focus on teaching algorithmic procedures causes students to develop the belief that learning mathematics means following rules that someone else presents (Cobb et al. 1992).
- In a traditional curriculum, many techniques taught for solving problems are based on surface characteristics of problems, like looking for key words that correspond to operations (Carpenter and Moser 1984). Such techniques encourage superficial analysis of problems, and their continued use leads to insufficient development of problem-solving skills.
- Classrooms that have adopted a constructivist approach to instruction have proved quite successful. For instance, when compared with students in traditionally taught classes, students in second-grade classes taking a constructivist approach were similar in computational skills but had higher levels of conceptual understanding, were more motivated by a desire to understand or to make sense of their mathematical experiences, and showed signs of developing intellectual autonomy (Cobb et al. 1991; Yackel, Cobb, and Wood 1991).

The role of units in elementary mathematics

Investigating the role and nature of *units* has been a recurrent theme in research on place value, multidigit addition and subtraction, and fractions. The notion is also essential in students' understanding of whole-number multiplication and division and of decimals, making it a natural idea with which to make *connections* across the mathematics curriculum, as suggested by the NCTM's *Curriculum and Evaluation Standards for School Mathematics* (1989).

Fuson (1990) and Baroody (1990) agree that to understand place value, students need to develop *multiunit conceptual structures.* It is not enough for them to think of 70 as a collection of seventy units; they need to form new units of ten so that they can also think of 70 as seven units, each of which is a collection of ten. That is, students have constructed a multiunit, or *abstract composite unit,* of ten when they can take sets of ten as single entities while simultaneously maintaining their "tenness." Baroody (1990) systematically describes a sequence of activities using different concrete embodiments that can help students develop multiunits of ten, hundred, and thousand: (1) using interlocking blocks to group 10 ones into a ten, (2) using base-ten blocks to trade 10 ones for a pregrouped ten, (3) using colored chips for trading in 10 ones for a different-looking ten marker, and (4) using trading boards to trade in 10 ones for an identical marker that represents ten because of its position on the board. Also see Cobb, Yackel, and Wood's (1992) discussion of "The Candy Factory" instructional sequence.

On the basis of studies, in particular Fuson and Briars (1990), Fuson (1990) recommends changes in the traditional textbook placement of place-value topics and in teaching practices. She thinks that multidigit addition and subtraction should first be introduced in second rather than first grade. From the beginning, students should be given multidigit prob-

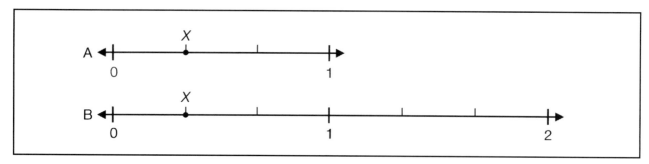

Fig. 1. What number is represented by X on these number lines?

lems with and without regrouping. Fuson and Briars (1990) report that using base-ten blocks to solve a wide variety of problems helped the second graders in their research study begin to develop the concepts of multiunits that are essential to understanding our numeration system. Once students have constructed these multiunits, they are better able to apply place-value concepts in such other circumstances as multiplication and division.

Another area in which "seeing" units is essential is in understanding fractions. Being able to identify the unit, or one, for any given fraction situation and recognizing how that unit is partitioned are crucial to developing an understanding of fractions as "numbers." For instance, fourth graders have little difficulty successfully partitioning a set into equivalent subsets, that is, sharing 28 cookies fairly among 4 dolls. Yet when trying to find 1/5 of 5 eggs, some of these students cannot employ their partitioning strategy because their concept of *whole* is one continuous object (Hunting 1983). To understand fractional parts of a set, students need to recognize that the collection is the unit, or one. That is, in all elementary school students' work with representing whole numbers, each object in a set is one, or a unit. However, when working with fractional parts of sets, students must think of the whole set as the unit. For example, a six-pack of cola comprises six cans; one-half of the *whole set* contains three cans.

The number-line model for fractions also requires students to refine their concept of unit, especially when the number line is longer than one unit. For example, in **figure 1**, students have more difficulty locating or identifying what number is represented by position X on number line B than on A (Bright et al. 1988). On number line B, some students associate "1/6" with point X because they assume that the unit is the total length of the visible portion of the number line rather than the segment from 0 to 1. The confusion seems to arise from students' thinking of the number line as a line segment model, for fractions, as shown in **figure 2**, causing them to lose track of the unit. In the line-segment model, the complete segment, regardless of length, is treated as the unit; in this example, each of the three subsegments is labeled 1/3. On the scaled number line, in contrast, 1/3 is the result of partitioning each *marked* unit into three equivalent lengths, with endpoints of segments of length 1/3 being labeled 1/3, 2/3, 1, and so on. Complicating matters even further, Bright and others note that "on the number line there is no visual discreetness between consecutive units" (1988, p. 215).

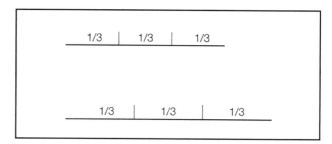

Fig. 2. Two line-segment models for one-third

Many other studies in *JRME* discuss in detail students' learning of specific topics. Readers are encouraged to examine *JRME* articles for details. As in the foregoing discussion of units, most of the studies illustrate the cognitive complexity of seemingly simple mathematical concepts and suggest that students require many diverse experiences before they can transfer learning from one type of situation, problem, or concrete model to another (e.g., Behr et al. [1984]; Kouba [1989]; Mack [1990]).

Classroom Application

In addition to the instructional hints given in our discussion of units, several practical suggestions for improving mathematics instruction can be derived from research.

- An essential ingredient for encouraging students' constructive activity in mathematics instruction is a set of properly chosen problem-solving tasks. To be effective, these tasks should represent genuine, but doable, problems for students who are functioning within a wide range of conceptual levels and should elicit variety in students' solution strategies and thinking (Cobb et al. 1991; Lamon 1993). For example., students could solve concretely or pictorially the aliens problem described earlier; they could use multiplication, reasoning that because 5 times 3 is 15, only 15 pellets are needed; or they could set up a proportion. All viable student strategies should be accepted as correct; adopting more sophisticated strategies is encouraged as students publicly discuss their strategies and solve additional problems.

- Instruction should promote a climate of sense making, inquiry, and reflection in the classroom. One way to accomplish this goal is to involve students first in small-group collaborative mathematical investigations, then in teacher-orchestrated class discussions in which students understand that their role is to make sense of, discuss, explain, critique, and justify mathematical ideas (Cobb et al. 1991).

- Word problems should not be treated as applications of previously learned computational procedures but as opportunities for problem solving and sense making. They should be viewed as a way to introduce new mathematical concepts. For example, students in the Lamon (1993) study were able to discover viable methods for solving the aliens problem before they had been formally introduced to ratio and proportion concepts.

- Mathematical ideas reside in students' minds, not in concrete materials or in mathematical symbols. These materials and symbols become meaningful when students can use them as tools to support their personal mathematical thinking (Cobb, Yackel, and Wood 1992). For example, because of their tendency to understand addition and subtraction word problems in terms of the action described, first graders were very successful in learning to solve such problems using open number sentences (Bebout 1990). Open sentences, such as $5 + n = 12$, permitted students to express meaningfully their intuitive understanding of such problems as "John had 5 marbles. After Mary gave him some more marbles, he had 12. How many marbles did Mary give to John?" Similarly, base-ten blocks become meaningful to a student as he or she learns to let a tens block represent the act of counting ten cubes and mentally grouping them into a unit.

- Students' informal mathematical ideas cannot be ignored. These informal ideas, along with previously learned formal ideas, form the current experiential mathematical reality of students. This reality must serve as the starting point for the construction of more sophisticated mathematical structures (Cobb, Yackel, and Wood 1992). Only after students have created these structures and developed meaning for symbols should they be practicing symbolic routines (Wearne and Hiebert 1988).

Conclusion

The striking similarity between the conclusions drawn from research and the basic tenets underlying the reform movement and the NCTM's *Curriculum and Evaluation Standards* (1989) is not a coincidence. Research in mathematics education has been essential in exposing shortcomings of past practice and suggesting avenues for improving mathematics instruction. The *Journal for Research in Mathematics Education* has played a leading role in supporting this endeavor.

References

Baroody, Arthur J. "How and When Should Place-Value Concepts and Skills Be Taught?" *Journal for Research in Mathematics Education* 21 (July 1990):281–86.

Bebout, Harriett C. "Children's Symbolic Representation of Addition and Subtraction Word Problems." *Journal for Research in Mathematics Education* 21 (March 1990):123–31.

Behr, Merlyn J., Ipke Wachsmuth, Thomas R. Post, and Richard Lesh. "Order and Equivalence of Rational Numbers: A Clinical Teaching Experiment." *Journal for Research in Mathematics Education* 15 (November 1984):323–41.

Bright, George W., Merlyn Behr, Thomas R. Post, and Ipke Wachsmuth. "Identifying Fractions on Number Lines." *Journal for Research in Mathematics Education* 19 (May 1988):215–32.

Carpenter, Thomas P., James Hiebert, and James M. Moser. "Problem Structure and First-Grade Children's Initial

Solution Processes for Simple Addition and Subtraction Problems." *Journal for Research in Mathematics Education* 12 (January 1981):27–39.

Carpenter, Thomas P., and James M. Moser. "The Acquisition of Addition and Subtraction Concepts in Grades One through Three." *Journal for Research in Mathematics Education* 15 (May 1984):179–202.

Clement, John. "Constructivism in the Classroom—a Review of *Transforming Children's Mathematics Education: International Perspectives,* edited by Leslie P. Steffe and Terry Wood." *Journal for Research in Mathematics Education* 22 (November 1991): 422–28.

Clements, Douglas H., and Michael T. Battista. "The Effects of Logo on Children's Conceptualizations of Angle and Polygons." *Journal for Research in Mathematics Education* 21 (November 1990):356–71.

Cobb, Paul, Terry Wood, Erna Yackel, John Nicholls, Grayson Wheatley, Beatriz Trigatti, and Marcella Perlwitz. "Assessment of a Problem-Centered Second-Grade Mathematics Project." *Journal for Research in Mathematics Education* 22 (January 1991):3–29.

Cobb, Paul, Terry Wood, Erna Yackel, and Betsy McNeal. "Characteristics of Classroom Mathematics Traditions: An Interactional Analysis." *American Educational Research Journal* 29 (1992):573–604.

Cobb, Paul, Erna Yackel, and Terry Wood. "A Constructivist Alternative to the Representational View of Mind in Mathematics Education." *Journal for Research in Mathematics Education* 23 (January 1992):2–33.

De Corte, Erik, and Lieven Verschaffel. "The Effect of Semantic Structure on First Graders' Strategies for Solving Addition and Subtraction Word Problems." *Journal for Research in Mathematics Education* 18 (November 1987): 363–81.

Fuson, Karen. "Issues in Place-Value and Multidigit Addition and Subtraction Learning and Teaching." *Journal for Research in Mathematics Education* 21 (July 1990):273–80.

Fuson, Karen C., and Diane J. Briars. "Using a Base-Ten Blocks Learning/Teaching Approach for First- and Second-Grade Place-Value and Multidigit Addition and Subtraction." *Journal for Research in Mathematics Education* 21 (March 1990):180–206.

Hiebert, James, and Thomas P. Carpenter. "Learning and Teaching with Understanding." In *Handbook of Research on Mathematics Teaching and Learning,* edited by Douglas A. Grouws, pp. 65–97. Reston, Va.: National Council of Teachers of Mathematics and Macmillan Publishing Co., 1992.

Hunting, Robert P. "Alan: A Case Study of Knowledge of Units and Performance with Fractions." *Journal for Research in Mathematics Education* 14 (May 1983):182–97.

JRME Editorial Board. "Information for Contributors." *Journal for Research in Mathematics Education* 24 (January 1993):3–7.

Kouba, Vicky L. "Children's Solution Strategies for Equivalent Set Multiplication and Division Word Problems." *Journal for Research in Mathematics Education* 20 (March 1989):147–58.

Lamon, Susan J. "Ratio and Proportion: Connecting Content and Children's Thinking." *Journal for Research in Mathematics Education* 24 (January 1993):41–61.

Mack, Nancy K. "Learning Fractions with Understanding: Building on Informal Knowledge." *Journal for Research in Mathematics Education* 21 (January 1990):16–32.

National Council of Teachers of Mathematics (NCTM). *Curriculum and Evaluation Standards for School Mathematics.* Reston. Va.: NCTM, 1989.

Wearne, Diana, and James Hiebert. "A Cognitive Approach to Meaningful Mathematics Instruction: Testing a Local Theory Using Decimal Numbers." *Journal for Research in Mathematics Education* 19 (November 1988):371–84.

Yackel, Erna, Paul Cobb, and Terry Wood. "Small-Group Interactions as a Source of Learning Opportunities in Second-Grade Mathematics." *Journal for Research in Mathematics Education* 22 (November 1991):390–408.

Research Resources for Teachers

Donald L. Chambers

A number of resources are available to teachers interested in further research-based information on mathematics teaching and learning in the elementary grades. Those included in the first group are developed with an audience of classroom teachers in mind. Those in the second group are written in a more technical style for an audience of researchers and scholars. Nevertheless, classroom teachers could benefit from reading and discussing articles on topics related to relevent aspects of their classroom teaching.

Nontechnical Resources

- Three or four times a year, *Teaching Children Mathematics,* a journal of the National Council of Teachers of Mathematics, contains a department formerly called "Research into Practice." That department was the source of many of the articles in this book. New articles will continue to appear several times each year in the department, which is currently titled "Research, Reflection, Practice." Articles in *Teaching Children Mathematics* are intended for readers who are elementary school teachers. Prior to September 1994, *Teaching Children Mathematics* was called *Arithmetic Teacher*. From 1983 to 1987, *Arithmetic Teacher* contained a department called "Research Report." "Research into Practice" first appeared in *Arithmetic Teacher* in 1987 and continued in *Teaching Children Mathematics* until it was renamed "Research, Reflection, Practice" in 2001. Available from NCTM, 1906 Association Drive, Reston, VA 20191, or by telephone at (800) 235-7566 (orders only).
- *Research Ideas for the Classroom: Early Childhood Mathematics*, edited by Robert J. Jensen, is one of the three volumes in the National Council of Teachers of Mathematics Research Interpretation Project. With an intended audience of early childhood teachers, the book summarizes and synthesizes research in mathematics teaching and learning and reports the findings in a straightforward manner. Each chapter presents classroom implications of the research and suggests small-scale investigations that teachers can conduct in their own classrooms. Available from NCTM, 1906 Association Drive, Reston, VA 20191, or by telephone at (800) 235-7566 (orders only).
- *Research Ideas for the Classroom: Middle Grades Mathematics*, edited by Douglas T. Owens, is a continuation of the National Council of Teachers of Mathematics Research Interpretation project. It is written for teachers of middle-grades mathematics and summarizes and synthesizes research in mathematics teaching and learning in a straightforward manner. Each chapter presents classroom implications of the research and suggests small-scale investigations that teachers can conduct in their own classrooms. Available from NCTM, 1906 Association Drive, Reston, 20191, or by telephone at (800) 235-7566 (orders only).
- *Research within Reach: Elementary School Mathematics*, by Mark J. Driscoll, is now out of print, but your school may have a copy, or you may find it in a university library. Much new research has appeared since this work was published in 1980, but the content is amazingly contemporary and extremely readable. The subtitle, *A Research-Guided Response to the Concerns of Educators*, indicates that the book uses a question-answer format in which the questions come from surveys of classroom teachers.

Technical Resources

- The research journal of the National Council of Teachers of Mathematics, *Journal for Research in Mathematics Education (JRME),* is a forum for disciplined inquiry into the teaching and learning of mathematics. Generally, each article reports on a single research study and includes a summary of related studies, a fairly technical description of the methodology, results, interpretations, and conclusions. Teachers who have enjoyed some of the research summaries written for teachers may want to explore a selection of the articles in *JRME*. Available from NCTM, 1906 Association Drive, Reston, VA 20191, or by telephone at (800) 235-7566 (orders only).

- ***Handbook of Research on Mathematics Teaching and Learning***, published in 1992, is the most comprehensive summary and analysis of research on mathematics teaching and learning now available. It does not require the reader to wade through complex statistical procedures; however, some of its technical language will require some effort on the part of readers. Available from NCTM, 1906 Association Drive, Reston, VA 20191, or by telephone at (800) 235-7566 (orders only).